Learning from Memory

Learning from Memory: Body, Memory and Technology in a Globalizing World

Edited by

Bianca Maria Pirani

CAMBRIDGE
SCHOLARS
PUBLISHING

Learning from Memory:
Body, Memory and Technology in a Globalizing World,
Edited by Bianca Maria Pirani

This book first published 2011

Cambridge Scholars Publishing

12 Back Chapman Street, Newcastle upon Tyne, NE6 2XX, UK

British Library Cataloguing in Publication Data
A catalogue record for this book is available from the British Library

ISBN (10): 1-4438-2884-X, ISBN (13): 978-1-4438-2884-0

TABLE OF CONTENTS

PREFACE

ROBERTO CIPRIANI

This volume is an important reference point in our field, addressing, as it does, three themes of vital importance to contemporary life (the body, memory and technology) and because it does so from a perspective which, some time ago, Roland Robertson aptly defined globalised. The topics and contents are of the utmost relevance because, in some cases, they anticipate future developments, which, at present stand half way between modernisation and virtuality, between continuity and innovation, between old and new frontiers.

The question underscoring the various articles (addressed to the discerning reader and the sociologist who wishes to go beyond the already consolidated) concern yet again the role of the social actor, faced not only with the problems of everyday life but also with the meta-generational dynamics involving our descendants.

What will happen, at planetary-market level, to the adolescents of today who invest in their future as protagonists, to the social and cultural structures of this new-born century, to currently accepted attitudes and patterns of behaviour, to individual and collective memory, to the physical self endowed already with all kinds of prostheses (from pacemakers to i-pods, from hand-held phones to personal electronic devices, myriad tiny supports, a part of the daily apparel we don as soon as we wake)?

This is no minor challenge. Are we slaves to the structures handed down to us by those who went before, designing our scenario, the kind of life, that is, we are obliged to experience, the conditions shaping our existence and survival? Or is there still some room for personal freedom, for the possibility to create an environment of one's own? Or is one obliged to opt for halfway measures to be found somewhere between upstream conditioning and downstream potential?

Within a similar scenario – where even capitalism is an unrecognisable, fleeting, polymorphous, invisible, anonymous, elusive, continuous flow chart devoid of clear outlines, which our person is obliged to deal with – what can our physical self expect? And what use can me make of the memory that belongs to us?

The editor of the book recalls, and rightly so, that "the 'passage' from biological to cultural, suggested by present day cyber-sociality, obliges us to 'rewrite the codes' both of the body and of the orientation system, in this computer society of ours. Information technology (ICTs) is changing the way people relate to and interact with each other and the objects surrounding them. The emergence of complex, compound, contemporary technologies involving virtuality, simulation and computer modelling (including tomography) has special implications for embodiment and perception. Technologies, inventions of the human brain, have, obviously, an immense impact upon the functioning of the brain, on the workings of the mind and on memory".

Furthermore, "the focus of technology is gradually shifting from the computer as such, to the user. Electronic communications – personal computers linked together by Internet – create a worldwide web of memory banks".

Given these premises, the present volume appears extremely promising, even provoking. Each chapter deals with a domain where acquired and new knowledge meet to be contrasted and compared, for the enhancement of scientific knowledge.

The common interpretative key to the various approaches represented here, is, perhaps, the notion of time: time can actually devour everything; time can reach beyond what has already been; time can remove itself from the past for good, without turning back. It grinds event after event, returning each one to the bowels of the earth from which it stemmed, condemning it to oblivion, to the river Lethe, that is, to forgetfulness. But above all, time is, by its very nature, un-stoppable. The very moment one tries to trace it, time escapes and seems to elude our clumsy efforts to block it. To be even more precise, the real attitude of time seems to be that of total indifference towards what it leaves in its wake.

But memory comes to the rescue and recuperates what times risks annihilating. So, not even the physical dimension is thwarted; to the contrary, it is reappraised thanks to memory and, more and more, thanks also to new technological devices (although there are doubts regarding their resistance in time: technical supports are still not totally trustworthy, their duration in time has not been scientifically tested and there are grave risks of sudden failure).

From another point of view, we see how increasingly fixable, modifiable, adaptable the body has become. Prostheses replace missing or malfunctioning parts. The *soma* can, therefore, be reprogrammed. Even death can be delayed, at least partially, thanks to intubation, life-support machines, oxygen supplies. The issue of "delayed death" has been eloquently illustrated thanks to the cases of illustrious personages like the Spanish dictator Francisco Franco and the Brazilian politician Tancredi Neves.

On the moral plane, however, questions concerning eugenics and euthanasia arise, that is, questions accruing to life and death, as focal aspects of the endless existential pathway. But the horizons also broaden to embrace less vital issues like cosmetics, diet, plastic surgery and ultraviolet sun beds for tanning out of season, as well as tattoos, piercing and mutilation.

One must also recognise the fact that contemporary society seems to have reached the acme of the glorification of the body. New aesthetic and reparatory surgery, neo-genetics, neo-dietetics and neo-beauty have all contributed to building an imposing, majestic monument to the ideology of wellbeing. Meanwhile alternative solutions providing relief and undue development of the body industry are being studied: the issues of abortion and contraception are now believed to have been superseded by questions like cloning, same-sex marriage, gene banks, artificial fecundation, modification of embryos.

Medicalisation has gone beyond bounds. The body is now considered a somewhat unique economic commodity, sold in parts for transplants, loaned to supply spinal marrow (at times even procreated for medical purposes). In actual fact, the body is becoming more and more a product among products, equally tradable. And although the body remains exclusively ours, nevertheless it seems to be giving rise to a new kind of capitalism, of private ownership of the physical self (with pricelists referred to single organs, a new butcher's shop window displaying human meat with its catalogue of prices and costs, like an insurance policy containing compensation-rates tables). The aim seems to be that of preserving as many individuals as possible of those no longer able to reproduce because of age. The quest for wellbeing at all costs continues unrelentingly.

INTRODUCTION

BIANCA MARIA PIRANI AND IVAN VARGA

1. Memory in the Real World

This book with contributions from international social scientists– presented at the program of Research Committee 54, "The Body in the Social Sciences" of the International Sociological Association's XVII. World Congress (Gothenburg, Sweden, 11–17 July, 2010) – focuses on the link between body and memory, with specific references to the use of computer technologies.

For neuroscientists it is a known fact that human beings automatically and unconsciously organize their experience in their bodies into spatial units the confines of which are established by changes in location, temporality and the interactive elements that determine it. Our memories might be less reliable than those of the average computer, but they are just as capacious, much more flexible and more user friendly. According to the current collective interactive elements enterprise of cognitive neuroscience – the systematic emphasis on the temporal dynamics of the neural networks that underlie learning – we consider the body as a spearhead of the social sciences, above all concerning the processes of memory.

For most people *memory* does not sound like a mysterious concept. In fact, the general public is often using the concept inclusively, thus meaninglessly, indiscriminately referring to all aspects of the mind. Ask ten people, what memory does, and the answers would be pretty uniform: learning names, telephone numbers, the multiplication table, and recalling for the final exam the dates of historical events one could do without. Memory is also among the most heavily studied aspects of the mind. In a typical study of the memory the subject is asked to memorize a list of words or a series of pictures of faces and then to recall or recognize the material under various conditions. However, both the perceptions of memory by the general public as well as the traditional ways of conducting research into memory have very little to do with the way memory operates in real life. In most real life situations we store and recall information as a prerequisite for solving a problem at hand. Here, recall is a means to an end, not the end itself. Furthermore, certain memories are accessed and retrieved not in response to an external command coming from someone else but rather in response to an internally generated need. Instead of being told, what to recall, I have to decide which information is useful to me in the context of my ongoing activities at the moment.

The kind of things people remember in everyday life include a great variety of different items such as, for example, remembering a shopping list or a recipe, remembering to telephone a relative or to fill up the car with petrol, recounting the arguments put forward at a meeting or the plot of a play seen on television, or remembering the amount of a bill that has to be paid. All these experiences are embedded in a rich context of ongoing events and scenes; they are influenced by a lifetime of past experiences, by history and culture, by current motives and emotions, by intelligence and personality traits, by future goals and plans. It is probably impossible to take all these factors into account, but everyday memory research does recognize the importance of the context in which an event occurs. The kind of things researchers exploit the way that reinstating the context can facilitate retrieval. Research into everyday memory also emphasizes the fact that remembering usually occurs in a social context. Most real life acts of memory recall involve deciding, what type of information is useful to us at the moment and then selecting that information from the totality of the available knowledge. People remember details of an event they witnessed when they are reminded of aspects of the context such as the scene, or the preceding or succeeding events. Memory is not just a private databank; it is shared, exchanged, constructed, revised, and elaborated in all our social interactions.

Memory in the real world (Cohen, 2008:1-20) is often known as *everyday memory*; it is concerned with the memory used as people go about their daily lives. Research puts its emphasis on the functional aspect of memory. Memory, then, is viewed as part of a repertoire of behaviour designed to fulfill specific goals. For example, autobiographical memory functions in order to build and maintain personal identity and concept of the self; prospective memory functions to enable the individual to realise plans and intentions; spatial memory functions so that the individual could navigate in the environment, etc., etc. Bruce (1985) stated that research into ecological memory must ask, how memory operates in everyday life, identifying causes and processes; what functions does it serve and why has it evolved both ontogenetically and evolutionarily in this way. Everyday memory is *context-bound* and not context-free.

Today, memory is conceived as a complex and diffuse mental faculty which does not reconstruct the past faithfully, but instead is responsible for a continuing process producing individual memories that depends on the meaning ascribed and the emotions linked to the embodied experiences of the individual. This is true above all when the events are important for one's sense of self (autobiographical memory), in conferring uniqueness to one's own life (Schacter 2001). Thus, we can state that the approach that is currently dominant in memory studies is "constructivist in nature" (Assmann A. : 2006), and that it has been supported by the evolution that has taken place in the neurosciences themselves, although substantial differences can be found when comparing the subjective and social levels of memory.

Memory takes many forms.

• Representations and reconstructions of the past, including autobiographical memories.

• Socially distributed and reconstructed memories, such as jointly reconstructed autobiographical memories.

• Collective memory, ranging from the stories a family tells about itself on family occasions to the culture of the past, including religious stories and myths, communal histories, and socially celebrated events and persons.

• And finally there is also memory so ancient that it is stored in our physiology and our genes, adaptations preserved by evolution as patterns in our behaviour and in the structure of our social networks, patterns that have augmented the adaptive fitness of our species. The chapters in this book will bring together reflections, by contributors from several research traditions, on the connections between electronic technologies and memory in all of these forms.

The first section of the book "Memory and the Enhanced Body", consists of contributions describing computer implementations to solve spatial problems, to simulate human or animal orientation and navigation behaviour, or to reproduce spatial communication patterns. The second section, "The Social Nature of the Body: Synchronization, Embodimemt and Technology", is an experimental performance aimed at fleshing out the *body in context* as a methodological standpoint for examining how the sense of our bodies, of our orientation in the world, and our everyday memory interact with information technologies. Each chapter brings together reflections by contributors from several research traditions on the connections between body, memory and electronic technologies. Contemporary understanding of these connections necessarily spans research in many disciplines, ranging from the social sciences to cognitive science, psychology, neuroscience and biology. We shall slight none of these perspectives, yet we propose a sociological analysis which benefits from recent advances in the study of social networks. In addition, we wish to link network analysis to emerging knowledge of synchronization in attachment behaviour and social interaction. We believe that an account at the sociological level will prove valuable in throwing light on accounts of human behaviour at the interpersonal and social level, and will play an important role in our ability to understand the neurobiological factors that underpin the various types of memory.

Research on social networks has taught the social sciences why so many of their canonical questions have seemed intractable using only traditional theoretical tools. Any list of the topics to which network research has given new understanding is impressive—the organization of economic markets, the spread of obesity, the sexual organization of cities, the structure of adolescents' friendship choices, health and recovery from illness, loneliness and the supports for happiness, the network advantages of brokerage and closure, constraints on reciprocity and altruism, patterns of entrepreneurial behavior, the organization of production networks, the structure of professional élite systems, rules governing the choice of sexual partners, the synchronization of religious communicants, moral integration, and many other topics. All are subjects that network research has opened up in new and productive ways.

What of memory and the body?

2. The Body Memory and Spatial Cognition

The starting point of the book is the hypothesis that the body is a spatial-temporal carrier of socially distributed processes that enable people, through links with others in social networks, to acquire, store, recall and manipulate information. It is common to characterize cognitive tasks in terms of "networks" of underlying brain structures derived by fMRI or other neuroimagining methods. These considerations imply that these "networks" change over time. Thus, it is impossible to characterize a task by assuming a single, static network. Often the data are averaged across the whole experimental sequence on the assumption that the underlying neural network is static. This is a big mistake, since the network is usually not static, and averaging across the whole sequence is like mixing data points representing different populations. Therefore one would be better served by segmenting the experimental sequence and averaging the data separately within each segment, thus expressing the findings as a *vector*, rather than a single state. One can design a suite of tasks activating cognition that would elicit the temporal dynamics of such spatial networks as a function of learning. The present book suggests this methodological approach because it is supposed to be more powerful in characterizing normal cognition, since a *vector* by definition contains more information than a single variable.

The spatial aspect of the body cannot be reduced to intersubjectivity only. Up until now the analyses have this in common: they treat the body as a "thematic object".

Up to this point, we have concentrated on the 'zero-point' of orientation as such, how it can be identified in various perceptual formations.

Research on spatial cognition is a rapidly evolving interdisciplinary approach to the study of cognitive spatial processes, be they real or abstract, originated by humans or machines. Without understanding human spatial cognition, we will not be capable to develop appropriate technology and interfaces for spatial information systems that communicate with humans by language and graphics in natural ways. Autonomous robots moving in an unknown environment require abilities to infer location of objects from incomplete and qualitative information from various sources and to follow imprecise instructions much like human beings. To use maps and other diagrams for communication with computers, we must understand, how people generate and interpret them. To fully exploit the potential of the technology of virtual reality, we must adapt its capabilities to human conceptions of space. In order to develop computers programmed by spatial structures rather than by sequential instructions, we have to fully understand the relevant aspects of space.

It is important to take into account the ways the body and objects relate to each other in the contemporary world that also change the old boundaries between the human subject and the material object. The world of the subjects and the world of the objects become increasingly intertwined.

Let us now turn to an examination of how the brain came to embed the properties of the external world, how it carries this out, and the evolutionary relationship of this phenomenon to the generation of such an amazing functional space as mindness. What we must stress here is that the brain's understanding of anything, whether factual or abstract, arises from the manipulations of the external world, by our moving within the world and thus from our sensory-derived experience of it. According to LLinas (2002: 59), "*the organization and function of our brains are based on the embedding of motricity over evolution*".

When we ponder the concepts of learning and memory, what seems most often to come to mind are the wonders of human capability. The immense amount of knowledge some people acquire from years of education, or the ability to recall a singular event from one's childhood decades later, as clearly as if one were living again, is what generally comes to mind when thinking of human memory. But one should keep in mind that the neuronal mechanisms subserving these fantastic capacities came to us, as do all things physiological, by the long evolutionary processes of trial and error. For our nervous systems,–for us– to be able to learn and to remember means that the evolution not only had to learn and remember, but that it had to learn and remember *how* to learn and to remember. What we, or any creature may learn, however, is a product of the myriad needs and events experienced during development. It was evolution's task to learn and slowly fine-tune the appropriate forms, the structural morphologies that added to the survivability of a given species. By so doing, it brought together the world of our external bodies with our brains. The result was the opposable thumb, the tail of a rat, the nose of a kitten, even the shape of the brain. This kind of memory would be considered *phylogenetic*. But the phylogenetic memory of structural forms that expresses itself at birth is not enough. We see interwined into the organ architecture (the plant) a second type of memory, as phylogenetically old. At this period in neuroscience, the issue of learning and memory are central. Indeed, the ability to learn is viewed as critical for bettering ourselves within the practical world in which we live.

Somatic sensibility arises from information provided by a variety of receptors distributed throughout the body. Somatic sensibility has four major modalities: *discriminative touch* (required to recognize the size, shape and texture of objects as well as their movement across the skin); *proprioception* (the sense of static position and movements of the limbs and the body); *nociception* (the signaling of tissue damage or chemical irritation, typically perceived as pain or itch) and *temperature sense* (warmth and cold). Each of these modalities are mediated by a distinct system of receptors and pathways to the brain. However, all share a common class of sensory neurons: the dorsal ganglion neurons. The somatic sensory stimuli we encounter in everyday life are complex, they cover large areas of the skin and have many characteristics. Each type of receptor is selectively activated by distinct spatial and qualitative properties of a stimulus. Different types of information about an object are transmitted by populations of different types of sensory neurons and conveyed in parallel pathways to the primary somato-sensory cortex where all the information is combined into a unified somatic percept. Our sensory systems are the way in which we perceive the external world, remain alert, form a body image and regulate our movements, Sensations occur when external stimuli interact with receptors. Sensory information is transmitted to the brain as trains of action potentials travelling along individual sensory neurons and by agglomerations of such neurons acting together.

The location and space of a stimulus are conveyed topographically, through each activated receptor's position in the sensory epithelium, called its receptive field. The complex qualities of sounds, visual images, shapes, textures, tastes and odours require the activation of large ensembles of receptors acting in parallel, each one signaling the attribute of a particular stimulus. For us to savour the richness and diversity of perception, the central nervous system must integrate the activity of an entire sensory ensemble.

The goal of the neuroscience is to understand the mind – how we perceive, move, think and remember. Molecular biology also has greatly expanded our understanding of how the brain develops and how it generates behaviour. In addition, the ability to develop genetically modified mice allowed us to relate single genes to signalling in nerve cells and to relate both of these to an organism's behaviour. Ultimately, these experiments will make it possible to study emotion, perception, learning, memory at both a cellular and molecular level.

Memory is not an exclusive property of *Homo sapiens*. However, human beings differ from animals inasmuch as their individual memory is influenced by – one could even state, dependent on – collective memory. The Freudian theory of psychoanalysis emphasized the role of instincts and drives in the foundation of the unconscious. Jung pointed out the importance of innate patterns of behaviour. In this aspect the ethology of Lorenz is similar, inasmuch as he also emphasizes the importance of innate patterns of behaviour amongst animals. (This, and some other examples, are derived from the article of Soren R. Ekstrom [2004] "The mind beyond our immediate awareness: Freudian, Jungian and cognitive models of the consciousness"). The early discoveries of Freud linked the unconscious to repressed memories, mainly about childhood sexual traumas. Jung's emphasis on archetypical factors, such as myths and beliefs, indicated that he assumed the existence of a common human constitution that is independent of historical conditions. It is, however, highly questionable whether certain similarities in, say, creation myths, etc., result from genetic predispositions. Ekstrom (2004) to the relatively new cognitive science and the works of George Lakoff and Mark Johnson, two cognitive linguists and philosophers who introduced a new approach to the unconscious in their attempt at overcoming the dualist conception of the body-mind relationship. The gist of threir argument is that the unconscious is operating beneath the level of cognitive awareness, thus doeIs not stem from repression in the Freudian sense. It has two areas: "The first has to do with all our automatic cognitive operations: visual and auditory processing, and motor operation among them. The other, particularly relevant when we deal with memory, is what they call 'our implicit memory' and they claim that 'all our knowledge and beliefs are framed in terms of a conceptual system that resides mostly in the cognitive unconscious" (Ekstrom, ibid.: 666). Highly complex computer simulations allow to conclude that the brain carries out two things at the same time: perception or motor control on the one hand, and conceptualizing, categorizing and reasoning, on the other. (Ibid.: 667). They conclude that humans' concepts relate to spatial-relations concepts, concepts of bodily movements and concepts indicating the structure of actions or events. What is highly important for our purposes, i.e. to understand and analyze the spatial aspects of the body's existence, these findings confirm the capacity of the brain to conceptualize the forms of our bodily existence.

Advances in artificial intelligence (AI) to simulate certain tasks performed by the brain rely on neural modelling. They are also relevant for rethinking the nature of the unconscious as well as of the mechanisms of memory. Roger Schank (1995: 669–70) of Northwestern University developed the theory that a wide range of stories or scripts serve in memory storage. They help to remember a great number of events and can be "indexed". These mechanisms ensure that many cognitive operations happen without our conscious participation. According to this theory knowledge has many features but only some of them are conscious. "There is the rational knowledge of acts necessary for logical thinking, and there is the emotional knowledge of being able to identify how we feel. But beyond these two, there is also [...] subconscious knowledge of which we are ordinarily unaware of and a physical knowledge that our body uses which is mostly unconscious. Finally, Schank assumes a non conscious knowledge which is being used in basic and ongoing mental activity. We are generally unable to articulate this activity as it occurs to us, but it can now be deduced from computer modelling." http://www.rogerschank.com

In modern times it was Maurice Halbwachs (1877–1945), the French sociologist belonging to the Durkheimian school, who first elaborated the characteristic features of collective memory. Already Durkheim (1912) mentioned the importance of the collectivity on laying the frameworks for the ideas that influence the individual's thinking of the past. Halbwachs, however, developed further the concept of collective memory. He claimed that the study of memory does not take into account the properties of the subjective, i.e. the individual, mind. Instead, memory is the result of minds working together in society and their operations are structured by social arrangements. He wrote (1912:38) "it is in society that people normally acquire their memories. It is also in society that they recall, recognize, and localize their memories". Halbwachs made a distinction amongst autobiographical memory, historical memory and collective memory. This distinction has its roots in Bergson's differentiation between memory of specific events and memory of enduring attitudes. Namely, Halbwachs analyzed the dreams and images of memory, the relationship between memory and language, the location of memories as well as the collective memory of the family and the religious collective memory. He also called the attention to the influence of the social classes and their traditions in building and maintaining collective memory. He considered that the social frameworks of memory include the temporal and the spatial frames, language and the dominant mode of thought. The temporal frame of memory relates to the historical process and to the actual social conditions within which the historical events are recalled or (re-)evaluated. The spatial framework refers to the position of the group in space. Objects, things, insisted Halbwachs, are parts of the society. Works of art and literature already suggest the social type or category of the human being or beings who are depicted or described. The group, say, the family, that is surrounded by a more or less stable and permanent surrounding (habitat, objects, etc.) develops an image of its external surroundings, the *milieu*. The individual experiences and evaluates it according to his/her belonging to the group. Buildings acquire meaning for the group, even if its members move elsewhere. (Namely, personal memories are intertwined with group memories, e.g. someone's memory of his/her university years are linked to memories of the university's physical and intellectual environment). Halbwachs attributes special importance to the city-space. It is structured according to functions (i.e. economic, entertainment, etc.) and customs. He also mentions that part parts of the city are distinct by social stratification: the streets inhabited by the rich are different from the ones where the poor live. (In the analysis of the city, its functions and stratification, Halbwachs anticipated and influenced Pierre Bourdieu's and his school's concepttions).An important element of Halbwachs' thought is the conception of the relation between the individual and the collective memory. Namely, the individual's memory is his/her own but as they are members of a group, they remember as group members. In other words, memories that the individual considers to be his/her own, are influenced, or even determined, by the collective memory, and change when the individual changes his/her relationship to other milieus.

Paul Ricoeur, in his monumental work, *Memory, History, Forgetting* (2004) acknowledges the seminal contribution of Halbwachs but complements it by introducing a phenomenological (as well as a hermeneutical-historical-epistemological approach to the subject and memory. His aim is to overcome the aporia that exists between the sociology of collective memory and the phenomenology of individual memory. In the segment *"Three Subjects of the Attribution of Memories: Ego, Collectives, Close Relations"* (2006:124-132), Ricoeur, analyzing the role and nature of language, elaborates on the relations between personal and collective memory. He says that "memory enters into the region of language; memories spoken of, pronounced are already a kind of discourse that the subject engages in with herself. What is pronounced in this discourse occurs in the common language, most often in the mother tongue, which is the language of others (Ibid.:139). A phenomenology of social world, as developed by Alfred Schütz, bridges the gap between the individual's memory and the memory of the others. As the individual never lives isolated from others, his or her memories are first of all influenced by the people who are closest to him or her – members of immediate family – and vice versa, the individual's birth and death affects them. Thus, there is a certain gradation in belonging to a group, from the intimacy within the family where emotions influence memories, to the common interest as manifested, say, in memories shaped by class interests. Solidarity with, or within, a group shapes memories and assessment of the importance of events that constitute particular memories. The problem of generations also emerges in this context. Ricoeur (1990) , in the third volume of *Time and Narrative* (1990) refers to the anonymous relationship between contemporaries, predecessors and successors as developed by Schütz. Ricoeur, then, goes beyond the Heideggerian analysis of the aporia of temporality that exists between the public (calendar) time and the mortal time, i.e. the life span of the individual and his/her generation. He wrote: "The notion of a succession of generations provides an answer to this antinomy by designating the chain of historical agents as living people who come to the place of dead people. It is this replacement of the dead by the living that constitutes the notion of a succession of generations." (Ricoeur, 1998:109.)

The problem concerning generations is important in two main respects: it is a pivotal aspect of carrying forth traditions and introducing innovation. Sociological approaches, especially Karl Mannheim's analysis, expanded the biological criteria of generations by introducing a common element: thinking, feeling and acting in similar ways. One can say that events experienced by people belonging to the same, or close, age groups, have been or are influenced by experiences they had in common. For example, World War II, the Vietnam War, the Holocaust, etc, mean something different for those who survived it in their home countries or in the battlefield or for those who learned of it from books or from accounts of survivors. These insights – the meaning and importance of which could be more deeply analysed from philosophical and sociological viewpoints – shed light to the importance of intersubjectivity and history in shaping collective and individual memory. The relationship to the past, be it an individual's to his/her own or a society's to its collective memory, is always shaped by the present. As Jacques LeGoff, (1992:54) the French historian wrote: "The study of social memory is one of the fundamental approaches to the problems of time and history, in relation to which memory is sometimes retreating, sometimes overflowing." (LeGoff, 1992:54.) With the arrival of written, in lieu of the oral, memory, the amount of preserved memories has immensely increased and has become independent of the individual's memory.

The arrival and expansion of computer generated and computer stored memory has in a significant way changed, how humans create, store and recall memory. One substantive difference is that memories of humans are unstable (depending on the individual's capacity of remembering and recalling), while the content of the computer memory remains stable and with a suitable programme can easily and quickly be recalled. Nevertheless, one has to bear in mind that the programming has been done by humans – applying their mind and body (hands) – and, as LeGoff remarks, "human memory preserves a large sector that cannot be reduced to 'information', and that, like all the forms of automatic memory that have appeared in the course of history, electronic memory is only an aid, a servant of memory and of human mind." (Ibid.: 92).

In human beings emotions can, and do, influence the capacity to recall memories and even their content. Collective memories involve assessments of past events as a society's influential forces define, evaluate them. Eviatar Zerubavel (2003:11) states that that memory is "patterned in a highly structured manner that both shapes and distorts what we actually come to mentally retain from the past" and "many of of these highly schematic mnemonic *patterns* are unmistakably social." (2003: 11). He argues that "human memory is our ability to mentally transform essentially unstructured series of events into seemingly coherent *historical narratives.*" Zerubavel concurs with most historians and sociologists who emphasize that past events are presented and evaluated according to historical situations, but adds that there are skriptlike *plotlines* in which we often remember (reconstruct) past events, "as we habitually reduce highly complex event sequences to inevitably simplistic, one-dimensional visions of the past." Instead, postmodern conditions foster the attitude of "presentism" (the expression used by Adam Schaff, the Polish philosopher) that diminishes the social and cultural frameworks in which social identity has been established. The element of *forgetting* already indicates the possibility of selective or blocked memory. At the individual level psychoanalysis investigates into repressed and suppressed memories, and one could eventually reveal the reasons of those removed from conscious memories. A person's relationship to his or her past – acceptance, denial, selection, interpretation or re-interpretation – as well as bearing the consequences from it has an important moral element that influences the individual's relationship to a group, be it a family or another collective. Jeffrey Blustein (2008) in his book, *The Moral Demands of Memory* (2008), develops a complex philosophical, ethical approach to individual and collective memory. One of the most important moral demands for historians is to discern the truth of past events and overcome the distortions imposed by the present. Temporal factors have to be taken into account in giving an accurate rendering of the past. If there are survivors of recent past (as mentioned above, the Holocaust, World War II, etc., etc.) it is easier to evoke a correct picture of events. In dealing with the remote past – especially if there is no written testimony left – there is always the moral problem for historiography, sociology and anthropology to establish not only their course but also their meaning for the present or else, give a reasoned account, why are they unimportant for the living members of a community or society.

There is no doubt that a great deal of our learning is implicit, in the sense that we can learn skills without being able to reflect and report on precisely what we know. By memory, we usually mean our capability to remember certain events of our past, or to retain and retrieve data and knowledge. However, the phenomena of memory are by no means restricted to that. As Descartes already noted, the lute player must also have a memory in his hands in order to play with such a skill. He would certainly be lost should he try to remember the single movements which he once learned deliberately. Obviously, there is a memory of the body apart from conscious recollection. Through repetition and exercise a habit has developed. Long-trained patterns of movement and perception have been embodied as skills or faculties that we practice as a matter-of-course in our everyday life – the upright gait, the ability of speaking, reading, writing and the handling of instruments, such as a piano or a bicycle or a keyboard.

Already conceptualized in French philosophy (Maine de Biran, Ravaisson, Henri Bergson), this kind of memory has come to be rediscovered and explored as *implicit memory* in the last two decades. Research into amnesic patients who may still learn simple motor skills though unable to retain new explicit recollections has demonstrated the existence of multiple memory systems. Above all, there ought to be a distinction made between the so-called declarative and implicit memory. Declarative or explicit memory contains single recollections or informations that may be reported and described; it may also be called a tacit "knowing that". In contrast, repeated situations or actions have melted into implicit memory, thus no more to be retrieved as single events. They have become a tacit "know-how", hardly to be verbalized – we would have some difficulty to describe, how to waltz. Thus, explicit recollection is directed from the present back towards the past; implicit memory, however, does not represent the past but *re-enacts* it in the course of the body's performance. What we have acquired as skills, habits and experience, has become what we are today; implicit knowledge is our *lived past.*

On the other hand, implicit memory is not a mere reflex programme realized by the body machine.

"Computer modelling helps to study the mind—brain connection. For example, the one carried out by Joseph LeDoux of New York University, focused on the neural aspects of emotions, in particular fear. He concluded that the importance of synaptic transmissions in brain function allows to demonstrate that the self is synaptic. Against the objections that such athesis ignores the psychological, social, moral, aesthetic or spiritual character of the self, LeDoux counters that his theory does not deny these features but asserts that the synaptic approach shows, how those features of the self are realized. (Ekstrom, 2004: 670).The importance of these discoveries is not only a more precise knowledge of the machanisms of the brain; it also sheds light on those processes that are not immediately available to consciousness.

We (our bodies) are performing functions of which we are not aware, such as "standard body maintenance like regulating heart rate, breathing rhythm, stomach contractions and posture; or many aspects of seeing, smelling, behaving, feeling, speaking, thinking, evaluating, judging, believing and imagining" (Ekstrom, *ibidem*). For LeDoux, synapsises hold the clues to the dilemma of nature versus nurture, genetics versus learning. Similarly, we become conscious of unconscious operations after the fact, when e.g. someone talks to us, we decode the sounds of words and the meaning of the sentence (phonology); assign meaning to the words (semantics); know the grammatical relations between words (syntax) and using our knowledge of the world (pragmatics). This distinct structure and functioning of the human brain has an important, unique impact on memory. The importance of these discoveries is not only a more precise knowledge of the mechanisms of the brain; it also sheds light on those processes that are not immediately available to consciousness. The study of brain circuits, combined with the focus on synaptic plasticity or learning as well as use of computer models allow to better understand the processes and working of the brain, the body—mind and the conscious—unconscious relationship. It also helps to get a more nuanced notion of the temporal element of the body. Neurological studies are ever more clarifying processes of the brain including the mechanisms of memory. The brain changes not only when we lose information but also when we are exposed to to excessive activation – for example when we practice a skill, such as learning to play a musical instrument with year-in, year-out, hour-after-hour drills. When scientists mapped the areas that receive sensory information for the left hand of string musicians, they found that the area activated by sensory impressions is larger than that in non-players. They also found that the area activated on hearing piano notes is roughly 25 per cent larger in pianists than in non-musicians, and the pathways conducting motor impulses are different. Juggling is not something that many people do on a daily basis. But if we were to start practicing, we would markedly improve in just a few weeks. In other words, it is an activity that lends itself to the study of what happens in the brain when a specific activity is learned. One study (Draganski, Gaser, Bush. 2004, *Nature* 427: 311–12) examined the structure of the brain in a group of subjects before and after a three-months course in juggling. What the scientists found was that an area in the occipital lobe specializing in the perception of motion grew over this period, but three months after the training stopped it had shrunk and lost roughly half of the increase previously induced by training. In other words, as little as three months' activity or three months' passivity had an immediate effect on the structure of the brain. Norman Doidge's book, *The Brain That Changes Itself* (2007), describes cases when the human brain did change itself. The author, a psychoanalyst and researcher at Columbia University and University of Toronto, documents the plasticity of the brain. During his contacts with eminent brain scientists he learned that "the brain changed its very structure with each different activity it performed, perfecting its circuits so it was better suited to the task at hand. If certain 'parts' failed than other parts could sometimes take over." (Doidge, 2007: xiv-xv.) This elasticity – a fundamental property of the brain – is called "neuroplasticity", and it refutes the long-held idea that the brain is "hardwired".

Research into training-based brain plasticity, such as the mentioned study of musicians, seems to confirm that the axiom "use it or lose it" is rather banal – at least for the brain researcher and the psychologist. True as it may be that the brain changes depending on *how* it is used but we should be careful not to generalize. The first question we should ask when we hear such assertions is, what "use" actually means. Are all kinds of activities equivalent? There are plenty of examples showing, how the brain can adapt to suit its environment and can be shaped by training.

3. Body and Memory in Digital Space

The "passage from biological to cultural", as suggested by the cybersociality, obliges us to "recodify", in the body, the orientation system in computer society. Information technology (ICT) is transforming the way people interact between themselves and with objects around them. The emergence of the contemporary complex, compound technologies that involve *virtuality, simulation* and *computer modelling (including tomography)* has special implications for embodiment and perception. Technologies, inventions of the human mind, obviously have a great impact on the functioning of the brain, on the working of mind and on memory. Kelly has coined a new term in his book, *What Technology Wants* (2010), that he calls the *technium.* Using this term he suggests that technology and life must share some fundamental essence, yet we have historically failed to understand and define, how technology is intertwined with biological life. Is technology a creation of human ingenuity or can it develop "wants" of its own? If one thinks about it, electronic networks exhibit near-biological behaviour in many respects. If you are still not convinced, then consider the PR2 research robot that has been programmed to look for power outlets and plug itself in, when it gets "hungry". Of course, we humans want certain things from the *technium* but at the same time there is an inherent bias in the *technium* that is outside our wants. Beyond our desires there is a tendency within the *technium* that – all other things being equal – favours a certain solutions.

The issues are complex and often confused, in part because there is overlap and loose usage concerning all the VSC technologies. In popular form, one can enter virtual raality in a virtual reality arcade. Unfortunately, the description and claims made for this arcade experience are usually cast in the now antiquated frame of early modern epistemology. According to Turkle (2005: 287), computers become "thinking tools" offering new models of what it means to know and to understand: computers, with their reactivity and interactivity, stand in a novel and evocative relationship between the living and the inanimate. They make it increasingly tempting to project our fee-lings onto objects and to treat things as though they were people—an impulse called by Turkle the "Eliza effect" after the early AI program that was designed to seem like a solicitous psychotherapist. Our connection with things, consequently, is becoming an "object relations" perspective “that takes *objects*” as its *subject*. Relational artefacts ask their users to see them not as tools but as “companions”, as “subjects” in their own right. The simple robots are marketed as toys and the more advanced robots remain largely confined to research settings or are used for complicated tasks, such as flight simulation or flying unmanned aircraft. In the presence of relational artefacts, people feel attachment and loss.

Technology's focus is gradually shifting away from the computer as such, to the user. Electronic communication – personal computers linked together by the Internet – creates a world-wide web of memory banks. It produced a new form of space: the cyberspace that is a ‘no-space space’ because it cannot be localized, and localization is unimportant or immaterial. Timelines (i.e. temporal models) are replaced by spatial models (hyperlinks amongst web sites). This change of paradigm aims at making communication and computer systems simple, collaborative and transparent to the user. A first sign of this change has been the creation of totally new interactive communication environments, such as Computer Mediated Communication (CMC) and Computer Supported Collaborative Work (CSCW). The final steps towards this vision will be made possible by three dominant trends: - The increase of *richness and completeness of human-computer interaction*, through technology extensions of the senses and of the human body;

- The relevant *role of mobility*, through the development of mobile communications and extended networks;
- The *pervasive diffusion of intelligence* in the space around us, through the development of advanced biosensors.

The merging of these trends allows the emergence of a new vision : *Ambient Intelligence* (AmI), *a pervasive and unobtrusive intelligence in the surrounding environment supporting the activities and interactions of the users*. Ambient Intelligence (AmI) is a new paradigm in information technology, in which people are empowered through a digital environment that is aware of their presence and context, and is sensitive, adaptive, and responsive to their needs, habits, gestures and emotions. The most ambitious expression of AmI is Intelligent Mixed Reality (IMR), an evolution of traditional virtual reality environments. Using IMR, it is possiible to integrate computer interfaces into the real environment, so that the user can interact with other individuals and with the environment itself in the most natural and intuitive way.
www.neurovr.org/emerging/volume.html

As psychologists studying the effects of cyberspace on people point out, the sensory impacts that one experiences in everyday life, in interactions with others in society, are seriously altered in cyberspace.

Some of our sensory experiences – smelling and touching – disappear; others – seeing and hearing – remain but in a somewhat changed form. John Suler (2002), in his essay *Presence in Cyberspace,* analyses the effects of cyberspace on the individual engaged in VR. He concludes that in spite of some shortfalls of sensory experience in smell and touch, cyberspace 'is becoming increasingly more sophisticated in the visual and auditory stimulation provided.' He also acknowledges that in 'the life span, especially during childhood, humans rely heavily on the close stimulation of touch and smell in developing awareness of, and intimacy with, significant others.' In this he follows the discoveries of neurosociology. The difference between the two approaches is that neurosociology emphasises the importance of the physical, dyadic experience that cannot be replaced by VR, in spite of the ever increasing sophistication of cyberspace. Adherents to the 'presence' approach in cyberspace draw a parallel between interactivity in society and in cyberspace. Suler, for instance, stresses that online environments – even though they are still limited compared with physical ones – allow us 'to express a wide range of thoughts, memories, emotions and motives' as well as make it possible to exhibit personal identity, in particular when the expressions of identity become interactive. The author acknowledges that some individuals prefer online expression of their identity, their self, and depending on the personality type, experience differently the presence of others whom they encounter in cyberspace. Professor Suler admits that despite the 'powerful possibilities for the presence online, we must remind ourselves that indeed our body sits in a room, in front of a computer, in a setting that is quite different from the online encounter' and we cannot immerse ourselves fully into cyberspace and in-person presence simultaneously.' He concludes this study by saying that the worlds and relationships of cyberspace remind us that the being, here and now of presence resides in the human mind.' One could say without exaggeration that the invention and proliferation of VR have created the most profound intertwining of body (mind) and technology. It, however, transforms a person into an image, lifts him or her out of his or her social environment and allows one to present oneself in a subjective way, according to the individual's 'wish. Further studies are needed to ascertain whether a deep immersion in cyberspace (i.e. in virtual reality) does induce changes in neural processes of the human brain. One could argue whether technologies could substitute for emotions or will. Some theorists (Levy: 2007) give a positive answer to this problem, but even those who deny this possibility acknowledge that microelectronic technologies could influence emotions and will, and often do. Therefore, the question arises: Does human memory change as 'mapped' by these mobile, shifting boundaries? The human brain—unique in the organic world—has the infinite capacity to adapt itself to the changing conditions of life. Over the last few years popular books about the brain have become a literary genre. As advances in information technology and communication supply us with information at an ever accelerating rate, the limitations of our brains become all the more obvious. The modern work situation, with its pace and simultaneous demands, often gives us the feeling of having attention difficulties. The torrent of information increases not only the volume of data we are expected to take in, but also the volume we need to shut out. New findings in psychology and brain research suggest that the difficulties we find with simultaneous performance and distractions converge onto one central limitation: *the ability to retain information.*

Namely, it is questionable whether the ever greater and intrusive role of machines (computers) in human cognition—the constantly more complex character of artificial intelligence'—contributes to the plasticity of the brain or to the nuanced cognition. This is an open question; it would be premature to take a definitive position, but one cannot exclude either a positive or a negative answer to it. Suffice it to say that at present brain scientists, psychiatrists and psychologists disagree about the problem. Nonetheless, the fundamental difference between the brain and even the most sophisticated machines producing artificial intelligence is that the brain's fashion of recognizing images, sound, etc. is 'assembling' by a very large number of parallel processors the visual, auditory, etc. patterns. By contrast computers separate the Central Processing Unit (CPU) – the control unit – and the memory unit. As Lynch and Granger (2008: 17–32) emphasize, the computer's processing time is much longer than the human brain's. Geoffrey Hilton, a distinguished researcher of the University of Toronto and a pioneer in artificial intelligence stated, in an interview in the daily *Globe and Mail*, Toronto, February 12, 2011. that 'it is very hard to get computers that have the same amount of processing power and particularly the same access to stored knowledge. The brain can access many gigabytes of knowledge in a tiny fraction of a second. Only the biggest supercomputers can do that kind of thing at present.' One can add to it that all computers that have been built the same way function identically. The differences are in the programming. Humans are different, unique in their intelligence, talents, interests and because of their sociability, react to their social surrounding. Even identical twins, if they find themselves in different social conditions, develop their individual personality traits.

Fortunately, the information–processing approach became increasingly influential. Two books were particularly important. Donald Broadbent's *Perception and Com-munication* (1958) developed and applied Craik's seminal ideas to a range of work carried out at the Medical Research Council Applied Psychology Unit in Cambridge, England. Some 9 years later, this growing field was then brilliantly synthesized and summarized by Ulric Neisser (1967) in a book whose title provided a name for this burgeoning field: *Cognitive Psychology*. Using the digital computer as an analogy, human memory could be regarded as comprising one or more storage systems. Any memory system– whether physical, electronic, or human– requires three things: the capacity to *encode*, or enter information into the system, the capacity to store it, and–subsequently–the capacity to find and *retrieve* it.

Neural science and cognitive psychology have now found a common ground, and we are beginning to benefit from the increased explanatory power thatresults from the convergence of two initially disparate disciplines. Recently, in large measure due to the work of the neuroscientists Patricia Goldman–Rakic (1987: 373–417) and Joaquin Fuster (1997), the concept of working memory has gained prominence. Much has been written on the subject of working memory over the last few years. A book by the Swedish cognitive neuroscientist Torkel Klingberg (2008), *The Overflowing Brain: Information Overload and the Limits of Working Memory*, stands out as a particularly lucid rendition of the subject. Working memory is closely linked to the critical role that the frontal lobes play in the temporal organization of behavior and in controlling the proper sequence in which various mental operations are enacted to meet the organism's objective. Today the concept of working memory is among the trendiest in cognitive neuroscience. As is the case with trendy concepts, it is often used arbitrarily and loosely, as times being rendered meaningless. This is why it is particularly important to discuss this concept carefully and rigorously. It is often said that "working memory is like short–term memory". Well, if it is so much like short–term memory, then why do we need a new term? Creating duplicate terminology without new meaning obfuscates things rather than clarifying them.

According to Goldberg (2009: 94), we define working memory as *the selection of task–relevant information.* The neural circuitry of working memory is the focus of intense research with both experimental and computational methods (Amit, Fusi and Yacovlev (1997:1071–92); Mongillo Machens, Romo, Brody, 2005: 1121:4); Mongillo, Barak, Tsodyks (2008: 1543–46). One of the defining characteristics of working memory is its very *limitation*. If you are told “Go straight ahead for two blocks and then left one block”, you will have no difficulty remembering where to go. The capacity limitation of working memory is one of the things that distinguish it from long-term memory. What makes working memory particularly interesting is that it not only retains instructions, numbers, and positions in the memory, but also seems to play a critical part in our ability to solve problems. It does not operate in a *vacuum*: it involves goals, plans, and beliefs and is about the stakes (active goals) and options a person has for managing person–environment relationships. Modern interest in unconscious processes, the concept that is sensory or perceptual is also perforce, motor, and what is called action theory, which emphasizes goals, intentions, and plans– all reflect a much freer search for understanding how the mind works than was possible in past decades. Although this freedom troubles many, we believe it has been a major stimulant for a renewed interest in the body. By inserting itself into every situation, the body carries its own past into the surroundings as a procedural field. Its experiences and dispositions permeate the environment like an invisible net that projects from its senses and limbs, connects us with the world and renders it familiar to us. Each perception, each situation is permeated by implicit bodily recollections. The memory of the body is an impressive refutation of the dualism of consciousness and the physical body. As an example one can mention that when o-ne is dancing, the rhythmic movements are released by the body, spontaneously, wi-thout the need to make them deliberately, and yet the dancer is guiding his/her movements according to the gesture and rhythm felt. The person is still dancing by him/herself, and is not a ghost in a body machine. Almost sixty years ago Sir Frederic Bartlett (1932) investigated the claim that the Swazi of South Africa had remarkable memory abilities. Bartlett’s findings are an example of one of the most important generalizations about memory, namely, information on any given topic will be much better remembered by individuals who have a great interest in it than by those who do not. For example, cattle are of great importance to the Swazi of South Africa, and so they remember cattle transactions in great detail. This is a typical exchange skill for the capacity of working memory. Effective motivation involves, indeed, implementation of intentions to minimize the chances of being distracted while working towards a goal.

It is up to us to control our environments and reshape the work we do to our abilities. The world around us is constantly changing and we have to react to the change through a plastic response allowing us to deal with new and unexpected experiences foreign to our individual past. In considering memory as a procedural faculty, we suggest therefore that rather than discarding the notion of culture, it should be reinvented, through the exploration of the place of culture in both the experiential and discursive spaces that people inhabit or invent.

The journey on which the book *Learning from Memory: Body, Memory and Technology* is about to take the Reader towards a new perspective on one of the biggest mysteries within the body: the social construction of working memory, a mystery that will fascinate not only professionals of social sciences but anyone interested in understanding the human memory.

References

Amit, D.J., Fusi and Yacovlev (1997), "Paradigmatic working memory (attractor) cell in IT cortex", *Neural Comput*, 9 : 1071–92); Mongillo G. Machens K, Romo, Brody C.D, (2005), "Flexible control of mutual inhibition: a neural model of two interval discrimination", *Science* 307: 1121:4); Mongillo G., Barak O., Tsodyks M. (2008), "Synaptic theory of working memory", *Science* 319: 1543–46).

Assman A. (2006). "The Printing Press and the Internet: from a Culture of Memory to a Culture of Attention." In *Globalization, Cultural Identities and Media Representations*, edited by Natasha Gentz and Stefan Kramer, 11-23. Albany: State University of New York Press.

Baddeley, A. Eysenck M.V., Anderson M.C., *Memory*, New York, Psychology Press.

Bartlett, F.C. (1932), *Remembering: a Study in Experimental and Social Psychology*, New York, Cambridge University Press.

Blustein, J. (2008), *The Moral Demands of Memory*. Cambridge–New York: Cambridge University Press.

Bruce, D. (1995) "The How and Why of Ecological Memory", *Journal of Experimental Psychology*, 77: 114: 78–90.

Cohen, G. (2008), "The Study of Everyday Memory", in *Memory in the Real World*, New York, Psychology Press, pp. 1–20).

Doidge, N. (2007), *The Brain That Changes Itself*. New York: Viking Penguin.

Draganski, B., Gaser C., Busch *et al.* (2004), "Neuroplasticity: Changes ijn Grey Matter Induced by Training", *Nature* 427.

Ekstrom, S. (2004), 'The mind beyond our immediate awareness: Freudian, Jungian, and cognitive models of the consciousness' in: *Journal of Analytic Psychology*, 2004, 49: 657-682.

Fuster J.M., (1997) *The Prefrontal Cortex: Anatomy, Physiology and Neuropsychology of the Frontal Lobe*, 3rd ed. Philadelphia: Lippincott–Raven.

Goldman–Rakic P.S. (1987), "Circuitry of primate prefrontal cortex and regulation of behavior by representational memory", in *Handbook of Physiology: Nervous System Higher Functions of the Brain*, Part 1, Bethesda MD: American Physiological Association, 373: 417.

Goldberg, E. (2009), *The Executive Brain: Frontal Lobes in a Complex World*, New York, Oxford University Press.

Halbwachs, M. (1992), *On Collective Memory*. Ed., transl. and with an Introduction by L.A. Coser. Chicago—London: The Chicago University Press

Kandel, E.(2007), *In Search of Memory: The Emergence of a New Science of Mind*, New York, W.W. Norton&Company, Inc.

Kelly, K (2010) *What Technology Wants*, New York, Viking Penguin;

Klawans, H.L. (1988), *Toscanini's Fumble and Other Tales of Clinical Neurology*. Chicago—New York: Contemporary Books.

Klingberg T. (2008), *The Overflowing Brain: Information Overload and the Limits of Working Memory*, New York: Oxford University Press

LeDoux, J. (2002) *Synaptic Self: How Our Brains Become Who We Are*, New York, Viking Penguin.

Le Goff, J. (1992), *History and Memory*. Transl. by S. Randall and E.Claman. New York: Columbia University Press.

LLinás, R.R. (2002), *I of the Vortex. From Neurons to Self* , Cambridge, Mass: the MIT Press.

Lynch, G.—R. Grange (2008), *Big Brain: The Origins and Future of Human Intelligence*. New York: Palgrave MMacmillan

Rabin, J.F. "Repression', in: *International Dictionary of Psychoanalysis*. http://www.enotes.com/psychoanalysis-encyclopedia/repression

Ricoeur, P. (2004), *Memory, History, Forgetting*. Transl. by K. Blamey and D. Fellauer. Chicago--London: The University of Chicago Press

—. (2008), *Time and Narrative*, vol.3. Transl. by K. Blamey and D. Fellauer. Chicago—London: The University of Chicago Press

Riva, G.—G.A. Waterworth, *Presence and the Self. A Cognitive Neuroscience Approach*. htpp://presence.cs.ucl.ac.uk/presenceconnect/articles/Apr.2003

Riva G.A. (ed. by) (2003), *Ambient Intelligence The Evolution of Technology, Communication and Cognition Towards the Future of Human-computer interaction* www.neurovr.org/emerging/volume.html

Rommeru, C. (1992), 'Statut du corps dans les différentes religions et les différentes civilisations' (The Status of the Body in Different Religions and different civilisations), in: Labesse, J. Ed. *Analyses et réflexions sur le corps* (Analyses and Reflexions on the Body). Paris: Editions Marketing.

Schacter Daniel L. 2007. *Alla ricerca della memoria. Il cervello, la mente e il passato*. Torino: Einaudi.

Schank, R.C. & Cleary. C. (1995). *Engines for Education*. Hillsdale, NJ: Erlbaum Assoc.

Suler, J. (2002) *Presence in Cyberspace*, in: http://www-usr.rider.edu/

—. *The Basic Psychological Features of Cyberspace: Elements of a Cyberpsychology Model.* http://usesr.rider.edu/~suler/psycyber/basicfeat.html

—. *Identity Management in Cyberspace,* http://users.rider.edu/~suler/psycyber/identitymanage.html

Taylor, L.M. (2005), *Introducing Cognitive Development*, Hove, UK: Psychology Press.

Zerubavel, E. (2003), *Time Maps: Collective Memory and the Shape of the Past.* Chicago—London: The Chicago University

PART ONE:

MEMORY AND THE ENHANCED BODY

CHAPTER ONE

KINAESTHETIC MEMORY

ANDREAS STASCHEIT

Abstract

The chapterr outlines aspects of an analysis of kinaesthesia, motion and embodied memory from the viewpoint of phenomenology. After briefly exploring the specific characteristics of the recollection of temporal objects, the discussion focuses on the topics simultaneity and synchronization. As embodied memory, the implicit or explicit embodied reference to the past, is not only based on what is given through experience but at the same time is itself a foundational moment of all forms and modes of experience, section 4 proposes to introduce the dimension of technology into the discussion of the bodily foundations of memory via an approach that starts from the question on how technology inheres in experience. The role and function of technology in the constitution of experience, the intertwining of the corporeal, technical and cultural dimensions of experience is introduced by referring to history, i.e. institutionalized memory. Thus the study of the history of technology provides the access to carving out the inherent technological character of lived animate-bodily experience. In order to present a first preliminary sketch of this approach, section 4 presents an outline of how experience shapes and configures technological inventions and development, followed by some remarks on the technological formation of experience and the methodological problems a study of this topic is confronted with. The concluding section 5 is dedicated to the analysis of the dimension complementary to embodied memory: the question of innovation and discovery in the kinaesthetic field.

1. Temporal Objects

In his famous paper "The Problem of Transcendental Intersubjectivity in Husserl" (1957) Alfred Schutz probed the limits of the transcendental approach with regard to the theoretical understanding of the genesis of social relationship and intersubjectivity, i.e. the genesis of the difference that separates and connects my life and the Other's life. At this point, in order to clarify the "immediacy of understanding by which the existence of the Other is apprehended in shared situations"[1], Alfred Schutz in the above paper proposed an approach grounded in the phenomenological analysis of temporality, referring to the "question of the simultaneity of the ego with Others, of the common Now as a presupposition for differentiating a Here and a There."[2] In the discussion that followed the presentation of the paper at Royaumont, Schutz therefore stated: "The problem of simultaneity, taken not merely as a common Now in objective time but also as a community of two inner flows of time – [as a community of] "durée" in Bergson's sense – seems to me to be of the greatest significance for the problem of inter-subjectivity, and that not only in regard to transcendental but also to mundane intersubjec-tivity."[3] Together with the concept "simultaneity", the notions "synchronization" and "tension" gain essential importance for Schutz' phenomenology of the social world. Although all of these are core concepts of Henri Bergson's philosophy, Schutz is far from performing a simple adaptation: Schutz is rethinking Bergson through the music-centered gaze.

Which theoretical aspects could be the motivation for taking "music matters" as a starting point for the inquiry into sociological questions? It is not just a marginal theoretical concern that fuels this approach, but a question that touches the very foundation of the social sciences: "The problem which has to be investigated", Schutz sums it up in the preliminary sketches to "Making Music Together", "is the following one: Is the communicative process really at the origin of social relationship or is there a preceding layer"?[4] In music inheres the potential to provide access to this preceding layer, that is: to the pre- and non-linguistic dimensions of meaning, together with and through the exploration of the temporal structures of experience and sociality. How does this work? How are these two aspects, the aspect of meaning and the aspect of temporality, screwed together?

This becomes understandable if we refer to Edmund Husserl's distinction of "two different modes in which the sense of past experiences can be 'grasped'":[5] Husserl first introduced the distinction between polythetic

---

[1] Alfred Schutz. Collected Papers III. Studies in Phenomenological Philosophy. Ed. by Ilse Schutz, with an introduction by Aron Gurwitsch. The Hague 1966, p. 85.

[2] Ibid. p. 88.

[3] Ibid. p. 88.

[4] *Alfred Schutz Papers*. Beinecke Library, Yale. Series I/ Box 6/ Folder 106, p. 3090.

[5] Alfred Schutz and Thomas Luckmann: *The Structures of the Lifeworld*. Transl. Richard M. Zaner and H. Tristram Engelhardt. Evanston 1973, p. 53.

and monothetic mode in volume 1 of *Ideas: General Introduction to Pure Phenomenology* in 1913 and took the topic up again 1938 in *Experience and Judgement.* In this work, Husserl analyzes experience as becoming originarily constituted step by step, resulting in a polythetic unity:

> "In streaming forth in a linear continuity, the act of contemplation would become a simple fixed view if it did not disengage itself and pass over into a chain of individual apprehensions, of individual acts, in a discrete succession of separate steps which, bound internally to one another, form a polythetic unity of the individual theses."[6] Many cases allow for a transformation of such a discrete succession of separate steps into a "one-rayed" object, which can be "grasped" monothetically: "Every such many-rayed (polythetic) constitution of synthetic objectivities – which are essentially such that "originally" we can be aware of them only synthetically – possesses the essential law conforming possibility of transforming the many-rayed object of awareness into one that is simply one-rayed, of "rendering objective" in the specific sense and in a monothetic act what is synthetically constituted in the many-rayed object." [7] But this transformation of a "many-rayed" into a "one-rayed" unity of experience is impossible in the case "of those experiences whose meaning is essentially contained in the polythetic structure of its elements, that is, experiences of so-called temporal Objects." Therefore, if I want to attempt to grasp the meaning of the experience of a temporal object in retrospect, I must "reflectively realize the polythetic building up of this experience".[8]

2. Simultaneity

In order to understand temporal objects, it is essential to relate to the experience of what is called 'rhythm' and 'tempo', two immediately interrelated dimensions of agency and experience that have central relevance for the temporal organization of moving and movement. The close interrelatedness of rhythm and animate motion can be demonstrated by distinguishing two regulating principles that guide the generation of rhythmic structures.

The first principle is based in the animate-bodily practice of walking and its different variations, resulting in a series of units of motion (also

---

[6] Edmund Husserl: *Experience and Judgment Investigations in a Genealogy of Logic*. Rev. and edited by Ludwig Landgrebe, transl. by James S. Churchill and Karl Ameriks. Evanston 1973, p. 112.

[7] Edmund Husserl: *Ideas: General Introduction to Pure Phenomenology.* Transl. W. R. Boyce Gibson. London / New York 1931, p.336.

[8] Alfred Schutz and Thomas Luckmann: *The Structures of the Lifeworld*. Op. cit. p. 53.

misleadingly called ‘beats’), organized in patterns of two: a binary rhythm. The old belief, that the reference for binary rhythms is provided by the throbbing heart has been widely criticized, e.g. by Curt Sachs in his book *Rhythm and Tempo*.[9] The second regulating principle in the generation of rhythmic structures is founded in the experience of breathing, with tension and relaxation alternating in mostly regular, but hardly equal, intervals. The result is a ternary rhythmic structure, which may be composed as a result of a rhythmical grouping (like 2 + 1 or 4 + 3 or 3 + 3 + 2).

The problem of “simultaneity” – i.e. the coherence of a multiplicity of distinct streams – can well be approached by analyzing the characteristics of polyphonic music in the perspective of the study of the temporality of experience. A reconstruction of this issue in the context of ‘making music’ could for instance refer to piano praxis, as the pianist finds himself confronted with the problem of the coherence of a multiplicity of distinct simultaneous streams as a part of everyday working practice. The practical implication of this problem is frequently called “independence of the hands”, programmatically demonstrated in Johann Sebastian Bach’s *The Well-Tempered Piano* and *Goldberg Variations*. This requirement of “independence of hands” pertains to the specific features of keyed instruments, the history of which being closely linked to the idea and art of polyphony in the sense of multi-part music. Although, as in praxis the so-called “independence of the hands” – i.e. polyphonic bodily movement – is not just naturally given to the pianist as a matter of course, it presents a fundamental problem and continuous challenge. Modeled by its physical and instrumental structure and typicalities of sound-production, the piano thus provides access to the experience of polyphony as experience of the moving animate body.

A first idea concerning the significance of the requirement “independence of the hands” might be conveyed by a simple musical experiment, performed while listening to a piece of music built upon the most elementary form of polyphony – dyaphonia. As musical material for this experiment, let us take the traditional French song “Sur le Pont d’Avignon”, arranged for two voices and played by a so-called “Spieluhr” or “musical clock”. This musical toy uses a technology derived from the 19th century “mechanical piano”, the piano-strings being replaced by chimes.

The experiment consists of three sections:

> In section 1, please listen to the song as you would in everyday life. [play music example]

---

[9] Curt Sachs: Rhythm and Tempo. A Study in Music History. New York 1953.

Now, in section 2, please try to take up a musician's perspective by centering the attention exclusively on one of the two parts, continuously tracking its flux without digression. Please try to protocol the specific motional characteristics that distinguish the parts, the different temporal qualities exhibited, and how these differences become manifest on the musical level as “this” or “that” melody, the respective musical meaning of either melody being demonstrated through its memorability and transposability. It may be helpful to assume that the music presented is being performed by two musicians. Within this scenario of an imagined duet, the experiment consists in taking up the perspective of exclusively one of the two musicians. In the course of such an imagined ‘Making Music Together’, the ‘I am doing’ becomes the focus of attention, together with the correlative dimensions of ‘*es zeigt sich*’, and ‘*ich erlebe mich*’, the pertaining temporal horizons and directions of description (*Beschreibungsrichtungen*).

While the musical example now is being played the second time, please try to focus exclusively on the upper voice. – [play music example] – Next, please try to center the attention on the accompanying lower voice without allowing any distraction by the simultaneous presence of the upper main melody. – [play music example]

What did the first two sections of the experiment reveal? – What presented itself as a unity in ‘naïve’ listening in section one is given as different meaningful processes in section two. This difference exists on the level of musical meaning (both processes appear as musically meaningful, but not in an identical way) as well as with regard to the dynamics of the musical movement and the corresponding temporal qualities: Either part is being experienced “as movement”, but as movement differing in terms of dynamics, mode and *Gestalt*.

Now follows the final section 3 of the experiment: Please try to alternate between the modes of listening practiced in section 1 (music listening in everyday life) and section 2 (focusing on one of the two voices). Please decide yourself – and ad libitum – when to change between the two modes. [play music example]

Let us now look at a passage in Bergson’s book *Durée et Simultanéité*, which can be read not only as a discussion of the concept of simultaneity, but also as a kind of instruction for the above experiment:

> “I call two flows "contemporaneous" when they are equally one or two for my consciousness, the latter perceiving them together as a single flowing if it sees fit to engage in an undivided act of attention, and, on the other hand, separating them throughout if it prefers to divide its attention between them, even doing both at one and the same time if it decides to

> divide its attention and yet not cut it in two. I call two instantaneous perceptions "simultaneous" that are apprehended in one and the same mental act, the attention here again being able to make one or two out of them at will".[10]

The concept of simultaneity as coined by Henri Bergson, which has nothing to do with temporal compatibility, synchronization and simultaneousness based in the notion of linear metric time, appears as the theoretical explication of the above experiment's essence.

3. Synchronization

When the problem of simultaneity is being discussed with regard to subjective and social dimensions of temporality, the second cardinal problem for philosophical analyses of the intertwining of animate-bodily Being-in-the-World, temporality and sociality becomes evident: the synchronization of motion and time-as-experienced (*temps durée*).

Motion essentially is qualified by being performed in a specific tempo. In order to understand the notion of tempo, it is helpful to reflect on how tempo can be distinguished from speed. As has often been argued, western traditions tend to think and speak about temporal phenomena in terms taken from the sphere of spatial experience. Duration is being equated with "extension" and events are conceived as single spots, "points in time". In this context, tempo can only exist as "speed", defining how fast I can get from one point in time to the next.

Whereas "speed" is defined as independent of the experiencing subject, "tempo" is one of the concepts that are grounded in everyday life and, at the same time, have a phenomenological nucleus, as a phenomenological analysis of the musical flux reveals that "tempo" has to be understood as the specific experience of "movement in relation-to-me". "Tempo as experienced" is based on structuring a succession due to a pace of reference, resulting in a specific modification of mood instead of being a mere matter of fast and slow. In other words, tempo is not at all "continuity", an homogenous flux, it requires the structure of a meter. And in the form of "tempo as mood", tempo represents the convergence of the temporal and the expressive dimension of human experience and self-experience.

As a matter of fact, synchronization of time-as-experienced is possible. This is revealed by practices that form part of everyday life, like

[10] Henri Bergson: *Duration and Simultaneity*. Transl. by Leon Jacobson, edited by Robin Durie. Manchester.

"Dancing", "Making Love" and "Making Music Together", to quote three of Alfred Schutz' favorite examples. Although, everyday practice also reveals that synchronization is not at all naturally given, but rather has to be established and maintained, which implies the possibility of failure.

How might the phenomenon of synchronized *temps durée* be productive in theoretical terms?

An exemplary situation for analyzing the problem of synchronization is provided by the situation of 'making music together' in a small chamber ensemble playing without a conductor (e.g. a string quartet). To start simultaneously and in a common tempo pertains to the elementary as well as demanding requirements of ensemble performance, and it is almost equally demanding to maintain a common, perhaps commonly modified tempo throughout the whole piece of music.

The synchronization of tempo is neither established nor maintained automatically with the flux of music. At the same time, arranging the attunement of tempo in the form of a *consensus* or a planned *project of action* is impossible. Also, the movements and gestures of the musicians cannot provide the foundation of synchronization. The cue of the primarius only presents a single temporal pulsation in spatial form, which consequently can only approximately indicate the definite tempo to be played *in futuro*. And finally, every attempt to respond or react as a result of communicative action embedded in the flux of music results in breaking the synchronization of tempo, as the reaction always comes "too late" (as musicians use to say).

Rather, the foundation and precondition for establishing a common tempo has to be specified as common level of *animate-bodily attention*, manifested as *tension*, the intensity of which determines the density of the intervals that function as meter. According to the given meter, the ongoing flux is structured and experienced as a succession of "steps". - With regard to the metrics of temporal structuring, the phenomenological study of making music together also corresponds with Bergson's philosophy of temporality. Let us have a look at the following passage in *Matière et mémoire. Essai sur la relation du corps à l'esprit*, translated by Nancy Margaret Paul and W. Scott Palmer (1911, p. 272): "The duration lived by our consciousness is a duration with its own determined rhythm [...], which can store up, in a given interval, as great a number of phenomena as we please."

When working with the concept of "tension" in the context of analyses of temporality, the history of philosophy suggests to go back to Augustine, who used several derivations of *tendere* und *tensio*, particularly in his reflections on time and time consciousness. In the field of musicology, the works of Ernst Kurth have to be mentioned, where tension is analyzed as constitutive with regard to "music as experienced" as well as to music as specific aesthetic form. Baumgarten adverts to the psychological concept of tension when discussing the foundations of the *impetus aestheticus*.

4. On the technological formation of experience as performed and performing

In the context of technological developments and innovations, the structures of sensory experience, bodily movement and adaptive behaviour often function as a *model*, e.g. in the fields of ergonomics and usability engineering, or as motivation and *starting point*, e.g. in cases, where the limits of human capabilities motivate the development of new so-called "solutions", i.e. technical compensations. The fact that all auditory phenomena fade away in the continuous flux of consciousness marks a prominent exemplary case of specific characteristics of human sensory capacities motivating the endeavor for technological innovation in the sense of a starting point. Before the phonograph and gramophone were invented in the last decades of the 19th century, documenting auditory results of human action – in particular documenting the spoken word or music performances –was impossible and an urgent desiderate. Due to the immediate, foundational and essential function of memory for auditory experience, this exemplary case is especially suitable for the discussion presented here.

Let us also mention an example of how experience may function as a model for technical inventions: We know as a fact from everyday life experience and specific practices like music, that the quality and volume of the sound produced by a vibrating object - like the violin's string or the membrane of a drum – is improved, the less vertical force is exerted on it. The less gravitation comes into play when producing vibrations using the violinist's bow or the drummer's sticks, the more subtle nuances of sound are made possible. The idea to record sound by using horizontal instead of vertical modulation on a cylinder and, after 1888, on a disc record, is one technical consequence resulting from these observations. To take the question of how experience configures technology as an approach to study and document the history of technology means to pursue a systematic reconstruction of the active role of experience and behaviour in the history of inventions.

Let us now go in the opposite direction and ask: How does technology shape and configure experience? – It is evident that technological evolution not only affects normative and institutional structures. Also, styles and modes of sense experience and understanding including their corporeal foundation are not left untouched. Because of this evolutionary dynamic, discussed e.g. by Bachelard, to speak of a "technological construction of experience" can – although this wording has the potential to mislead because of the term "construction" – open up an interesting discussion. The allusion to Peter Berger's and Thomas Luckmann's famous book *The social construction of reality*[11] is quite intentional. This approach has been highly influential in many areas including science and technology studies, where it provided the basis for what was discussed as "social construction of technology". But let us go one or two steps back in history and look at the foundations of Berger's and Luckmann's thought. In the introduction to the German translation, Helmuth Plessner pointed out that for Husserlian phenomenology, and so through Alfred Schutz for Berger's and Luckmann's approach, Schelling's notion of a "working knowledge" ("werktätiges Wissen") was very important. If – following this thread of *experience as performed and performing* – we raise the question of the genealogy of technological implications of experience, two different modes of "doing experience" have to be distinguished. First: the implicit technological formation of experience via experience we just *go through*, in contrast to – second – experience we *produce* in a more or less intentional process.

When we "go through" or even find ourselves "exposed to" "what happens", the inherent subjective activity remains non-thematic. In this way, and as a result of the mere ubiquitous presence of technological devices in everyday life, technology and the "physiognomic structure" of technical products are introduced *tacitly* into the subject's realm as results of long-term, slow and often clandestine processes.

How can this process of a "technological formation of experience" be studied in detail? Let us sketch in a few words a simple experiment which everybody can try out some day. A part of the lab for this experiment is provided by the technologically based reproduction of recorded music at home. It is safe to say that disc records and cd's, together with the turntable or, nowadays, its digital counterpart, in some way pertain to and are part of everybody's life. Thus, the use of the turntable or cd player belongs to the taken-for-granted structures in and on which our everyday life is grounded.

Think of your collection of sound recordings at home. Among them, take out one of your favorites you have listened to very frequently. The

[11] Berger, Peter L., and Thomas Luckmann. *The Social Construction of Reality. A Treatise in the Sociology of Knowledge*. London : Allen Lane, The Penguin Press, 1967.

music presented by the recording may come from any epoch of musical history including contemporary Jazz and Popular Music. For sure you know very well every part of the work of music presented, and you are very much acquainted with the style and atmosphere of the interpretation, the acoustics and the specific sound of the instruments on the recording, as well as the musicians' style of playing.

As the second step of the experiment, go and attend a live performance of that particular work of music. If you have chosen a recording of a work of music that has been created und handed down in written form, the work should be performed on stage by another artist or ensemble as on the recording. – What will happen? – The style and atmosphere of the interpretation on stage will be different from the recording, the performers' style of playing and the sound of the instruments will be different as well. Most probably, you will find yourself tacitly or even explicitly involved in an ongoing process of comparing and distinguishing the live performance to what you have got acquainted with via listening so many times to the recording of the respective work of music at home, in your car or while travelling by aeroplane.

The technological heritage in and alteration of auditory experience we are confronted with through the second step of this experiment, has led to the inversion of what has been named "High Fidelity", beginning as an advertisement slogan in the 1960s and defined as a technical standard later on. Initially, "HiFi" meant the endeavour to make the reproduction of a music recording evoke an auditory experience as close as possible to the experience one would have while listening to this music in a concert hall. We do not have to go further into the details of the two sound transformation processes, the microphone at the beginning and the loudspeaker at the end of the long chain of audio signal processing. The many variables and potential bias-inducing factors along this chain have kept whole industries alive for decades. What is interesting in our context is, that as a result of a slow but steady development, the meaning of "High Fidelity" has changed to its opposite. In the meantime, it is the musician on stage, from whom the audience expects to "emulate", not a specific recording, but "recorded music", i.e. technologically grounded music. These expectations reference terms of playing style, level of perfection, tempo, articulation, and other aspects of music performance.

Now follows the third and final step of the experiment: While listening to the music in the concert hall, try to assume an attitude that permits you to temporarily suspend the sedimentations and memories of the previous listening experiences, striving for a mode of listening that does not let the sediments stemming from listening to technically produced and reproduced music interfere with the present experience in the concert hall. – What will happen? Most probably you will find this difficult and you will succeed only to a certain extent, not in a pure and "perfect" way. This third step of the experiment is just a short exercise in what has been discussed e.g. by Bernhard Waldenfels, referring to the procedure of "phenomenological reduction", as "technological reduction".[12]

If we now take a look at the second variant of "Doing Experience" mentioned above – experience *we produce* in a more or less intentional process – a specific way of understanding is provided if we examine what happens when *new* technologies are introduced into what Schutz called *paramount reality*, i.e. into everyday life.

New technologies always open up new potentials of "we can". At the same time, they pose new obligations and limits to action and behaviour, which appear on a global level, on the level of societies as well as on the level of the individual actor. This represents a second way in which technology *alters and re-configures* habitual and potential repertoires of action and experience.

But *before* new technologies can open up new potentials of "we can", they are terra incognita for us. To be confronted with the inaccessibility of new technical apparatus or a renewed and thus unfamiliar user interface design, pertains to the basic experiences of contemporary everyday life. As new or modified technologies require new or modified ways of *handling*, *access* to an apparatus has to be opened up when we use it for the first time. In today's "digital age", this is a very frequent challenge, and for sure everybody will be able to give wonderful examples.

Besides, in order to lack accessibility, the respective technology does not necessarily have to be new. The same occurs with historical tools and apparatus, like today the turntable player, the descendant of Emile

---

[12] In the chapter entitled *„Technische Eingriffe in die Erfahrung“* (*„Technical Intrusions into experience“*) of his book *Bruchlinien der Erfahrung. Phänomenologie, Psychoanalyse, Phänomenotechnik* (Frankfurt : Suhrkamp, 2002: 374f.) Bernhard Waldenfels says: "Um einen Zugang zu diesem Grenzbereich zu eröffnen, bedarf es einer technologischen Reduktion, die es einerseits vermeidet, erlebte Phänomene in einer funktionierenden Technik aufgehen zu lassen, die sich andererseits nicht damit begnügt, technische Prozesse und Gegenstände als Phänomene zu betrachten. Letzteres würde eine Distanz zur Technik voraussetzen, die gerade durch die Annahme einer genuinen Phänomenotechnik in Frage stellt wird. Phänomenotechnik besagt nämlich, dass technische Momente im Bereich der Phänomene eine Rolle spielen, so daß diese gleichsam technisch eingefärbt sind. Die technologische Reduktion wäre somit ein Bestandteil der phänomenologischen Reduktion, vergleichbar der eidetischen beziehungsweise der strukturalen Reduktion, die allgemeine Erfahrungsstrukturen ans Licht hebt.“

Berliner's famous invention, the gramophone. Probably, not many of us still really know how to configure the turntable player's crucial part, the pickup-system.

Three ways to open up access to something still not accessible have to be distinguished: to train, to play and to practice.[13]

In the case of technical apparatus like the turntable, face-to-face communicated instructions or written documents and the other varieties of "technical documentation" provide – or rather: *can and should provide* – the necessary explicitly defined basis *to train* the usage or application of the respective technology. Consequently, educational measures concerning the introduction of new technologies into business and production processes are usually called "training".

Playing around and experimenting with technical apparatus is another way to gain access to it. In contrast to training, this way is neither based on nor does it make use of written documents or guidelines distributed by word of mouth. What is essential in order to make use of an apparatus in *this subject-relative way*, is that the apparatus is being explored via an intuitional style of experimenting, a kind of "learning by doing", attentively responding to the observed "behaviour" of the apparatus.

This second way to open up access to something up to now inaccessible shows many links and transitions to the third – i.e. to practice, as we use to call the primary working method in the context of the arts. When approaching a technical device via practicing, the apparatus is transformed into an instrument. In the case of the turntable, the result of this process is the turntable's conversion from an apparatus dedicated to the *reproduction of music recordings to a musical instrument*, itself now being involved in the production of music, like e.g. in "turntable-jazz".

5. Innovation and Discovery

When in the *Phenomenology of Perception* Merleau-Ponty introduces the case of playing a musical instrument in order to show, how "habit has its abode neither in thought nor in the objective body, but in the body as mediator of a world" [14], he sketches with the following words the character of the musician's involvement in the process of 'music in the

[13] See Andreas Georg Stascheit. "Doppelgriffe. Phänomenologische Motive vom Gesichtspunkt der Musikarbeit." In *Phänomenologie und soziale Wirklichkeit: Entwicklungen und Arbeitsweisen*. Ed. Ilja Srubar and Steven Vaitkus. Opladen: Leske & Budrich, 2003, 215–38.

[14] Maurice Merleau-Ponty: *Phenomenology of Perception*, transl. Colin Smith. London and New York 1962, p. 167.

making': "Between the musical essence of the piece as it is shown in the score and the notes which actually sound round the organ, so direct a relation is established that the organist's body and his instrument are merely the medium of this relationship."[15] Although, in praxis, establishing so direct a relation between musician and instrument, between musical essence and actually sounding tones, cannot just naturally be done. It turns out to be an unsolvable problem for the beginner, transforms into a mystery for the advanced student and continues to pose a continuous open task to the most proficient performer.

However, as musical practice demonstrates, extending the horizon of the "I can", thus providing access to something up to now inaccessible, *can* be achieved by a specific kind of practice: Very commonly, the verb 'to practice' and the noun 'practice' are used to designate the heuristic method aiming at the *intended development* of new potentialities of agency and the incorporation of new *ways and means* into one's repertoire of animate-bodily expression.

In genetic phenomenology, Husserl no longer conceives the "I" as "empty pole of identity" [16], but as defined by capabilities, by positions taken, by convictions assumed, by the pregivenness of the world as the horizon of an "I can". As Alfred Schutz has pointed out in his review „Edmund Husserl's Ideas, Volume II": "The I as a unity is a system of faculties of the form 'I can'",[17] the horizon of "I can" itself developing over time, pointing back to earlier experiences.[18]

"Now what does all that mean? What I can do, what is in my power, what I know myself capable of and am conscious of as such, that is what a practical possibility is."[19] With these words, Husserl starts the section dedicated to the analysis of the "I can" in the second book of *Ideas*, titled "The 'I can' as logical possibility, as practical possibility and impossibility, as neutrality modification of practical acts, and as original consciousness of abilities (subjective power, faculty, resistance)." And a few sentences later he continues: "Then just what sort of modification is the 'I can,' 'I

---

[15] Ibid. p.168.

[16] Rudolf Bernet, Iso Kern, Eduard Marbach: *An Introduction to Husserlian Phenomenology*. Evanston 1993, p.8.

[17] Alfred Schutz: „Edmund Husserl's *Ideas*, Volume II". In: Alfred Schutz: *Collected Papers III. Studies in Phenomenological Philosophy*, ed. Ilse Schutz. Den Haag 1966, p. 32.

[18] See ibid. p. 199 ff.

[19] Edmund Husserl: Ideas Pertaining to a Pure Phenomenology and to a Phenomenological Philosophy. Second Book. Studies in the Phenomenology of Constitution, transl. Richard Rojcewicz, André Schuwer. Dordrecht 1989, p. 270.

have the power to,' 'I am capable of'? In experience, the 'I can' is distinct from the 'I cannot' according to their phenomenological characters. There is a resistanceless doing of things, i.e., a consciousness of an ability that meets no resistance, and there is a doing as an overcoming of resistance, a doing that has its 'against which,' and a corresponding consciousness of an ability to overcome the resistance".[20]

In the following six sections, this paper sketches a phenomenological analysis of practicing in the sense of – to put it in a Husserlian expression – the specific "mode of the 'I do'"[21] that provides access to intentionally extending, modifying or restructuring the "horizon of ability".

First contributions to this analysis may be obtained by an indirect approach, for instance via the distinction of "practicing" ("Üben") and "training" ("Einüben"). Both modes of activity show similarities, as they share a common starting-point: practicing as well as training commence at a point, where something cannot just naturally be done, but turns out to be a problem. However, practicing and training differ with regard to their intention and motivation: The objective of training essentially is to achieve assertiveness by developing the most perfect control and self-control. Where training seeks for achievements, practicing is done for the sake of its effects on the subject of practice and implies an attitude that is characterized by the simultaneity of intentional activity and, to put it in a word of Ludwig Landgrebe, "perceiving without intention to control".

References

Aristoxenus T.. (1990), *Melik und Rhythmik des Classischen Hellenentums*. 2 vols. Trans._R. Westphal. Leipzig, 1883; *Elementa Rhythmica: The Fragment of Book II and the Additional Evidence for Aristoxenean Rhythmic Theory*. Ed. and Trans. Lionel Pearson. Oxford: Clarendon Press.

Barber, M. D. (2004), *The Participating Citizen: A Biography of Alfred Schutz*. Albany: State University of New York Press.

Berger, P.L, and Luckmann T.(1967), *The Social Construction of Reality. A Treatise in the Sociology of Knowledge*. London: Allen Lane, The Penguin Press..

Bergson, H. (1970), *Essai sur les données immédiates de la conscience* [1888]. Paris: Presses Universitaires de France,

[20] Ibid. p. 270 f.

[21] Edmund Husserl: *The Crisis of European Sciences and Transcendental Phenomenology*, transl. David Carr, Evanston 1970, p. 106.

—. (1910), *Time and Free Will: An Essay on the Immediate Data of Consciousness.* Trans. F. L. Pogson. London: George Allen and Unwin.

—. *Durée et simultanéité: à propos de la théorie d'Einstein.* 2nd enl. ed. Paris: Félix Alcan, 1923; *Duration and Simultaneity: Bergson and the Einsteinian Universe.* Ed. Robin Durie. Trans. Leon Jacobson, with supplementary material trans. Mark Lewis. Manchester: Clinamen Press, 1999.

Fink, E. (1980), "Operative Begriffe in Husserls Phänomenologie." *Zeitschrift für philosophische Forschung* 11 (1957), 321–37; "Operative Concepts in hs Phenomenology." Trans. William McKenna. In *Apriori and World: European Contributions to Husserlian Phenomenology.* Ed. William McKenna et al. The Hague: Martinus Nijhoff, 1981, 56–70.

Georgiades, T. G. (1955), *Greek Music, Verse, and Dance.* New York: Merlin Press.

Husserl, E. (1975),. *Experience and Judgment Investigations in a Genealogy of Logic.* Rev. and edited by Ludwig Landgrebe, transl. by James S. Churchill and Karl Ameriks. Evanston.

Kersten, F., I. (1976), "Preface" to Alfred Schutz, "Fragments on the Phenomenology of Music." Ed. Fred Kersten. *Music and Man* 2.

Schopenhauer, A. (1859), *Die Welt als Wille und Vorstellung.* Vol.1. 3rd enl. ed. Leipzig: Brockhaus.

Schutz, A.. (1956), Papers. General Collection, Beinecke Rare Book and Manuscript Library, Yale University.

—. (1951),"Making Music Together: A Study in Social Relationship." *Social Research* 18 (1951), 76–97; rpt. in his Collected Paper II. Ed. Arvid Brodersen. (1956), "The Hague: Martinus Nijhoff," 1964, 159–78.

—. "Mozart and the Philosophers." (1964), *Social Research* 23 (1956), 219–42; rpt. in his Collected Papers II. Ed. Arvid Broderson. The Hague: Martinus Nijhoff, .

—. (1962),*Collected Papers I. The Problem of Social Reality.* Ed. Maurice Natanson. The Hague: Martinus Nijhoff.

—. (1964), *Collected Papers II. Studies in Social Theory.* Ed. Arvid Brodersen, The Hague: Martinus Nijhoff.

—. (1966), *Collected Papers III. Studies in Phenomenological Philosophy.* Ed. Ilse Schutz. The Hague: Martinus Nijhoff, 1966.

—. (1944),"Fragments on the Phenomenology of Music." [1944] Ed. Fred Kersten. *Music and Man* 2 (1976), 5–71; rpt. in *In Search of Musical Method.* Ed. F. Joseph Smith. London: Gordon & Breach, 1976, 5–72; rpt. as "Fragments Toward a Phenomenology of Music." In his

Collected Papers IV. Ed. Helmut Wagner and George Psathas with Fred Kersten. Dordrecht: Kluwer Academic Publishers, 1996, 243–75.

—. (1982), *Theorie der Lebensformen. Frühe Manuskripte aus der .Bergson-Periode*. Ed. Ilja Srubar. Frankfurt am Main: Suhrkamp, *Life Forms and Meaning Structure*. Trans. Helmut R. Wagner. London: Routledge & Kegan Paul..

—. "A 'Construction of Sociological Aspect of Literature.'" Ed. Lester Embree. In Alfred Schutz (1988), *"Sociological Aspect of Literature": Construction and Complementary Essays*. Ed. Lester Embree. Dordrecht: Kluwer Academic Publishers..

—. (2003), *Alfred Schütz Werkausgabe*. Ed. Richard Grathoff, Hans-Georg Soeffner, and Ilja Srubar. 9 vols. Konstanz: Universitätsverlag Konstanz, 2003ff.

Schutz, A., and Luckmann Th.. (1975), *Strukturen der Lebenswelt*. Vol.1. *The Structures of the Lifeworld*. Vol.1. Trans. Richard M. Zaner and H. Tristram Engelhardt. Evanston, IL: Northwestern University Press,..

—. (1984), *Strukturen der Lebenswelt*. Vol.2. Frankfurt am Main: Suhrkamp.; *The Structures of the Lifeworld*. Vol.2. Trans. Richard M. Zaner and David J. Parent (1989),. Evanston, IL: Northwestern University Press.

Schutz, A., and Gurwitsch A.. (1985), *Briefwechsel, 1939–1959*. Ed. Richard Grathoff. Munich: Wilhelm Fink, 1985; *Philosophers in Exile: The Correspondence of Alfred Schutz and Aron Gurwitsch, 1939–1959*. Trans. J. Claude Evans. Bloomington: Indiana University Press, 1989.

Schutz, A. and Eric Voegelin E..(2004), *Eine Freundschaft, die ein Leben ausgehalten hat. Briefwechsel 1938–1959*. Ed. Gerhard Wagner and Gilbert Weiss. Konstanz: Universitätsverlag Konstanz.

Stascheit, A. G. (2003),. "Doppelgriffe. Phänomenologische Motive vom Gesichtspunkt der Musikarbeit." In *Phänomenologie und soziale Wirklichkeit: Entwicklungen und Arbeitsweisen*. Ed. Ilja Srubar and Steven Vaitkus. Opladen: Leske & Budrich, 2003, 215–38.

—. (2007), "Polyphonie und Kontrapunkt im Kontext einer Phänomenologie der sozialen Welt." In *Modell Zauberflöte: Der Kredit des Möglichen. Kulturgeschichtliche Spiegelungen erfundener Wahrheiten*. Ed. Mathias Mayer. Hildesheim: Georg Olms., 97–113.

—. (2008), "Béla Bartók und die Milman Parry Collection der Harvard University Volksliedforschung im amerikanischen Exil." In *Komponisten im Exil: 16 Künstlerschicksale des 20. Jahrhunderts*. Ed. Ferdinand Zehentreiter. Leipzig: Henschel, 2008, 155–64.

Ungeheuer, G. (1990),. “Nietzsche über Sprache und Sprechen, über Wahrheit und Traum.” In *Kommunikationstheoretische Schriften II. Symbolische Erkenntnis und Kommunikation*. Ed. H. Walter Schmitz. Aachen: Alano.

Winternitz, E. (1970), “The Role of Music in Leonardo’s Paragone.” In *Phenomenology and Social Reality: Essays in Memory of Alfred Schutz*. Ed. Maurice Natanson. The Hague: Martinus Nijhoff 270–96.

Zeeb, Tanja. (2004), “Die Wirkung Nietzsches auf die deutsche Gesellschaft der Jahrhundertwende im Spiegel der Tagespresse.” *Nietzsche-Studien* 33 :278–305

CHAPTER TWO

APPEARANCE AND GENDER: THE HISTORIES OF THE BODY IN THE CONTEXT OF CLOTHING

DAMAYANTHIE ELUWAWALAGE

Abstract

This chapter will examine gender in attire as a dichotomy of social agenda and cultural phenomenon; how and in what sense fashion and clothing are sociological phenomena, with different theoretical approaches to each standpoint. The study of clothing is interdisciplinary as scarcely any single theory or field of knowledge is able to explain the entire concept. Gender and related issues are too complex to understand in a solitary discipline. Biological gender explanation for instance, basically enhanced our comprehension of the sexes, nevertheless it scarcely explained the relations and evaluations of the sexes in contemporary forms of life. Therefore, the sexual asymmetry of society requires to be examined in the socio-cultural context, as a part of an anthropological scholarship. The dominant ideological beliefs, such as, the male public sphere and the female domestic sphere; masculine intellect and feminine emotionalism; male activism and female passivism; and masculine physical and mental strength and feminine physical and mental weaknesses, must have prompted the prominent gender differentiation of male dominance and female submission throughout the centuries. The questions: Why did men and women dress distinctively? Why did women's dresses conceal their bodies? And, Why were women's dresses decorative?, were directly associated with gender issues such as social and cultural differentiations of gender, and different social roles and behaviours expected from each gender in society.

Introduction

The body is our general medium for having a world. Sometimes it is restricted to the actions necessary for the conservation of life, and accordingly it posits around us a biological world; at other times, elaborating upon these primary actions and moving from their literal to a figurative meaning, it manifests through them a core of new significance: this is true of motor habits [sic] such as dancing. Sometimes, finally, the meaning aimed at cannot be achieved by the body's natural means; it must then build itself an instrument, and it projects thereby around itself a cultural world.[1]

> Know, first, who you are; and then adorn yourself accordingly.
> —Epictetus, Discourses 3.1

Attire, throughout history, performed a vital function as a form of non-verbal communication. Clothing and its decorative version called 'finery' is an effective symbol which signifies the exhibitors' or wearer's societal standing, occupation and gender. 'Gender' is a widely applicable subject, which could be analysed in many different facets. To examine only within one discipline is to isolate it from the other disciplines, of which it is a part. Therefore to comprehend 'gender' appropriately, this study endeavoured to weave the work of sociological, psychological and anthropological work into the historical material.

How does clothing express gender? This paper will investigate gender in attire as a dichotomy of 1) social agenda and 2) cultural phenomenon; how and in what sense fashion and clothing are sociological phenomena, with different theoretical approaches to each standpoint. The study of clothing is interdisciplinary as scarcely any single theory or field of knowledge is able to explain the entire concept. Gender and related issues are too complex to understand in a solitary discipline. Biological gender explanation for instance, basically enhanced our comprehension of the sexes, nevertheless it scarcely explained the relations and evaluations of the sexes in contemporary forms of life. Therefore, the sexual asymmetry of society requires to be examined in the socio-cultural context, as a part of an anthropological scholarship.

The term 'culture' derived from the Latin word 'colere', meaning to inhabit, to cultivate, to protect and to honour with worship. The word 'cultura', meaning the idea of cultivation, was developed from the former. The utilisation of the English term 'culture' was initiated in the early fifteenth century.[2] According to the *Oxford English Dictionary,* gender is a grammatical classification corresponding roughly with the two sexes and

---

[1] Merleau-Ponty, M. (1974 [1945]), *Phenomenology of Perception,* London: Routledge & Kegan Paul (pp. 146).

[2] Barnard, M. (1996). *Fashion as Communication* (pp. 32). London, New York: Routledge.

sexlessness. Nevertheless, as Ann Oakley defines, the term 'gender' is more than an identification with a biological sex, it is a cultural concept which refers to the social classification into 'masculine' and 'feminine'.[3]

The questions: Why did men and women dress distinctively? Why did women's dresses conceal their bodies? and, Why were women's dresses decorative?, were directly associated with gender issues such as social and cultural differentiations of gender, and different social roles and behaviours expected from each gender in society. Writers, such as, Dentith, Barnes, Eicher and Davies, analysed gender in clothing as a cultural phenomenon, because clothing is a matter of custom and a form of communication.[4]

The dominant ideological beliefs, such as, the male public sphere and the female domestic sphere; masculine intellect and feminine emotionalism; male activism and female passivism; and masculine physical and mental strength and feminine physical and mental weaknesses, must have prompted the prominent gender differentiation of male dominance and female submission throughout the centuries. Any breach of the 'great law of subordination', especially between husband and wife, was regarded as a sort of petty treason.[5] The phenomena of male dominance could be analysed in many theoretical facets, such as sociological, psychological, economical and biological. According to these arguments, patriarchal male dominance and female dependence was caused by dichotomal rationales, such as their respective biological roles related sexual differences and social inequalities of gender.

> "The position of women rests, as [does] everything in our complex society, on an economic base."
> —Eleanor Marx

According to the economic argument in relation to male dominance, as some writers argue, women's subordination and male domination is economically established, determined and developed. "Labor is the source of all wealth", wrote Karl Marx. In a Capitalist society, where value was measured in terms of monetary worth of commodities and market production, women in domestic labour, traditionally considered as valueless

---

[3] Oakley, A. (1972), *Sex, Gender and Society* (pp. 16). South Melbourne: Sun Books.

[4] Dentith, S. (1998). *Society and Cultural Forms in Nineteenth Century England* (pp. 128-129). London: Macmillan. Barns, R. & Eicher, J. B. (eds.) (1992), *Dress and Gender, Making and Meanin.* New York: Oxford University Press. Davies, F. (1992). *Fashion, Culture, and Identity.* Chicago: The University of Chicago Press.

[5] Perkins, H. (1969). *The Origin of the Modern English Society 1780-1880* (pp. 37). London: Routledge and Kegan Paul.

and non-waged labor, were not regarded as part of the market economy. Therefore, in the economic sense, women were subjected to submission to men, who were the primary contributors of accumulation of capital.[6] Marxist feminism perceives capitalism as the basis of women's oppression, social gender differentiation and segregation. It attempts to analyse the relationship between the subordination of women and the organisation of various capitalist modes of production. Engels in *The Origin of the Family* argued, the emergence and ascendance of the class society; the introduction of commodity production; women's status as privately owned property and women's scarcity of productivity in the economic sense; were the primary rationales for women's inferiority in the social sphere.[7] As Beauvoir explained, this social arrangement befitted the economic, social and political interests of men.[8]

In terms of attire in the nineteenth century, as a symbol of their economic standing, fashionable women wore decorative and expensive dresses designed to restrict movement, which made them incapable of working ("The role of the lady of the house is to exhibit her master's wealth and his pecuniary strength to remove her from the workforce."[9]), to justify their male counterparts' status and wealth. "The unchanged splendour of women's toilettes and the opulence of their flesh signified the social status and monetary power of their fathers, husbands or lovers."[10] Women's fashionable finery in the nineteenth century was regarded as an indication of women's social subordination, as women adorned in fashion were observed as show-pieces of their families and were symbolic of their role as a 'man's chattel'.[11]

According to the biological argument in relation to male domination, there are many theories of gender inequality established on biological aspects of sexes. According to Freud, the anatomical distinction between

---

[6] Marx, K. (1930). *Capital. Vol. 1* (pp. 3-193). London: Dent. Evans, M. (1982). *The Woman Question: Readings on the Subordination of Women* (pp. 195). London: Fontana Paperbacks.

[7] Engels, F. (1986). *The Origin of the Family, Private Property and the State* (pp. 102-106 & 195-217). Harmondsworth: Penguin.

[8] Beauvior, S. De. (1972). *The Second Sex* (pp. 159). London: Jonathan Cape.

[9] Veblen, T. (1899). *The Theory of the Leisure Class*, London: Unwin Books.

[10] Perrot, P. (1994), *Fashioning the Bourgeoisie: A History of Clothing in the Nineteenth Century* (pp. 35). Princeton: Princeton University Press.

[11] Veblen, T. (1899). *The Theory of the Leisure Class*, London: Unwin Books. Crane, D. (1933). *Fashion and its Social Agendas: Class, Gender, and Identity in Clothing*. Chicago & London: University of Chicago Press. Davies, F. (1992), *Fashion, Culture, and Identity* (pp. 41). Chicago: University of Chicago Press. Lurie, A.(1981). *The Language of Clothes* (pp. 220). London: William Heinemann.

the sexes resulted in clear physical differences. These differences may create distinctive psychological characteristics in the two sexes, which determine male and female behaviour.[12] Along similar lines, Hutt argues that physical and psychological abilities are characteristically different in men and women. The male is physically stronger, less resilient, more independent, adventurous, ambitious, aggressive, competitive, and more likely to construe the world in terms of objects, ideas and theories. The female is more nurturing, affiliative, matures more rapidly physically and psychologically. According to Hutt's argument, psychological and behavioural differences between sexes were based on their biological dissimilarities.[13] Contrary to the above argument, Maccoby and Jackline claimed, the sexes are psychologically much alike in many aspects. The greater majority of men and women are very similar in terms of their intelligence, spatial and verbal aptitude and aggression. Sexes are very similar in relation to their innate characteristics, yet men and women are socially subject to different kinds of expectations.[14]

In relation to gender inequality in behaviour, and male dominance in particular, some socio-biologists assume a biological and genetical base in male authoritative behaviour. The male command is an outcome of genetically programmed and evolved male societal behaviour, which is primarily derived from the species' hunting heritage. In hunter-gatherer societies, a sexual division of labour would be highly functional, as in most primitive societies and modern industrial societies, man's hazardous surviving mechanisms developed into male autocratic demeanour.[15] In concurring with the above theory, anthropologists Parker and Parker asserted, primitive men acquired their power and dominance as a reward for their risky activities such as hunting and warfare, which has been reinforced throughout civilisation.[16] Contrary to the above argument that, 'male dominance and female dependence are part of our genetic primate heritage', Coontz and Henderson stated male domineering behaviour is not determined by our genes.

Perhaps clothing was designed to focus on these physical and biological

---

[12] Freud, S. (1977). *On Sexuality* (pp. 331-343). Harmondsworth: Penguin.

[13] Hutt, C. (1972). *Males and Females* (pp. 132-133). Harmondsworth: Penguin.

[14] Maccoby, E. & Jacklin, C. N. (1974). *The Psychology of Sex Differences* (pp. 373). California: Stanford University Press.

[15] Wilson, E. O. (1980). *Sociobiology* (pp. 271-299). Cambridge: Harvard University Press. Rosaldo, M. S. & Lamphere, L. (eds.) (1974). *Women, Culture and Society,* Stanford: Stanford University Press.

[16] Parker, S. & Parker, H. (1979). *American Anthropologist Vol. 81, No. 1-2* (pp. 289-303). WA: American Anthropological Association.

inequalities of men and women. According to James Laver, "Women's clothes are governed and denoted by the erotic or seduction principle, by contrast men's clothes are directed by the hierarchical principle of class or status."[17] In the context of Victorian dress, the biological and physical gender differentiation has been described as: "men were serious (they wore dark clothes and little ornamentation), women were frivolous (they wore light pastel colours, ribbons, lace and bows); men were active (their clothing allowed them movement), women were inactive (their clothes inhibited movement); men were strong (their clothes emphasised their broad chest and shoulders), women were delicate (their clothing accentuated tiny waists, sloppy shoulders and a softly rounded silhouette); men were aggressive (their clothing had sharp definitive lines and a clearly defined silhouette), women were submissive (their silhouette was indefinite, their clothing constricting).[18] As these descriptions suggest, the genders' biological distinctiveness was uniquely illustrated in their attire.

Throughout the centuries, gender inequality and segregation has continued to be passed from one generation to another, primarily because of the assistance from cultural symbol systems such as religion and language. Language, for instance, was a stereotypical cultural phenomenon which denigrated women as inferior to men and second-rate in social stratification. Religion, on the other hand, in the cultural perspective, not only recognised female submission, but also agreed with it via its beliefs and principles.

Cultural Structure, Symbolism and Gender Asymmetry of Language

The habits of the body 'do not just vary with individuals and their imitations, they vary especially between societies, educations, properties and fashions, prestiges. In them we should see the techniques and work of collective and individual practical reason rather than ...merely the soul and its repetitive facilities.[19] Every society has been established along gender lines in all facets. Expressions, communications and verbalisations, for instance, convey the oppositional message of masculinity and femininity

---

[17] Laver, J. (1948). *Taste and Fashion* (pp. 199-202). London: George G. Harrap. Laver, J. (1988). *Costume and Fashion: A Concise History* (pp. 268-269, 272-273). London: Thames and Hudson.

[18] Roberts, H. (1977). The Exquisite Slave: The Role of Clothes in the Making of the Victorian Woman. *Signs*, 2(3), 554-569.

[19] Mauss, M., (1973 [1936]) Techniques of the Body, *Economy and Society,* 2:75.

in everyday life. 'Words and the ideas that they signify, are elements in the shared cultural heritage of social groups'.[20] Language as a communication system embodies gender inequality and symbolises male dominance. Some languages such as Japanese are used differently by male and female speakers. In relation to the English language, females and female versions are often described in terms of males. For instance the female version of: actor is actress, doctor - lady doctor, lawyer - woman lawyer etc. It also illustrated the demarcation and segregation of women in relation to gender. In addressing a female, the terms Mrs. or Miss indicate a woman's marital status, which is an automatic sign of her personal identity. It also suggested the inferior social status of women as 'chattels of men'. In contrast, when addressing a male, Mr. has no signal of his personal status, which suggested male independence and female dependence.

There are as many as one thousand English words and phrases which describe women in sexually derogatory ways; over five hundred synonyms for prostitute, whereas there are only about sixty five for the masculine term of whoremonger.[21] Historically, women were to be linguistically educated for the dichotomical purposes of: to fulfil the role of the mother, passing on pure language to the child, and to act as companion to the male in the public sphere.[22] The language also indicates gender differentiation in the manner in which the male and female learn to speak, as often female children converse in more mellow tones than male children. It reflects the cultural domination of the male gender and female submission in cultural aspects such as communication.

The similar phenomenon is observable in relation to clothing related words. According to Norbert Elias' Symbol Theory, language symbolically represented the social function and objects.[23] Particular vocabulary for gendered clothing in the nineteenth century specifically expressed the tone of the particular gender. The word 'blouse' for instance, automatically referred to women, whereas 'the suit' alluded to men. The words such as 'corset', 'petticoat', 'dress' and 'skirt' were firmly associated with femininity, and 'knickerbockers', 'shirt' and 'trouser' expressed masculinity, because they were strictly reserved for that particular gender. The idea of

---

[20] Scott, J. (1995). *Sociological Theory: Contemporary Debates* (pp. 102). Brookfields: Aldershot.

[21] Stochhard, J. & Johnson, M. M. (1992). *Sex and Gender in Society* (pp. 4-6). NJ: Prentice Hall. Boudreau, F. A., Sennott, R. S. & Wilson, M. (1986). *Sex Roles and Social Patterns* (pp. 52-56). NY: Piaeger.

[22] Crowley, T. (1996). *Language in History: Theories and Texts* (pp. 90). NY: Routledge.

[23] Elias, N. (1991). *The Symbol Theory.* London: Sage.

gendered fabrics also evolved through the above phenomena.[24] The delicacy and the lightness of silks, satins and laces were primarily regarded as feminine and wool was regarded as least feminine, while fabrics such as linen and cotton had a neutral image.

Culture and Clothing: 'The meaning of clothing is culturally defined'

The body is not an object to be studied in relation to culture, but is to be considered as the subject of culture, or n other words as the existential ground of culture.[25] Historically, cultural patterns connected with etiquette are morally associated with attire. As these cultural patterns are distinctive from society to society, each society classifies its own clothing behaviour in accordance with its moral and ethical standards. Clothing, as an aspect of culture, is an important element of creating masculinity and femininity, and the way in which bodies are gendered in society.

"An article of clothing has no inherent meaning. Trousers do not have the idea of masculinity built into them, nor does a skirt automatically signify that the wearer is either female or feminine. The meaning of clothing is culturally defined."[26] The narrative of the trouser provided ample evidence to strengthen the above statement. The modern version of the man's suit, consisting of a coat, waistcoat and trousers, is considered to have evolved by the last quarter of the seventeenth century. Although it has been reformed and transformed over the centuries, the fundamental form remains intact and un-mutated.

Trousers were apparently invented in Persia in the late pre-historic period, and then adapted by northern Europeans and central Asians. In Medieval Europe, male aristocracy experienced a series of trouser variations. During the fifteenth-century, elaborate and very short robes were worn over tight stockings. During the sixteenth-century, these were converted into a doublet and puffy bloomers, which were remodelled into knee-breeches in the seventeenth-century. During the early nineteenth-century, knee-breeches were replaced by trousers.[27] The phenomenon of

---

[24] Sahlins, M. (1976). *Culture and Practical Reasons* (pp. 182-183). Chicago & London: The University of Chicago Press.

[25] Cordas, T. J. (1990)Embodiment as Paradigm for Anthropology, *Ethos,* 18 (1), (pp. 5).

[26] Kidwell, C & Steel, V. (eds.) (1989). *Men and Women: Dressing the Par* (pp. 6). WA: Smithsonian Institute Press.

[27] Byrde, P. (1979). *The Male Image: Men's Fashion in Britain 1300-1970* (pp. 26-96). London: B.T. Batsford. Kidwell, C & Steel, V. (eds.) (1989). *Men and*

the disappearance of knee-breeches and the emergence of trousers could be explained using the Social Role Theory, which discussed changes in appearance as changes in the social role and identity in any given society. It could be argued that the western male dress reformation in the early nineteenth century reflected the transformation of social roles in contemporary European society. It also indicated the changes in social identities, primarily because of the social and economical modification.

A trouser, in Western cultures, especially in the nineteenth-century, was associated with masculinity and considered 'improper' for women to wear until the twentieth century, whilst historically a 'Turkish trouser' was a female accessory. In Imperial China, a trouser was demarcated as inferior clothing and was restricted to working-classes.[28] It suggested an article of attire was exhibited and represented distinctly in different cultures; either as gender differentiation or as social class differentiation. Historically, a skirt is familiar and customary with the Western femininity. In contrast, a kilt; a Scotsman's skirt, associated Highland dress and the tartan, are among the most powerful of all symbols of Scotland. According to Scottish belief, wearing of the kilt confers extra stature on its wearer. "A man in a kilt is a man and a half."[29] The gender in clothing, either as masculinity or femininity, is characterised differently in various cultures and is ultimately determined by the society.

The anecdote of the Western hat strengthens the argument that gender in clothing is a cultural phenomenon. Formal head-wear, especially hats in the pre-nineteenth-century, expressed a social hierarchy and authority. It also demonstrated the gender hegemony. As McDowell explains, despite its needlessness, the hat presents a powerful prerogative apparatus. Hats proclaimed man's status, attitudes and beliefs, and woman's class, breeding and even matrimonial state.[30]

It is significant to notice the application and practice of hat wearing and the demarcation of gender and social status in different social institutions. The utilisation of hats in churches for instance, was controlled and authoritated by gender superiority and power. On entering a church, men removed their hats, whilst women's head remained covered. "Every

---

Women: Dressing the Par (pp. 14). WA: Smithsonian Institute Press. Boucher, F. (1966). *A History of Costume in the West,* London: Thames and Hudson.

[28] Kidwell, C & Steel, V. (eds.) (1989). *Men and Women: Dressing the Par* (pp. 6). WA: Smithsonian Institute Press.

[29] http://members.aol.com/-ht-a/sconemac/kilt.html. N.A.MacCorkill, 22/05/2002

[30] McDowell, C. (1992). *Hats, Status, Style and Glamour* (pp. 97). London: Thames and Hudson. Crane, D. (2000). *Fashion and Its Social Agendas* (pp. 82-83). Chicago: The University of Chicago Press.

man praying or prophesying, having his head covered, dishonours his head [who is Christ]. But every woman who prays or prophesies with her head uncovered dishonours her head [who is her husband] ...."[31] At the other end of the spectrum, the soldier retains his head-dress even in a Christian church. According to Flugel, similar to woman's dress and her head-wear, a military uniform is a whole, and the removal of any part of it gives the wearer an undressed, and it could be viewed as immodest appearance. It is connected with the different forms of respect which social convention prescribes for the two sexes. As he explains, the assumption is that clothes are a sign of disrespect in man, while nakedness is a sign of disrespect in woman.[32] Therefore, the meaning and interpretation of the hatted-head fluctuated and varied in relation to the wearer's gender. The woman's hat exhibited her fashionability, rather than coded signs of class and status. Significantly, the expression of the hatted male was contrastive with the declaration of hatted females, a cultural-based ideology, which dominated the pre 1900 societies.

Culture, Gender Symbolism and Color

Humans possess an individual body being, constituted by drives, appetites and sensory impressions which are 'necessarily egoistic', yet also possess the capacity to transcend themselves and develop on the basis of social categories and emotions.[33] Historically, the color-coded attire demarcated differences, for example, differences in social position and differences in gender etc. As explained in previous chapters, color-coded attire designated the social station of the wearer in the antiquity. As Geoffrey Chaucer's *The Canterbury Tales* explains, distinctive colors are worn by different classes in medieval England, as knights wore embroidered garments, yeomen in green, merchants in motley and doctors in purple and light blue.[34]

Is pink naturally a feminine colour? By the twentieth-century Western society, certain colors were firmly associated with the specific gender; as pink for girls, and blue for boys. The color-coded gender phenomenon was a novelty at the turn of the twentieth-century and became widely practiced after World War II. Prior to that era, there were no distinctions or

---

[31] *Corinthians* 11: 3-5, 7-12

[32] Flugel, J. C. (1930). *The Psychology of Clothes* (pp. 103-104). London: Hogarth Press.

[33] Durkheim, E., (1995 [1912]). *The Elementary Forms Religious Life,* (pp. 151) New York: Free Press.

[34] Chaucer, G. (1980). *The Canterbury Tales.* London: Arnold.

differences in the dress of younger children, in relation to their gender. Children under five years of age wore similar clothing, often short frocks with low necklines and short sleeves. Sometimes pleating of the skirts distinguished the gender.[35] Pink for men nowadays is regarded as improper and unsuitable. Nevertheless, in the nineteenth century, pink suits were highly acceptable for gentlemen's attire.[36] As Chenoune explains, Frenchmen in the eighteenth-century wore colors with an infinite range of hues like pink, plum, white, cream, blue, yellow, puce and sea green,[37] which nowadays is unquestionably associated with femininity. Nineteenth-century European male austerity, primarily caused by the usage of sombre colors such as black, white and grey, has been described as an element for dignity, control and morality.[38]

Amongst the homogeneous female gender in the nineteenth century, there were many color-associated distinctions, as certain colours were reserved for the unmarried, the single or the aged. Two luminous colours in one dress was 'vulgar' and yellow was generally regarded as unladylike. (Yellow was the colour utilised for most correctional attire.) Orange was prohibited for the unmarried.[39] As the above statements suggested the gender differentiation through color, which could be regarded as a cultural phenomenon, was practiced diversely in different cultures and epochs.

In analysing nineteenth-century attire, as many historians argued, fashionable dressing in the nineteenth century was more gender-differentiated than in previous centuries.[40] The significant difference between eighteenth and nineteenth century attire was, contrary to the eighteenth century, the noticeable austerity of nineteenth-century male dress and the extravagance of female dress. Masculine austerity, as Flugel described, emphasised a shift in class relations. As a result of that shift, male attire became more uniform, homogenous and integrated (amongst

---

[35] Cunnington, P. & Buck, A. M. (1965). *Children's Costume in England from the 14th to the 19th Century* (pp. 199). London: Adam and Charles Black.

[36] Kidwell, C. & Steele, V. (eds.) (1989). *Men and Women: Dressing the Part* (pp. 6 & 22). WA: Smithsonian Institute Press.

[37] Chenoune, F. (1993). *A History of Men's fashion* (pp. 10). NY: Flamarian.

[38] Perrot, P. (1994). *Fashioning the Bourgeoisie: A History of Clothing in the Nineteenth Century* (pp. 32). Princeton: Princeton University Press.

[39] Cunnington, C. W. (1948). *The Art of English Costume* (pp. 135). London: Collins Clear Type Press.

[40] Byrde, P. (1979). *The Male Image Men's Fashion in Britain 1300-1970* (pp. 11). London: B.T. Batsford. Davies, F. (1992). *Fashion, Culture, and Identity* (pp. 33-39). Chicago: The University of Chicago Press. Flugel, J. C. (1930). *The Psychology of Clothes* (pp. 103-121). London: Hogarth Press. Laver, J. (1937). *Taste and Fashion*, London: George G. Harrap.

the classes), than its traditional hierarchal approach, which he calls 'The Great Masculine Renunciation'.[41]

This distinct divergence in relation to finery between the pre-nineteenth centuries and the nineteenth-century was argued and examined by many writers.[42] The rationales suggested were: the descendancy of European feudal aristocracy, the ascendancy of mercantilism and bourgeoisie-class, the collapse of the class system, and the transformation of social structure; primarily because of the Industrial Revolution.

The precise reformation in relation to clothing in the Victorian era was the austere appearance of male attire. Contrary to previous practice, the colorfully ornated and embellished male attire evolved into a tedious and sombre appearance. The fragmentary mutation of male attire in Britain occurred gradually as swords were transformed into walking-sticks; the tax on flour for hair powder during the war accelerated its decline; the drive for clean linen made ruffles and laces prohibitive and breeches were replaced by trousers.[43]

During the mid nineteenth-century, in English and French cities, gloomy and dark colors, especially black became standard.[44] Dark colours began appearing during the third quarter of the eighteenth-century and were gradually accepted and adopted.[45] Men in every social stratum during the nineteenth-century, regardless of their societal standing and class; the aristocracy, the gentry, the bourgeoisie and working-classes, attired in black on Sundays. By the early twentieth-century, men's suits had settled into a convention of black, dark grey or dark blue for formal day wear and patterned clothes or tweeds of brown or greenish hues for informal or country clothes.

Male austerity in the nineteenth-century requires explanation. Although male clothing designs transformed into simplicity, the variety of styles and fashions relatively flourished. Styles and variations of men's coats, jackets

---

[41] Flugel, J. C. (1930). *The Psychology of Clothes* (pp. 114-119). London: Hogarth Press.

[42] Davies, F. (1992). *Fashion, Culture, and Identity* (pp. 37-39). Chicago: The University of Chicago Press. Kidwell, C & Steel, V. (eds.) (1989). *Men and Women: Dressing the Part* (pp. 14-25). WA: Smithsonian Institute Press.

[43] Davidoff, L. & Hall, C. (1987). *Family Fortunes: Men and Women of the English Middle Class 1780-1850* (pp. 411-412). London: Hutchinson.

[44] De Marly, D. (1985). *Fashion for Men: An Illustrated History* (pp. 94-98). London: B. T. Batsford. Byrde, P. (1979). *The Male Image: Men's Fashion in Britain 1300-1970* (pp. 72). London: B.T. Batsford.

[45] Chenoune, F. (1993). *A History of Men's Fashion* (pp. 17). NY: Flammarian.

and neck-wear increased. Men's coats for instance varied, in terms of length, lapel size and function, as numerous developments such as tail coats, frock coats, and morning coats were utilised.

Varieties and specific design attire, for example, coats/jackets, for leisure and formal wear indicated the entire male population was scarcely dressed uniformly, rather they attired fashionably. "Elegant dress serves its purpose of elegance not only in that it is expensive, but also because it is the insignia of leisure. It [is] not only shows that the wearer is able to consume a relatively large value, but it argues at the same time that he consumes without producing."[46] At the other end of the spectrum, as Veblen designates, these fashionable masculine fineries especially reserved for the upper-classes, indicated the wearers' status and social position. This transformation of adopting simpler attire, also portrayed the new dimension and identity of masculinity, characterised by activity, fitness and strength. Unlike previous societies in which leisured men wore ruffs, wigs, jewellery and furs[47], in the new industrial society masculinity was identified by industriousness, which required less complicated attire. "In fact, bourgeoisie men displayed their glory or power in an oblique way, not through what they were, but through what they owned."[48]

Costume writers forwarded different perspectives and reasons in relation to this significant evolvement of male attire. Berger wrote: "Men act, women appear."[49] According to Perrot's theory, the switch of elegance from male attire to female's illustrated the novel way to display affluence. The absence of sumptuousness from male attire also signified the transformed social responsibility.[50] Perrot's argument was confirmed by Laver describing Victorian man as a dowdy bird, displaying wealth and status vicariously through the exotic plumage of his women, children and servants.[51] In concurring with the above assertion, as Flugel analyses, the sociability of men and their greater social participation influenced the uniformity and lesser decoration of their clothes.[52] The nineteenth-century

[46] Veblen, T. (1899). *The Theory of the Leisure Class* (pp. 121). London: Unwin.

[47] Boucher, F. (1966). *A History of Costume in the West,* London: Thames and Hudson.

[48] Perrot, P. (1994). *Fashioning the Bourgeoisie: A History of Clothing in the Nineteenth Century* (pp. 34-35). Princeton: Princeton University Press.

[49] Berger, J. (1972). *Ways of Seeing.* NY: Viking Press.

[50] Perrot, P. (1994). *Fashioning the Bourgeoisie: A History of Clothing in the Nineteenth Century* (pp. 34-35). Princeton: Princeton University Press.

[51] Laver, J. (1929). *English Costumes in the Nineteenth Century.* London: Adam and Charles Black.

[52] Flugel, J. C. (1930). *The Psychology of Clothes* (pp. 116). London: Hogarth

economic historian Thorstein Veblen proposed a divergent thesis that women's elegant and elaborate dressing (in the nineteenth century) was a requirement demanded by men who commanded the social structure as an element of a vicarious consumption, "At the stage of economic development at which the women were still in the full sense the property of the men, the performance of conspicuous leisure and consumption came to be part of the services required by them."[53]

In the economic perspective, historically, the triumph of Capitalism, a system in which trade and industry are controlled by private owners, was one of the contributing factors for mutating the masculine ideal. Man, who produced labour and capital in the rapidly transformed industrial capitalist society, required more practical clothing suitable for his public and functional obligations, which resulted in the gloominess of male attire.

By contrast, Chenoune, in his *A History of Men's Fashion,* claims a divergent ideology as; "a masculine image of reserve, austerity and gravity [was] stemmed from the ethical Puritan conceptions, which distincted from the ethic of magnificence attributed to the Catholic aristocracy."[54] Contrary to the above arguments, Crane claims, the exquisiteness of female attire was a form of social control which contributed to the maintenance of women in dependent, and subservient roles.[55] Crane's argument also suggests the gender paradigm of patriarchal power.

As explained previously in sociological, anthropological, psychological and cultural contexts, patriarchal associated male governing society was supported and reinforced by statutory legislations, as married women were denied access to property or any form of legal institutions, without first consulting their husbands. Blackstone in *The Commentaries on the Laws of England* stated; "By marriage, the husband and wife are one person by law: that is, the very being or legal existence of the woman is suspended during the marriage, or at least is incorporated and consolidated into that of the husband."[56]

Symbolic of man's authority and control of the household is the surname. Although English law does not require a woman to procure her husband's surname on marriage, traditionally it was a customary and

---

Press.

[53] Veblen, T. (1899). *The Theory of the Leisure Class* (pp. 64-65 & 126). London: Unwin Books.

[54] Chenoune, F. (1993). *A History of Men's Fashion* (pp. 17). NY: Flammarian.

[55] Crane, D. (2000). *Fashion and Its Social Agenda* (pp. 112). London: The University of Chicago Press.

[56] Cited in: Evans, M. (1982). *The Woman Question: Readings on the Subordination of Women* (pp. 346-351). London: Fontana Paperbacks.

appropriate exertion. Since most societies have accepted and expected woman's domestically confined social space, her submissiveness to men as their chattel or property (according to Veblen's theory as an indication of men's conspicuous leisure and consumption[57]), her economic dependence, her biological fragility and her subordinative social status, the phenomenon of patriarchy is strengthened.

> She is my goods, my chattels; she is my house,
> My household stuff, my field, my barn,
> My horse, my ox, my ass, my anything.
> —William Shakespeare[58]

The strict gender segregation through attire maintained firm boundaries between masculine and feminine, which preserved and protected the male dominance and governance.

> Fashion was presented as evidence of feminine vanity, falsehood, extravagance, conformity, ignorance and stubborn silliness.[59]

The image of the Victorian woman portrays sexually repressed and socially oppressed characteristics in every facet, as her attire manifested social, sexual and economical subordination. The Theory of the Shifting Erogenous Zone,[60] which explains clothing's desire to heighten sexual allure, could be analysed as another facet of male dominance in the context of attire. Decolletaged female dresses were permitted in the courts and among ruling aristocratic circles as a sexual display.[61] In the early 1930s, features of women's clothing emphasis shifted from the legs to the back. Backs of dresses were bared, and had been designed to be seen from the rear. Even day dresses had a slit up the back and the skirt was drawn tightly over the hips so as to reveal...,[62] which suggested the attention on female appearance, body and dress was determined and focused with men's sexual desires and references. In examining the relationship

---

[57] Veblen, T. (1899). *The Theory of the Leisure Class* (pp. 41-80). London: Unwin.
[58] Shakespeare, W. (1928). *The Taming of the Shrew* (3.2.230) (pp. 175). Cambridge: Cambridge University Press.
[59] Steel, V. (1985). *Fashion and Eroticism* (pp. 175). NY: Oxford University Press.
[60] Flugel, J. C. (1930). *The Psychology of Clothes*, London: Hogarth Press. Davies, F. (1992). *Fashion, Culture, and Identity*. Chicago: The University of Chicago Press. Laver, J. (1988). *Costume and Fashion: A Concise History*. London: Thames and Hudson.
[61] Kunzle, D. (1982). *Fashion and Fetishism* (pp. 1). NJ: Rowman and Littlefield.
[62] Laver, J. (1988). *Costume and Fashion : A Concise History* (pp. 241). London: Thames and Hudson.

between the meaning of the woman's physical body symbolised by the dress and male autocracy, the significant features, such as tiny waists and bursted breasts caused by corsets (the corset has been interpreted as an instrument of physical oppression and sexual commodification[63]), and petite feet of Chinese women as a result of footbinding[64] for instance, has been interpreted as a male fascination for the woman's body as a part of male domination.

The transformation process of female dress was less dramatic than males, nevertheless significant. As the gulf between the gender differentiated attire became extensive, the extremely elaborate and detailed female attire emerged and developed from the 1820s. By the mid-century to 1860, the extravagant female fashions; crinolined, multi-petticoated, corsetted and bustled attire, appeared and were universally utilised in Western countries. Sumptuous modes such as voluminous underwear/ petticoats, tight-fitting bodices, corsets and armholes/sleeves, gowns embellished with lace, ruffles, bows, ruching, gathers, braids, frills and beads; which often weighed ten to twenty pounds, and skirts with long trains, restricted the woman's body movement. The standard dress required approximately twelve yards or eleven meters of fabric.

According to Veblen's theoretical analyses, the aristocratic noble women, who were exempted from toil, could afford to wear elegant bonnets, heavy jewellery, high-heeled shoes, cosmetics, corsets and elaborate dresses. This attire, which was arduous to wear, was especially designed for conspicuous display rather than practicality. The corset, bonnet and wide-skirt, according to Veblen's economic theory, was designed for the purpose of lowering the subject's vitality, and rendering her incapable for all productive employment.[65]

For Durkheim, society is built upon the basis of an enduring tension between our individual and social existence, and our nature as both egoistic and moral beings.[66] Clothing demarcated the society, class and gender, as well as negotiated the boundaries and limitations of the social structure. The sartorial extravagance indicated the class, power and prestige. The gender hegemony was evident in the aristocratic fashionable arena in many forms and facets throughout the centuries. Gender in attire could also be classified into dichotomous conception; social and cultural.

---

[63] Steel, V. (1985). *Fashion and Eroticism.* NY: Oxford University Press.

[64] Dorothy K, (1999). 'Bondage in Time: Footbinding', *Fashion Theory Vol. 1, Issue 1* (pp. 3-38).

[65] Veblen, T. (1899). *The Theory of the Leisure Class* (pp. 121). London: Unwin.

[66] Durkheim, E., (1995 [1912]). *The Elementary Forms Religious Life,* (pp. 151) New York: Free Press.

Historically, gender did not mediate through vestmentary codes throughout Western history, as gender-segregated attire was visible from the nineteenth-century. Noticeably, female attire transformed into extremely elaborate, whilst male dress developed into uniformity and sobriety. The universality of female submission was emphasised by the patriarchal-related sexual and societal hierarchy. Male authority was considered a form of social control mechanism, in the economic and spiritual context. In Western civilisation, clothes performed the dual function of designating social position and gender, while fashionable clothes operated as a mechanism of social control, directing social distinctions.

> Everyday we make decisions about the social status and role of the people we meet based on what they are wearing; we treat their clothes as 'social hieroglyphics', which conceal, even as they communicate, the social position of the wearer.
> —Malcolm Barnard

References

Barnard, M. (1996). *Fashion as Communication,* London, New York: Routledge.

Barns, R. & Eicher, J. B. (eds.) (1992), *Dress and Gender, Making and Meaning.* New York: Oxford University Press.

Beauvior, S. De. (1972). *The Second Sex,* London: Jonathan Cape.

Boucher, F. (1966). *A History of Costume in the West,* London: Thames and Hudson.

Chaucer, G. (1980). *The Canterbury Tales.* London: Arnold.

Crane, D. (1933). *Fashion and its Social Agendas: Class, Gender, and Identity in Clothing.* Chicago & London: University of Chicago Press.

Crowley, T. (1996). *Language in History: Theories and Texts,* NY: Routledge.

Davies, F. (1992), *Fashion, Culture, and Identity,* Chicago: University of Chicago Press.

Dentith, S. (1998). *Society and Cultural Forms in Nineteenth Century England,* London: Macmillan.

Durkheim, E. (1915). *The Elementary Forms of the Religious Life,* London: George, Allen & Unwin.

Elias, N. (1991). *The Symbol Theory.* London: Sage.

Engels, F. (1986). *The Origin of the Family, Private Property and the State*, Harmondsworth: Penguin.

Evans, M. (1982). *The Woman Question: Readings on the Subordination*

of Women, London: Fontana Paperbacks.
Flugel, J. C. (1930). *The Psychology of Clothes,* London: Hogarth Press.
Freud, S. (1977). *On Sexuality,* Harmondsworth: Penguin.
Hutt, C. (1972). *Males and Females,* Harmondsworth: Penguin.
Kunzle, D. (1982). *Fashion and Fetishism,* NJ: Rowman and Littlefield.
Lurie, A. (1981). *The Language of Clothes,* London: William Heinemann.
Maccoby, E. & Jacklin, C. N. (1974). *The Psychology of Sex Differences,* California: Stanford University Press.
Marshall, T. H. (1950). *Citizenship and Social Class,* Cambridge: Cambridge University Press.
Marx, K. (1930). *Capital. Vol. 1,* London: Dent.
Perkins, H. (1969). *The Origin of the Modern English Society 1780-1880,* London: Routledge and Kegan Paul.
Perrot, P. (1994). *Fashioning the Bourgeoisie: A History of Clothing in the Nineteenth Century,* Princeton: Princeton University Press.
Steel, V. (1985). *Fashion and Eroticism,* NY: Oxford University Press.
Veblen, T. (1899). *The Theory of the Leisure Class*, London: Unwin Books.
Wilson, E. O. (1980). *Sociobiology,* Cambridge: Harvard University Press.

CHAPTER THREE

RECEDING HORIZONS: UNSTABLE BODIES AND THE TOPOGRAPHY OF MEMORY

ALAN DUNNING AND PAUL WOODROW

Abstract

The recent advent of digital technology, and its accompanying hybrid systems of communication, display and representation, has radically transformed the way in which we presently think about the notion of memory. Given that the body can now be characterized as distributed phenomena, hyper-spatialized and atemporal, memory is less than ever a fixed, graspable and retrievable entity. This chapter discusses the work of the Einstein's Brain Project, to examine issues surrounding the limitation of the technologies of representation in the processing of bodily perception and transformation. In particular, it explores the generative systems the art and science collaboration has developed to use Electronic Voice Phenomenon and Closed Eye Visualization as a means to investigate ideas about memory, hallucination and machine perception.

The work of the art and science collective the Einstein's Brain Project has long been closely associated with issues surrounding memory, time and the means by which we access the past. In particular, works that have been engaged with the deep analysis, through pattern and feature recognition, of information accreted over long periods of time, have contributed to a core understanding of memory as a shared construction between machine and human. In these works the task of archiving the data has been consigned to machine memory, but analysis and interpretation has remained with the human element in the symbiotic, cyborgian systems that constitute the works. The relationships between time, human interpretation and distributed and archived memory, together with the irresistible shifts that occur in the false continuities of reconstruction, have suggested new ways to consider memory and the body as its instantiating agent, as is increasingly distributed throughout a diffused and expanding world. They suggest that memory may not only be discrepant - dynamic and neomorphic - but also more purely hallucinatory and more bound up in the complex processes of remembering that lay beyond both the recall of persistent memory, and the paroxysm of involuntary memory.

As artists we use critical theory rather than empirical methodologies to engage with the notions of perception, memory and technology. But our productions are complex physical systems of consumption, representation and interpretation, and as such our works act as both thought and physical experiment; as models for new technologies of perception and reception.

Within our processes and systems, we have observed directly that new hybridized technologies of perception may not only be unreliable, but unavoidably false and delusional. This applies generally to memory as it is sustained through the new social constructions emerging from the conflation of human and machine generated fictive and non-fictive space, and specifically to the hallucinatory tendencies of memory machines.

> "The wish for memory is that it be machinic: external, vigilant, irreproachable. The truth may be that memory is inextricably cathected in the forces of interiority, desire, and identification."[1]

What we have observed is that the machines in our systems produce a particular kind of false memory that arises out of their inner workings. Far from being vaults for information kept intact and incorruptible, the machine's memory is built on parasitized probabilities rather than certainties:

> "This is the essential structure of the memory machine: it is a host apparatus that has been invaded by another figure, which is neither inside nor outside the apparatus-a phantastic projection like James Clerk Maxwell's thermodynamic agent."[2]

In Shapes of Thought (2006 – 2010)[3] streams of data are formed into objects containing all the information gathered in an EEG recording, and in works such as Ghosts in the Machine (2008)[4], and ColourBlind (2010)[5]

---

[1] Lippit, Akira M., Martin Arnold´s Memory Machine. In: Afterimage.
The Journal of Media Arts and Cultural Criticism. Vol 24 No.6, Rochester, NY 1997

[2] Ibid

[3] http://people.ucalgary.ca/~einbrain/new/text/the_shapes_of_thought.html

[4] http://people.ucalgary.ca/~einbrain/new/text/ghosts_in_the_machine.html

[5] http://people.ucalgary.ca/~einbrain/new/text/colourblind.html

very large amounts of data culled from noisy colour streams are processed by algorithms in attempts to perceive pattern and form in apparently random streams, or in undifferentiated monochrome fields of colour. Extremely large amounts of data (in excess of 5 terabytes) are written to disk and analyzed both in real time and after the fact. In each work, the task of archiving data is consigned to a machine, with analysis occurring in the interstice between machine and human as both attend to the task of searching for pattern. What the analysis reveals is that patterns and forms assert themselves that are neither properly in the data nor indexed elsewhere. While representing the data exactly, these forms arise through simple differences that recalls Deleuze's notions of difference intensity:

> "Every phenomenon refers to an inequality by which it is conditioned. Every diversity and every change refers to a difference that is sufficient reason. Everything which happens and everything which appears is correlated with orders of differences: differences of level, temperature, pressure, tension, potential, difference of intensity."[6]

Simple differences in tone, or colour or hue result in a morphogenesis based on emergence, on becoming and driven by intensity differences. The machine archive – machine memory – contains its own mechanisms for generating form from within.

Participants approach these works and are bombarded with a vast stream of data passing rapidly in front of their eyes. Visible for only a 60th of a second the stream passes by too fast to be properly seen. Yet out of these streams emerge recognizable patterns and forms. They are not indices of existing objects. They are not a record of observation. They are instead a function of an observer's memory of a similar form, and as much imagined as revealed, and as such are not exactly memories, though they may trigger memories.

University of Pennsylvania researchers have determined that false memories and true memories produce distinctly different brainwave patterns, particularly in the moments just before the retrieval of a memory occurs.[7] What is of particular importance is that the difference seems to be in the recall of contextual information associated with the past experience. If this recall is complicated by rapidly changing past and present contexts, – as is common in electronic remix environments – a memory act becomes

---

[6] Gilles Deleuze, *Difference and Repetition*, Columbia University Press, New York, 1994

[7] Sederberg Per B., Schulze-Bonhage, Andreas, *Gamma Oscillations Distinguish True From False Memories,* Psychological Science, November, 2007, Sage Journals Online

an isomorphic act. There is some degree of congruence with a memory, but it is not a memory.

Perhaps these are new forms that must emerge in electronic space; a conflation of the imagined, the recognized, and the misinterpreted, taking place in a constantly reconfigured space. Not a memory, but a new creation, bearing a similarity to a memory, that is confidently built on the recognition of extant forms, in an expanding present. Unlike a photograph, in which the object is the very record of the act observation, the record of salient forms in these works is always receding, and it is this recession that gives rise to false memory, to false certainty of a future past.

Machine memory suddenly appears less finite, less closed to discrepancy than first imagined. Forms and patterns are generated through the very intensities and complexities of difference that are the quintessence of code - ones and zeros. It is as if human algorithmic access to an archive – machine memory - is, and must be, fundamentally hallucinatory.

The electronic hallucination has been with us for some time. According to Nielsen Media Research's report, in 2009 the average American 153 hours of TV every month. Some 131 million Americans who watch video on the Internet watch on average about 3 hours of video online each month at home and work. 13.4 million Americans who watch video on mobile phones watch on average about 3 ½ hours of mobile video each month. Watching only one movie a month would amount to more than 840 films in a lifetime.[8] Photography changed our relationship to memory forever. Cinema and television conflated the viewer and image and simulated interior states – visualizing dreams, memories and lived experience. New technologies of display and dissemination have moved us closer to an even more complete simulation. Alison Landsberg has suggested that mass media's lasting legacy will be to provide what she terms prosthetic memory:

> "the technologies of mass culture and the capitalist economy of which they are a part open up a world of images outside a person's lived experience, creating a portable, fluid, and nonessentialist form of memory."[9]

As mass media shapes and creates our individual and collective memory so the possibility of memories that have not been lived can be experienced,

---

[8] http://blog.nielsen.com/nielsenwire/online_mobile/americans-watching-more-tv-than-ever/ accessed March 2009

[9] Landsberg, Alison, *Prosthetic Memory: The Transformation of American Remembrance in the Age of Mass Culture*, Columbia University Press, New York, 2004

and can be newly lived. No longer limited to lived experience, the simulation of memory and its generative process is actual memory.

If new media has had an impact on our collective memories then the development of sophisticated digital recording and storage has also changed the actual way we remember. High resolution digital recording devices, social networks and online archives have archived, catalogued, and dematerialized the indices of our memories, to such an extent, that it has obviated the need for us to engage in mental recall. Freed from the individual, the local need to remember, and combined with super-distributed public dissemination, atomized memories are released as to recombine and remix in social space.

Our memories have always been susceptible and impressionable. Jacquie Pickrell and Elizabeth Loftus at the University of Washington have shown the ease with which a false memory can be created.[10] Their study convincingly planted false memories through the use of a fake advertisement that showed the unlikely event of Warner Brothers' Bugs Bunny appearing at Disneyworld, causing participants to remember, falsely, meeting him. Prosthetic memory and remix culture combine to create an endless variety of new memories, new hallucinations, existing without regard for lived experience.

New technology has made commonplace constructed mnemonic spaces that are fundamentally different from the earlier technologies of memory, promoting new emergent entities, unattached to lived experience. This brings into focus acts of cognition that are inextricably linked to the building of meaning, the understanding of narrative, and, in turn, to the techno-subjective restructuring of the body.

It is the process of accessing of memory that is at the heart of memory itself. Memory is non-existent until enacted – that is, it is nowhere until actually accessed, and then it is instantaneously everywhere. Bodies are increasingly defined by their infinitely extensible, and hence infinitely porous, boundaries. This super-distributed body is also the instantaneous body – with the ability to be everywhere at all times. The delocalized body becomes just one more self-aware entity in a sea of dematerialized objects and processes arising out of random fluctuations in a chaotic system – a Boltzman brain[11] if you will. Memory here is a function not of the body/mind but of the distribution process itself.

---

[10] *False memories*, http://www.unexplainedstuff.com/Mysteries-of-the-Mind/The-Mechanics-of-Memory-False-memories.html, accessed February 2010

[11] Named for the physicist Ludwig Boltzmann (1844 – 1906), who advanced an idea that the known universe arose as a random fluctuation, a Boltzmann brain is self-aware entity that arises due to random fluctuations out of a state of chaos.

Preceding decades have witnessed a rapid emergence of technological devices and programs that have brought into question the fundamental status of the body that is now perceived as being in crisis. Not only is the body being challenged from a physical standpoint but also there have been radical developments and transformations in the field of social perception. Theorists have characterized this shift as a move from a photo-based mimetic to a sensibility anchored in a world of virtualization. Jonathon Crary suggests in Techniques of the Observer[12] that vision has been relocated to a plane severed from a human observer. Central to this shift is the role played by speed and the subsequent emergence of a hyperreal world of immediacy. Paul Virilio's conception of the 'Dromosphere'[13], expresses his fear of the automatization of perception and the disappearance of historical or chronological time transforming us into victims of spatio-temporal compression. The depth of time with its sense of a past and a future is increasingly eclipsed by an eternal present.

Taking his impetus from Neitchze, Arthur Kroker has described the present situation vis-à-vis technology as "the will to virtuality". When he says that data has become flesh and

> "to become data to such a point of violent implosion that the body finally breaks free of the confining myth of "wired culture" and goes wireless. The wireless body? That is the floating body, drifting around in the debris of technotopia: encrypted flesh in a sea of data."[14]

Kroker's prediction is for a future where the wireless body combines the velocity of virtualized substitutes into its cellular organization.

> "Not a body without memory or feelings, but the opposite. The wireless body is the battleground of the major political and ethical conflicts of late-twentieth-and early-twenty-first-century experience."[15]

What does it mean to be wired? Machines wired to bodies; bodies wired to machines? There are of course the literal meaning of being physically attached to a machine, and we talk about brains as being hard wired, but today's usage more normally functions as a metaphor – as a reference to the wired generation and to the more recent wireless generation – as

---

[12] Crary, Jonathon, *Techniques of the Observer : On Vision and Modernity in the 19th Century*, Cambridge MIT Press, 1992

[13] Virilio, Paul, *The University of Disaster*, Polity Press, Cambridge, 2010

[14] Kroker, Arthur and Weinstein, Michael, *Data Trash: the theory of the virtual class*, New World Perspectives, Montreal, 1994

[15] Ibid

plugged in or connected to the invisible worlds of technoculture and electronic communication. We are all familiar with technologies of distribution and the network, with on-demand information, instantly uploaded video and image, with touch-pads, cellular phones, speed, and through these familiar with the core ideas of the remix: fragmentation, instability, emergence and hybridity. We are all connected extensively through digital and electronic devices to the technosphere. As literal extensions of our body these devices and their networks make old conceptual models that separate us from a world linguistically framed through subjects and objects increasingly redundant. It is as if the role of technology is now perceived as a means by which we interface with a new expansive, immaterial world, in which there exists self-aware entities and other worlds constantly remixed and reborn.

A succession of different theoretical models that center on the notion of the self in conscious experience has been formulated by Thomas Metzinger.[16] According to his *self-model theory of subjectivity*, there are no such things as 'selves' in the world. In its place, there exist *phenomenal self-models.*

Metzinger's hypothesis is that nobody ever had or ever was *a self.* All that exists are conscious *self-models* that cannot be acknowledged as models.

> "The phenomenal self is not a thing but a process and the subjective experience of being someone emerges if a consciousness information processing system operates under a transparent model. You don't see it, you see with it –as you read these words you constantly confuse yourself with the content of the self model activated by your brain. Neither the object component nor the physical body carrying the human brain has to exist, in principle, to phenomenally experience yourself as being related to certain external or virtual objects…"[17]

In the conclusion to *Being No One* Metzinger invokes Plato's metaphor of the Cave to exemplify our present condition. He asserts that our phenomenal model of reality comprises a cave in which we live our conscious life. However the cave is only a model, "…the experience of which could be generated by internally stimulating the brain, independent of the outside world."[18] He goes on to suggest that a brain in vat "…could if stimulated, activate the conscious experience of being a self in attending

---

[16] Metzinger, Thomas, *Being No One: The self model theory of subjectivity*, MIT Press, Cambridge, 2004

[17] Ibid

[18] Ibid

to the colour of this book in its hands."[19] The distinction between the observer in Plato's cave however, is that in our case, there is no self present in the cave. What we take as our 'self' is just part of the model, part of the cave.

> "The experience of *being someone* is just a way of encountering the world. Metzinger's theory explains how and why we engage in the phenomena of being someone, yet it suggests that there is no-one (self) taking part in this experience. As we shall see later, the implications of this for memory is startling."

But how does the introduction of technology transform bodily processes and systems? Technological systems and operations do not appear to function in the same manner as humans, that is, biologically nor evolution. Artificial systems as they presently exist are not directly reflected in the bodily processes. But conversely technological extensions reconfigure the body as a techno-phenomenal processer. The substantial self has disappeared, the mind is both extended and embodied wherein embodiment becomes a feature of a system that is linked to the on-going world of action, other bodily systems, other brains, other worlds, other technologies.

This situation is complicated when contextualized by time, place and memory. When subsumed by technology, mind/body processes such as perception and memory take on a different character. Our understandings of place, time and memory and our processes of perception are framed by an historical view of technology, that insists it is separate from ourselves, and not directly grounded in the body as a function of evolution. The new technologies rewrite the historical view as the body is more and more enmeshed in electronic communities, and in doing so reframe technology as an implicit component of human evolution. Time, place and memory are reconfigured as functions of the body, as techno-biological processes: atemporal, non-local and amnemonic.

The effect of cyberspace technologies on time, place and memory is pervasive. One of the most profound consequences rests in its ability to create instantaneous connections. Through the compression of time and space a vast range of information from many worldwide sources is available. Instant retrieval practices have supplanted aspects of memory, transforming experience into information, reducing it to data and phenomenological despatialization.

Virilio characterizes the new found freedom proposed by cyberspace as an accident of knowledge - even going so far as to propose that its

---

[19] Ibid

practitioners be hospitalized. He imagines a hospital for scientists and technologists that need to be treated like patients suffering from a long-term illness. The principal malady of cyberspace is the deprivation of depth and its replacement by surface.[20] He sees communication technologies as a type of pollution. Moving across the surface of informational objects, selecting or acknowledging bits of information that might be of interest or use in the future - like the acquisition of trivia. We have the illusion that we are somehow knowledgeable about the world and about events and happenings that are occurring elsewhere. The critical distance that is needed to engage in meaningful activity has been eliminated by the compression of time and space. The information that we encounter tells us what is happening. There is little need to remember as long as we have access to total recall -a most attractive feature of cyberculture.

As biological and technological systems entwine it becomes increasingly difficult to view one without the other. In the face of technology, biological matter is subject to such redefinition and reconfiguration that it becomes part of data retrieval systems.

> "With today's technology it is possible to download the entire three billion digits of your DNA onto about four CDs A three – gigabyte genome sequence represents the prime coding information of a human body -your life as numbers. Biology, that pulsing mass of plant and animal flesh, is conceived by science today as an information process."[21]

Memory is virtual, subject to imperfection and discrepancy, and unreliable, subject to invention and to fictions. In the context of technology and cyberculture, the phenomenal self model and matter and memory, the notion of body is transformed into what appears to be an indivisible and incomprehensible substance, an unyielding non-spatial element comprising reality. What we consider as the world is a model of which we are a part. In regards to ourselves there exists nothing apart and technological engagement makes these notions more evident.

Recent work of the Einstein's Brain Project has developed generative systems to use EVP (Electronic Voice Phenomenon) and hallucination to examine ways in which we construct worlds, and bodies in worlds, through machine human interfaces. It explores how pareidolia (the psychological phenomenon involving a vague and random stimulus, often an image or sound, being perceived as significant), and apophenia (the

---

[20] Ibid

[21] Kelly, Kevin, Wired #10.12, Conde Nast Publications, San Francisco, 2002

seeing of connections where there are none) contribute to our creation and generation of space and memory.

This work has used, amongst other strategies, face tracking and feature recognition, to explore the felt presence of absent bodies, using intelligent symbiotic systems, comprising both machine and human vision and analysis, to reveal patterns – the shapes of faces, the sounds of voices, geometric clusters - in apparently random visual and audio noise. These works explored the construction of a world delineated by presence and absence, and pattern and randomness, relocating the mind/body through a construction that is both machine and human.

In each of the works there is a symbiotic relationship between machine and human. In the cyborg embrace, the tasks of memorizing data archiving and cataloguing and the initial search for pattern are assigned to the machine, but the role of the human increases as chaos coalesces into recognizable form. Images and sounds act as mnemonics. Shapes and voices are formed through the remembrance of other forms.

The act of recognition is neither fully the task of the machine nor of its human collaborator. The computer does the hard work of scanning gigabytes of data looking for likely candidates for analysis, and the human component fixes the meaning of the form as it is manifested, but neither is able to work alone. The impossibility of a human analyzing huge streams of data passing only briefly across a visual field is certain. The machine must play its part in the construction of the remembered form. In recognizing form machine and human are engaged in similar tasks. It might seem that the machine would have less fragile memory, one less susceptible to the vagaries of discrepant memory and invention, but the reality is that the analysis throws up all sort of false positives, and in doing so creates a false continuity of reconstruction. Over long periods of time the accretion of false positives and continuities creates a machinic version of what Deleuze terms a Time-Image: an unrecognizable image which disrupts the present logic. This time-image creates a moment in which the machine is forced to go into its own memories to construct meaning for itself. The act of memory is instantiated through interpolation and invention.

The work produces images that look to all intents and purposes like faces and sounds that resemble human voices. Where are these faces and voices in the world, since they can only index non-existent objects? The processes that brought them into being – into a form of existence that is recognizable by humans is one that appears to be purely technological. They are invented forms born of the distortion of a memory task assigned to a machine. In this case, machine memory is as fragile and susceptible as that ascribed to humans.

A new work *Machine Imagination: Closed Eye Hallucination and the Ganzfeld Effect* (2010) explores the internal workings of a machine through an implementation of the Ganzfield Effect and Closed Eye Visualizsation insofar as they relates to ideas about hallucination in human and machine hybrids. The work explores ideas about machine vision and how hybrid interpretation gives rise to unbidden and unexpected images and patterns in streams of unstructured data, and how colour affects interpretation and imagination. Through an examination and analysis of visual system noise expressed as spatio-temporal voxel volumes, the work explores the product of these investigations as machinic hallucinations.

The Ganzfeld (complete or open field) effect[22] is a phenomenon of visual perception caused by staring at an undifferentiated and uniform monochrome field of color. Usually this is accomplished by the subject wearing tight fitting goggles that block out all but one colour of the spectrum. In the 1930s psychologist Wolfgang Metzger established that when subjects gazed into a featureless colour field they were unable to see anything after even a few seconds. In further experiments subjects that were immersed in the monochrome field for extended periods of time, consistently hallucinated and recorded distinctly different EEG patterns of activity and change, similar to those of sensory deprivation.

The effect is a well-known phenomenon with historical precedents in the followers of Pythagoras entering dark caves to gain wisdom through visions[23], and in reports of trapped miners hallucinating and seeing ghosts. Similar experiences are often cited by Arctic and Antarctic explorers who report altered states of mind while traveling across large featureless landscapes. It is thought that the hallucinations in extended Ganzfeld experiments are the result of the brain amplifying neural noise in order to look for the missing visual signals, and it is this noise that is interpreted in the higher visual cortex, giving rise to hallucinations.

In Closed Eye Hallucination an individual sees blobs and colour motion even though the eyes are closed. Closely linked to the experiences of subjects in Ganzfeld experiments there are often patterns discovered within the blobs - most cited are webs, grids, honeycombs and other geometric and repeating structures.

CEH experiences can take a number of forms: visual noise - seemingly random noise of pointilistic light/dark regions with no apparent shape or

---

[22] Ramesh B., *The Ganzfeld Effect*, http://www.shvoong.com/exact-sciences/biology/1671321-ganzfeld-effect/ accessed March 2010

[23] Ustinova, Yulia, *Caves and the Ancient Greek Mind: Descending Underground in the Search for Ultimate Truth*, Oxford University Press, 2009

order; light/dark flashes – regions of intense black or bright white that appear in the noise; patterns – highly organized motion and color forming complex geometric patterns and shapes; and finally objects and things.

A camera is turned on, and covered with a single Ganzfeld goggle and bathed in a pure monochrome light. The video stream is sent to a computer where the input is cropped and adjusted for fall off at the edges of the camera so that the monochrome colour field is undifferentiated by tone or hue. The camera image is processed to construct a voxel volume that is analyzed for optical features within a specified region of interest. Tiny inconsistencies in the colour field, invisible to the human eye, false positives if you will, are tracked recognized and sampled as they pass across planes in the x, y and z dimensions of the volume.

Over time patterns gradually emerge as the sampled frames are accumulated and are multiplied together. What starts as pure noise gradually resolves itself into patterns with structure and form. Sometimes these form rapidly, but usually these are the results of layers of noise blended together over long periods of time. The patterns that emerge bear a striking resemblance to patterns that are normally associated with those seen in closed eye hallucination and Ganzfeld effect experiments in which subjects stare at monochrome uniform fields of colour.

Participants in the work are acted upon by two systems. One is a large, very bright colour field that fills the periphery of the vision of the observer and physically creates the Ganzfeld effect. The other is the machine's analysis of the inconsistencies in the colour field, and their projected accretions. Presented with these two systems participants come face to face with the real and the imagined. But, oddly, the imagining is left to the machine. The effect of the colour field on the participant is physical, acting on the optic nerve and, in turn, on the brain. The patterns from the machine are imaginary, relying on recognition by an observer for their existence. What participants report experiencing are two different memory effects – one that is internal that relies on the require recognition of a form within something that is essentially formless, and one that directs the observation to the (re)cognition of a predeterminded form.

The inconsistencies in the field surely come from the image sensor itself, in part thermal noise, and in part amplified background electrical noise that is present in the system, but the work suggest some other possibilities. It may be that these patterns are nothing more than the chance interactions and collisions of one another, but they hint at an analytic that exists in the interstice between machine and body that exists in pattern and information flows.

The patterns that emerge from the random flickerings in a machine hint at an immaterial hybrid body that exists in the pattern and information flows that were fusion of body and machine, suggests that there might be real information contained within the random noise of the work, but its lasting impression is to acknowledge an ontological anxiety that imagines a remembering body so inextricably enmeshed with its surroundings and the technologies that support it, that it becomes indistinguishable from the processes, systems and mechanisms of storage and distribution.

Viewers consistently describe their physical and mental experiences through the machine's vision of the effect. How a viewer remembers, and how that memory is expressed, produces, in this instance, a new memory. Each individual experience is difficult to describe, and the images and forms of such a fleeting nature, that individuals turn to the projected machine generated patterns as mnemonics to recall the experience and provide a descriptive syntax. As they do so, and as new patterns emerge, memories conform to the machine's algorithms.

> "But the real question is whether machines can imagine too, whether in the darkness there is anything that might cause the machines to see as humans do- to see things where there are none and to recognize within these visual mistakes optical and aesthetic possibilities?"[24]

Memory can be seen as the recognition of patterns and their repetition, the future as a function of past experience, and in some instances merely a copy. But according to Ilse Walker, it is found that asymmetry and irreversibility as a consequence of complexity are the basic principles of memory function.

> "The immediate and inevitable consequences of memory function are shown to be irreversible expansion and irreversible, unpredictable diversification, leading to growing complexity and size on all but the sub-molecular levels of biological structure."[25]

We can be certain that this characterization applies to human behavior. But when human memory is linked to machines that are capable of almost instant retrieval there is no need to remember everything. Where there are gaps in memory it is possible to invent new memories that replace, complete or improve on the missing ones, believing them to be true. We witness to this all the time in the re-writing of histories, in all manifestations

---

[24] Heibert, Ted, in conversation with Alan Dunning and Paul Woodrow, Calgary, April 2010

[25] Ibid

of revisionist thinking and the reconfiguration of our own personal memories.

The definite uncertainty about our memories - the instability that filters into every corner of our lives - is mediated by technology and its accompanying systems. In the work of Einstein's Brain Project we have explored the machine vision and computational analysis to examine the machine's interiority - its phenomenal self-model. The question remains: where is this model to be found? It is not located within the artifacts themselves and neither is it located within the person. It is the product of a relation.

Our past work has investigated the relation between perception and memory where memory is characterized as an index. New work suggests is possible to use machines and their interiorities (subjectivities) in a phenomenologically driven investigation to discover hallucinatory tendencies that can create false memories. Accessing memory is at the heart of memory itself. Memory is non-existent until enacted. It is nowhere until recalled, and then it is instantaneously everywhere, creating and occupying a completely fictive space.

Memory is a constantly receding event horizon. Even as it springs into existence it is dispersed through its own act of discovery and recovery. Ungraspable and unfixed, its dynamic is such that it can never be never be here and now. Memories are always memories however poignant, however disruptive. Memory events are atemporal and hyper-spatialized, but are always moving away. There will never be a past or future in the sense that this was then, this is to be - but an imagined future past that reconstitutes itself as the present.

References

Ackerman, D. (1990) *Natural History of the Senses,* New York: Random House.

Ascott, R. (ed.) (2005) *Engineering Nature: Art & Consciousness in the Post-Biological Era*, Bristol: Intellect Books, 2005.

Becker, R. O. and Gary Selden G. (2008), *Body Electric,* New York: Morrow, 1985.

Bishop, Ch. M. (2008), *Pattern Recognition and Machine Learning*, New York: Springer, 2008.

Bossis B (2005), *La Voix et la Machine: La Vocalite Artificielle dans la Musique*, Rennes: Presses Universitaires de Rennes..

Crary, J. (1992), *Techniques of the Observer: On Vision and Modernity in the 19th Century*, Cambridge: MIT Press.

—. (1992), and Sanford K., *Incorporations*, London: Verso, 1992.

Critical Art Ensemble, (1998) *Flesh Machine; Cyborgs, Designer Babies, Eugenic Consciousness*, Williamsburg: Autonomedia, 1998.

Damasio, A. (1999), *The Feeling of What Happens: Body and Emotion in the Making of Consciousness*, New York: Harcourt Brace and Company,

Danto, A. (2002), *The Body Problem: Selected Essays,* Berkeley: Univ. of California Press, 2002 .

Deitch, J. (1992), *Post Human*, New York: DAP.

Deleuze G. (1994), *Difference and Repetition*, New York: Columbia University Press.

Edelman, G. (1992), *Bright Air, Brilliant Fire: On the Matter of the Mind*, New York: Basic Books.

Eder, K. (1996), *Social Construction of Nature*, London: Sage.

Evernden, N. (1992), *Social Creation of Nature*, Baltimore: Johns Hopkins Press.

Forsyth, D. A., Ponce, J (2002) ., *Computer Vision: A Modern Approach*: Upper Saddle River, Prentice Hall.

Garnet ,E. (2003), *Narrative Body*, Williamsburg: Autonomedia.

Geary, J. (2002),. *The Body Electric: An Anatomy of the New Bionic Senses*, New Brunswick: Rutgers Univ. Press.

Haraway, D. Simians (1991), *Cyborgs, and Women: The Reinvention of Nature*, London: Free Association Books.

Harrison, J. (2001), *Synaesthesia: The Strangest Thing*, Oxford: Oxford University Press.

Hayles, N. K. (1999), *How We Became Posthuman: Virtual Bodies in Cybernetics, Literautre, and Informatics*, Chicago: University of Chicago Press.

Herman, D. (Ed.) (2003), *Narrative Theory and the Cognitive Sciences,* Stanford: CSLI Publications.

Ihde, D. (2002), *Electronic Mediations Series, Vol. 5: Bodies in Technology*, Minneapolis: University of Minnesota Press.

Iser, W.(1974) *The Implied Reader: Patterns of Communication in Prose Fiction from Bunyan to Beckett,* Baltimore, Johns Hopkins University Press,

Kevles, D. & L. H. (1992), *Code of Codes*. Cambridge: Harvard University Press..

Kroker, A. and Weinstein, M. (1994), *Data Trash: theTtheory of the Virtual Class*, Montreal: New World Perspectives.

Landsberg, A.,(2004) *Prosthetic Memory: The Transformation of American Remembrance in the Age of Mass Culture*, New York: Columbia University Press.

Lippit, A. M. (1997), *Martin Arnold´s Memory Machine,* in Afterimage, The Journal of Media Arts and Cultural Criticism. Vol 24 No.6, Rochester.

Mandoki, K. (2007),, *Everyday Aesthetics. Prosaics, the Play of Culture and Social Identities,* Williston: Ashgate.

Marcuse, H. (1964), *One-Dimensional Man*, Boston: Beacon Press.

Martindale, C., Locher, P.aul & Petrov, V.r (eds.),(2007), *Evolutionary and neurocognitive approaches in aesthetics, creativity, and the arts*, Amityville: Baywood.

McConachie, B. & Hart, E. (eds.) (2006), *Performance and Cognition: Theater Studies and the Cognitive Turn*, New York: Routledge.

Metzinger, T.(2004), *Being No One: The self model theory of subjectivity*, Cambridge: MIT Press.

O' Mahoney, M. (2002),, *Cyborg: The Man-Machine,* London: Thames & Hudson.

Opdahl, K., (2002), *Emotion as Meaning: the Literary case for how we imagine*, Lewisburg: Bucknell University Press,

Pal S. K, and Pal A, (2008) *Pattern Recognition*, Hackensack: World Scientific Publishing.

Prince, S. (1996), *True Lies: Perceptual Realism, Digital Images, and Film Theory*, Film Quarterly, Vol. 49, No. 3 (Spring, 1996), Berkley: University of California Press.

Ramachandran, V.S. and S. Blakeslee S. (1998), *Phantoms in the Brain: Probing the Mysteries of the Human Mind,* New York: William Morrow.

Richardson, A. and S.polsky, E. (ed.) (2004), *The Work of Fiction: Cognition, Culture, and Complexity*, Burlington: Ashgate.

Ricoeur, P. (1985), (K. McLaughlin & D. Pellauer, Trans.), *Time and Narrative*, University of Chicago: Chicago Press.

Ripley, Brian, D., *Pattern Recognition and Neural Networks*, Cambridge: Cambridge University Press.

Sederberg, P. B., Schulze-Bonhage, A. (2007), *Gamma Oscillations Distinguish True From False Memories,* Psychological Science, November 2007, Sage Journals Online.

Shaviro, S. (1998), *The Cinematic Body,* Minneapolis: University of Minnesota Press.

Sheehan, J. J., and Sosna, M., (eds.) (1991), *The Boundaries of Humanity: Humans, Animals, Machines,* Berkeley: University of California Press.

Sontag, S. (1966),, *Against Interpretation and Other Essays*, New York: Farrar, Strauss & Giroux.
Theodoridis, S. Koutroumbas, K., (2006), *Pattern Recognition*, Amsterdam: Academic Press, Elsevier.
Todes, S., (2001), *Body and World*, Cambridge: MIT Press.
Turner, B. S., (1992) , *Regulating Bodies*, London: Sage.
Ustinova, Y.(2009), *Caves and the Ancient Greek Mind: Descending Underground in the Search for Ultimate Truth*, Oxford: Oxford University Press.
Van Oostendorp, H., Goldman, S. (eds.), (1999), *The Construction of Mental Representations During Reading*, Mahwah: Erlbaum.
Virilio, P, (2010), *The University of Disaster*, Cambridge: Polity Press.
—. (1984), *The Vision Machine*, Indiana University Press, Bloomington.
Wegenstein, B. (2006), *Getting Under the Skin: Body and Media Theory*, Cambridge: MIT Press.
Whitelaw, M., *Metacreation: Art and Artificial Life*, Cambridge: MIT Press, 2004
Wollen, Peter, *Signs and Meaning in the Cinema*, Bloomington: Indiana University Press, 1972.

CHAPTER FOUR

AN INTERNET OF OLD THINGS

ANGELINA KARPOVICH AND CHRIS SPEED

Abstract

This chapter reflects upon the temporal characteristics of the emerging phenomenon known as the Internet of Things, which refers to the addition of small electronic chips or bar codes to objects, enabling the exchange of object-related data across a network. In the Internet of Things, objects would record and contain the histories of their interaction with people and with other objects. The advent of this ever-growing catalogue of histories means that every object will be 'in touch' with its current and previous owner at all times, and suggests that whilst owners might like to 'forget' about an object, we will never truly be detached from them. However the authors suggest that there exists a social and cultural inertia that is tied to a teleological perception of time and that the weight of this is hampering opportunities for the Internet of Things to embrace old things. The paper uses a series of cultural coordinates to outline our relationship with personal and social histories before exploring the potential for digital technology to network the past and develop an Internet of Old Things.

Things and Linearity

The 'Internet of Things' refers to the technical and cultural shift that are anticipated as society moves to a ubiquitous form of computing in which every device is 'on', and every object is connected in some way to the Internet. The specific reference to 'things' invokes the idea that every new object will also be able to part of this extended Internet, because it will have been tagged and indexed by its manufacturer during production. For example, the technology has enabled supermarkets to track the temperature of consignments of prawns from the fishing boat that caught them, to the in-store freezers. A product's life cycle can be traced from cradle to grave, shelf to landfill. Tracked and monitored as they move around the world, objects are becoming networked and 'always-on' (Greenfield, 2006), a condition that means it will become harder to disassociate an object from its memories. The Internet of Things is related to the concept of ubiquitous computing (ubicomp) refers to the integration of information technology into everyday objects and processes, to such an extent that the end-users are often unaware of the technology. According to Greenfield, ubicomp has significant potential to alter not only our relationship with technology, but the very fabric of our existence:

> "A mobile phone […] can be switched off or left at home. A computer […] can be shut down, unplugged, walked away from. But the technology we're discussing here–ambient, ubiquitous, capable of insinuating itself into all the apertures everyday life affords it–will form our environment in a way neither of those technologies can." (2006).

Greenfield's ideas are neither hypothesis, nor hyperbole. Ubicomp is already a reality. Dodson notes,

> "Ubicomp isn't just part of our [...] future. Its devices and services are already here. Think of the use of prepaid smart cards for use of public transport or the tags displayed in our cars to help regulate congestion charge pricing or the way in which corporations track and move goods around the world." (2008).

The Internet of Things advances the ubicomp notion of objects embedded with the capacity to receive and transmit data and anticipates a move towards a society in which every device is "on" and in some way connected to the Internet; in other words, objects become networked. Information contained within and transmitted among networked objects becomes a "digital overlay" (Valhouli 2010) over the physical world. Valhouli explains that objects, as well as geographical sites,

> "become part of the Internet of Things in two ways. Information may become associated with a specific location using GPS coordinates or a street address. Alternatively, embedding sensors and transmitters into objects enables them to be addressed by Internet protocols, and to sense and react to their environments, as well as communicate with users or with other objects." (2010).

The Internet of Things is not a theoretical paradigm. It is a framework for describing contemporary technological processes, in which communication moves beyond the established realm of human interaction, to enable a whole range of potential communications: "person-to-device (e.g. scheduling, remote control, or status update), device-to-device, or device-to-grid" (Valhouli 2010).

The current applications of ubicomp's are largely functional, used in transport, security, and stock control. Industry is not geared up to handle the histories of objects, it is focused on the production of new objects. The innovative use of linear models of time underpinned the development of manufacturing and distribution systems throughout the twentieth century. Since the industrial revolution, time and space have been treated as discrete units in order to develop more and more innovative means of accelerating production processes. From Ford's development of the production line to the Toyota Production System, time and space have been compressed to develop increasingly flexible forms of accumulation (Harvey 1990). In each case, time has been imagined as travelling in a unilateral direction: from a cradle to a grave, choosing to pay little attention to looking backwards.

The Internet and other comparatively recent technologies have the potential to reconfigure and recontextualise conventional understandings of time and space. Of course, before the establishment of computer-mediated communication (CMC), we already had multiple means of connecting people commonly separated by space (Gitelman and Pingree 2003). The Internet and related technologies have allowed us to see each other whilst separated by great distances, to share stories, images and other media online, to co-construct content and to do all this within groups, rather than merely between individuals (Weinberger 2002). Yet while the boundaries and limitations of physical space have been reconfigured by the new processes of communication and sharing, the same processes have not had a parallel impact on the limitations of conceptualising time in unilateral linear terms. This 'forward-looking' mode of production of things' is epitomised in the technological determinist slogans from recent UK history: 'We're getting there' (British Rail 1980s), 'Where do you want to go today?' (Microsoft 1990s) or 'The future's bright the future's Orange' (Orange mobile network 2000s). A characteristic of this condition is the assumption that we are able to relinquish ourselves from any past profile and that the things that are new are best.

Tags on current brand-new consumer artefacts tend to fall into three categories along a linear timeline:

1. Past - Ability to identify where something was or what condition it was in, in the past (often used in surveillance or evidence to justify the integrity of something that is fresh or new, or alternatively confirm that it is outdated),

2. Present - Ability to check on the status of something, its condition now,

3. Future - New opportunities to connect to related products or services, for future consumption (offers printed on to products or found on posters).

These three temporal perspectives offer the primary benefits for tagged products as they emerge into mass circulation, with different parties able to access different databases in order to promote the consumption of new things. Whilst the consumer landscape is kept 'fresh' with information about the new, the network properties of the Internet of Things offer other opportunities that do not adhere to a linear model of time, and instead offer value to objects through the recovery and retention of information from the past. However the question that this paper proposes is whether the Internet of Things can escape the inertia of society that uses a teleological basis for many of its approaches to production and manufacture?

Valuing the future and forgetting the past

In the summer of 2009 the UK artist Jasper Joffe staged the sale of everything that he owned at the Idea Generation Gallery in London. Everything, including his paintings, drawings, teddy bears, and rare books, was grouped into 33 different lots, each on sale for £3,333. Part of the publicity for the show involved a short interview on BBC Radio. During this interview Joffe described how the installation / performance offered him an opportunity to "re-think everything" and to overcome a tendency of "getting stuck with old habits".

> "My emotions exist I guess in my brain, not in the stuff that I own, the things that I feel... the things I do, don't relate to the photos I've got in a box or an object that I keep at home, or you know, an antique teddy bear." (Joffe 2009).

Joffe's rejection of intrinsic meaning, or emotional value, in his possessions may of course be part of his overall performance. Certainly, the seemingly-arbitrary assignation of identical monetary values to seemingly-arbitrary 'lots' initially suggests not only a lack of emotional value, but also that the individual objects, as a result of this lack, are interchangeable and entirely unremarkable. At the same time, the actual context of the performance gives the objects additional value and meaning as both 'works of art' and 'an artist's belongings'. The first of these meanings invokes the connotations of a museum piece or an otherwise culturally-significant object. Along with works of art, such objects include monuments, the possessions of historical figures, religious artefacts, and archaeological finds. At the same time, even while Joffe claims that his possessions have no emotional value, they acquire another layer of cultural cachet as the authentic belongings of an artist. Authenticity is something that the public actively seeks in its interactions with culturally-significant objects. As MacCannell (1999) points out in his seminal study of the production and consumption of tourist attractions, "The rhetoric of tourism is full of the manifestations of the importance of the authenticity between the tourists and what they see: this is a typical native house; this is the very place where the leader fell; this is the actual pen used to sign the law; this is the original manuscript". Thus, while Joffe's belongings may not, as he claims, have intrinsic emotional value, they are, by virtue of being part of his performance, certainly somewhat extraordinary.

The current value system which defines an object's cultural significance appears to replicate Bourdieu's assessment of the hierarchies which define aesthetic concepts such as taste (1984). In both cases, the popular, everyday, or otherwise mundane is deemed to possess less cultural capital than that which is less accessible or otherwise associated with the social elites. As a result, objects whose histories are well-known are mostly found in museums, untouchable and unused, whereas objects which are within reach, all around us, tend to travel from owner to owner without anyone considering what histories they might contain. Joffe's apparent ease in detaching himself from an object, and the memories associated with it, is a Cartesian trait, in which the breaking down of systems into discrete units, in particular those of subject/object and time/space, support a position of control. The linear model of time that production models have inherited tends toward an industrial interpretation for the Internet of Things that, like Joffe, disassociates itself from the past and is interested in only producing the new.

Creative opportunities for the Internet of Things

It has been suggested that people surround themselves with between 1,000 and 5,000 objects (Waldner 2007). Of those thousands of objects many of them are probably not truly cared for and end up in rubbish bins or in storage. But for every owner, in almost every household there are a selection of objects that hold significant resonance, and will already connect them to an 'Internet' of memory and meaning. An intrinsic human trait is the process of imbuing meaning onto objects so that they provide connections to people, events and environments. Artefacts across a mantelpiece become conduits between events that happened in the past, to people who will occupy the future. These objects become essential coordinates across families and communities to support the telling of a stories and passing-on knowledge.

Developed in collaboration with the Oxfam charity shop in the student quarter of Manchester, a creative/technical intervention entitled 'RememberMe' explored how memories that are attached to objects can affect consumer habits. Oxfam are a charity that has 700 shops across the UK. The shops receive donations of clothes and artefacts from the public, and sell them on to new owners as second-hand goods. A research associate worked for one week in the Oxfam shop in Manchester and asked the people who dropped off their object donations to tell a brief story about the object into a microphone (e.g. where they acquired it, what memories it brings back and any associated stories). These audio tracks were then uploaded to the Audioboo service (http://www.audioboo.com) and linked to newly created stories on the Tales of Things website (see section 3). One week later, with the permission of people involved, this audio track was linked to two-dimensional barcodes and RFID tags that were attached to the objects in the shop with a custom RememberMe label. Two dimensional barcodes, commonly known as QR codes (Quick Response) are printed paper barcodes which can contain an Internet address, and like RFID Tags can easily be associated with information or data files. Figure 1 illustrates a screenshot of a tag attached to an artefact in the Oxfam shop.

Visitors to the shop could use bespoke RFID readers and the Tales of Things iPhone and Android phone applications to scan the labels. Once triggered, speakers located in the shop played back the audio stories associated with the label. Although the team anticipated an interest in the stories, we were surprised at how affective the very individual voices were upon visitors to the shop. The actual sound of somebody's voice associated with an object offered an almost-supernatural extension to handling an artefact. People visiting the shop, browsing the objects and scanning the tagged donated items spoke of the "personal connections" made as artefacts conjured an actual voice that gave the object additional meaning. The red silk toiletries bag that had no history or geography was transformed into an object loaded with place and personality as the story of its previous owner described a shopping trip in Bangkok that involved a near death experience in a tuk tuk (Figure 2).

Figure 1: Tale View in Tales of Things

> "Well my item is the little red silk make up toiletries bag its from a place called Narai in Bangkok and it was one of the very first things that I bought when I went to visit my uncle and his wife Noi who lived just outside Bangkok themselves and I believe if this is the shopping trip that I'm thinking of, I believe its also one of the very first times that I got a tuk tuk and nearly fell out, on the middle of the motorway, on the way back which I'm pretty certain it is actually so yeah that's my story and I risked life and limb to get that toiletries bag." Red Toiletries bag, Anonymous donor.

The project's emphasis upon personal stories and not quantitative data such as price, temperature or other logistical data, offered a rich immaterial dimension to each object's material instantiation. The result of this supplementary information meant that every object (approximately 50 in total) was sold, even the types that are notoriously hard for a second hand shop to sell.

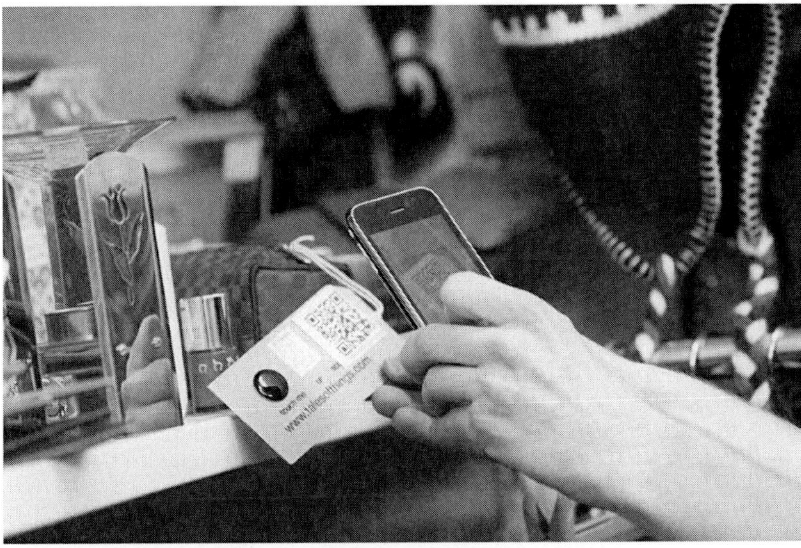

Figure 2: Scanning the red silk toiletries bag

Tales of Things

The RememberMe project relies upon a technical framework: Tales of Things (http://talesofthings.com). Tales of Things consists of a web application that provides a backend service to connect different media to a unique two-dimensional barcode. Users who register for a free account at Tales of Things can add new objects to a user-generated object database via a web browser interface. During this process they provide some (optional) information (e.g. name, keywords, location) and a story (tale) about the object (thing). Tales can be told using text and any additional media that can be referenced via a URL (Unified Resource Locator). The system is capable of analyzing provided URLs and rendering media from services such as YouTube, Flickr and Audioboo in an integrated media player interface.

When a new object is added, the service creates a unique two-dimensional barcode (QR Code) that can be printed out and attached to the object. People are also able to link the objects using RFID tags. The web interface provides additional functions such as a commenting system, a view of the location of things and tales on a map, search, user profiles, email and Twitter notifications, and groups. A public API are currently under development. We offer downloadable bespoke mobile clients that can read Tales Of Things QR Codes and that provide additional functions such as adding of new tales when a barcode has been scanned for the iPhone and Android platform. Further custom Bluetooth RFID Readers have been developed, which read Tales of Things-registered RFID tags and can communicate to other devices via Bluetooth.

The aim of Tales of Things is to provide both a context and a mechanism for enabling individuals and community groups to share object-related stories and memories through digital media, and to make these object tales persistently accessible in the off-line world through the objects themselves. The online platform will serve two primary functions. It will become an archive for object memories and thus grow to become what we term an "archaeology for the future". We hope that future generations will be able to return to this repository and learn about the things that are meaningful to groups and individuals right now, and to access user-generated personal narratives of significance told by 'ordinary' people in their own words. The platform also serves as an arena for contemporary communication. As the project develops, more and more object memories become directly accessible through tagged artefacts, as well as through browsing and keyword searches on the project website. Users are already able to communicate via the TOTeM platform, through commenting on object stories and forming and joining groups with a thematic or geographical focus. For example, the Tales of Wartime Things group brings together memories of wartime objects from the residents of the London borough of Hillingdon, collected in the process of our research (although many of the groups are user-created, the group functionality also allows the research team to bring together and easily showcase object stories collected by us during fieldwork, focus groups, and workshops). While the individual objects in the group are diverse (they include both 'typical' wartime objects such as a ration card and pieces of shrapnel and perhaps-unexpected wartime objects such as a magazine, a piano, and a teddybear, as well as contemporary objects which serve as metonyms or conduits for wartime memories, such as a map and a banana), the group provides a thematic, geographical, and temporal context for considering the objects, and associated object stories. The group functionality within the Tales of Things site is one of the ways in which points of commonality can emerge between seemingly disparate objects and object-related narratives.

While the Tales of Things site can allow for both casual browsing and a relatively structured and systematic exploration of object tales, the increasing number of objects 'in the wild' which carry tags linking them to tales about them (as in the RememberMe project) simultaneously allows for a haphazard and serendipitous exploration of objects in their real-world, in situ, context. This presents new opportunities for considering the conceptual and spatial relationships between our objects, our environments, and our selves. Objects tagged with object tales are, in a sense, an instance of augmented reality (AR). However, while most current AR applications suffer from the same unidirectional future-oriented approach to temporality which we outlined at the start of this chapter, Tales of Things-tagged objects are augmented with references to the present (the moment in which the tale is created) and the past (the history associated with the object). The present is, of course, always-already history-in-waiting; a concept implicit in our conception of Tales of Things as an 'archaeology for the future'. The potential of object tales serving as ephemeral 'time capsules' was highlighted to us, anecdotally, by one workshop participant reviewing his video object tale several months after recoding it. The participant had undergone chemotherapy in the months after recording the tale, and reported that one of the pleasures in reviewing it lay in having a record of himself not only recounting a happy memory but also looking his healthiest.

As the project develops and the number of tagged objects 'in the wild' grows, we anticipate not only that our relationship to objects might change (through adding a layer of 'emotional value', as demonstrated during RememberMe), but also that memory and the past might gain greater prominence in people's everyday interactions with their environment and with each other. Whereas previously, the history, or provenance, of an object, was relevant mostly in commercial terms, in the context of an auction or a museum, Tales of Things brings ideas of provenance, of emotional rather than material value, into a do-it-yourself context in which everyone's tales are equally valuable. The hierarchies previously inherent in designating some, but not all, objects as culturally-significant, highlighted in our discussion of Jasper Joffe's performance, do not apply to user-generated object stories focused on the past, created in the present, and preserved for the future.

Conclusions

This chapter has emphasised the teleological tendencies of contemporary society and its propensity to concentrate upon the new and detach itself from the past. Through a brief analysis of the artist Jasper Joffe and his motives for the work 'Sale of a lifetime', we established a basis upon which to explore how easy it might be for people to ignore their past because of their orientation towards the future.

The RememberMe project was used to evoke an alternative temporal economy, one in which the arrow of time reflects back to use memories and history as a means of adding value to artefacts as they pass through society. And, as objects become tagged and catalogued within networks, the Tales of Things project offers a 'bottom-up' approach allowing the public to tag objects and ensure that the Internet of Things is not just focusing upon new items, but identifies the value of old things. This temporal 'turn' offers a significant shift in the linear cradle-to-grave production and consumption path that has underpinned the twentieth century, one that in contrast offers a 'memory economy' in which value isn't predicated on the idea of the 'new' and which overcomes the assumption that we can detach ourselves from things in order to move into the future without ties to the past.

Acknowledgments

The Tales of Things project is supported by a Digital Economy, Research Councils UK grant, and made 'real' by our team: Barthell, R., Blundell, B., Burke, M., De Jode, M., Hudson-Smith, A., Leder, K., Karpovich, A., Manohar, M., Lee, C., Macdonald, J., O'Callaghan, S., Quigley, M., Rogers, J., Shingleton, D., Speed, C.

References

Bourdieu, P. (1984) Distinction: A Social Critique of the Judgement of Taste. London: Routledge.

Dodson, S. (2008) "Forward: A Tale of Two Cities" in van Kranenburg, R. The Internet of Things: A Critique of Ambient Technology and the All-Seeing Network of RFID. Amsterdam: Institute of Network Cultures, Network Notebooks 02, 2008. pp. 5-9. Available from http://www.networkcultures.org/_uploads/notebook2_theinternetofthings.pdf, accessed 20 November 2010.

Cox, B. (2009) Inglourious Basterds is cinema's revenge on life. Guardian Film Blog. Available from: http://www.guardian.co.uk/film/filmblog/2009/aug/20/inglourious-basterds-tarantino-change-history accessed 20 November 2010.

Castells, M. (1996) The Rise of the Network Society (Second Edition). Oxford: Blackwell.

Gitelman, L., and Pingree, G. B. (eds.) (2003) New Media: 1740-1915. Cambridge, MA: MIT Press.

Greenfield, A. (2006) Everyware, The Dawning Age of Ubiquitous Computing. Berkley: New Riders

Haneke, M. (2009) Every Film Rapes the Viewer. Spiegel Interview with Director Michael Haneke. Spiegel Online International. 21st October 2009. Available from: http://www.spiegel.de/international/spiegel/0,1518,656419,00.html accessed 20 November 2010.

Harvey, D. (1990) The Condition of Postmodernity. Cambridge: Blackwell.

Joffe, J. (2009) Interview on BBC Radio FiveLive, Breakfast Show. 30th June 2009.

MacCannell, D. (1999) The Tourist: A New Theory of the Leisure Class. Berkeley, CA: University of California Press.

Valhouli, C. A. (2010) The Internet of Things: Networked Objects and Smart Devices. The Hammersmith Group Research Report, available from http://thehammersmithgroup.com/images/reports/networked_objects.pdf, accessed 20 November 2010.

Waldner, Jean-Baptiste (2007). Inventer l'Ordinateur du XXIeme Siècle. London: Hermes Science.

Weinberger, D. (2002) Small Pieces Loosely Joined: How the Web Shows Us Who We Really Are. Oxford: Perseus Press.

CHAPTER FIVE

MOTORCYCLES, BODY AND EMOTIONS: THE MOTORCYCLIST'S SOCIAL CAREER

GABRIEL JDERU

Abstract

In this chapter, I approach motorcycling as a social phenomenon. The main concept used is that of social career, as advanced by Erving Goffman and David Matza. I highlight the specific bodily practices and emotions which occur in the successive stages of a motorcyclist's career. Throughout their social career as bikers, they learn how to manage emotions generated by this social practice. I then describe each stage of the social career. The study is based on data gathered from the main motorcycling on-line forum in Romania, participant observation carried out among bikers in Romania in 2008 and 2009, conversations and interviews with bikers. This work was supported by CNCSIS-UEFISCSU, project number PN II-RU 68/2010, *Modernity and Mobility in Romania: Motorcycle, Body and Emotions of Automobility*.

Introduction

Social identities are subject to learning processes. Unlike other social identities, such as sex or ethnic group, which are imposed to individuals from outside, the biker identity is taken on deliberately, emerging at the intersection of the *in-group* (motorcycle community) and the *out-group* (society as a whole), that is in relation with both bikers and non-bikers because "a community does not exist in a social vacuum but forms part of a wider society" (McDonald-Walker 2000, 14).

Even if the practice of using a motorcycle as a social learning process has already been analyzed in the sociological literature (McDonald-Walker 2000; Wolf 2000), the role played by body techniques and emotions in the development of the biker's social identity has not been studied yet.

Therefore, I will demonstrate that to become a motorcyclist requires an automobility which is socially situated, embodied and emotionally expressive.

The theoretical framework of this research is based on Goffman's work (1961) on the social career, Matza's theory (1969) on *affinity, affiliation and signification* and Sheller's concept (2004) of automotive emotions. I will also refer to Mauss' analysis (1979) of body techniques, Hochschild's theory (1983) of emotional work and Merleau-Ponty's phenomenology of the body (1962).

Methodology

The research data used in this text are part of an ongoing personal ethnographic research whose topic is a *Biker's Social Career in Romania.* I carried out and analyzed 24 semi-structured in-depth interviews with bikers and ex-bikers aged between 24 and 84 year old, I collected information by participant observation and informal discussions with bikers at their weekly meetings in Bucharest (March-October). I also attended 6 national meetings in 2008 and 2009, as well as regularly consulted motorcyclists' discussions on the most significant forum in Romania, www.motociclism.ro. These meetings allowed me to select individuals for in-depth interviews. The main topic of the in-depth interviews was the story of their first motorcycle acquisition. The reason I chose this stimulus-topic was to avoid a collection of *motorcycle myths.* Starting from this true *leit-motif,* I tried to identify the stages of a motorcyclists' social career. In this chapter I will present aspects bearing on the bodily practices and emotions experienced by motorcyclists.

Social Research on Motorcycling

Alford and Ferris (2006) classify motorcycle-related studies into several categories: 1) memoirs; 2) sociological studies; 3) travel books; 4) books on famous motorcycle brands focused on different historical aspects or information about the mechanics of certain classical motorcycle models; 5) manuals on driving techniques; 6) motorcycle stories and movies; 7) albums presenting motorcycle as an aesthetic object.

Sociological studies have been rather rare. Thus, Wolf (2000) has studied a group of Canadian outlaw bikers and McDonald-Walker (2000) analyzed motorcycling as a social movement. The recent volume edited by Veno (2007) gathers articles written by journalists, social researchers and motor bikers on American outlaw bikers. Bourne (2007) approached analyses motorcycle riding in philosophical manner. Truitt (2008) analyzed this phenomena within the sociology of mobility framework, approaching mobility as a social stratification variable in urban Vietnam. The author analyzed motorcycle as a factor of social stratification in the urban areas of Vietnam. Mellstrom (2002; 2004) and Nyazi (2008) discussed the relationship between masculinity and motorcycles, whereas Natalier (2001) described the the risk perception of motorcyclists. The reality show *American Chopper* was analyzed by Carroll (2008), who describes it as a perfect expression of the contemporary working class masculinity. An interesting study regarding the relationship between body and motorcycles from the perspective of the consumption theory is that of Halnon and Cohen (2006). Libret (2008) carried out a comparative analysis along several dimensions (structure, rituals, socialization processes, clothing etc) of the American outlaw bikers' clubs and the motorcycle police officers' organizations.

The motorcycle may be approached as a type of technology of mobility. Modernity creates new forms of sociability. Urry (2001) argues that, in modern societies, sociability may be linked to mobility. Moreover, both people and objects can be characterized by mobility, while movement involves certain "mobile technologies". The train, the airplane, the car, the motorcycle, the bicycle etc. are as many automotive technologies and movement "expert systems" (Giddens 1990). Within the vast field of sociology of mobility, automobility represents a distinct multidimensional subfield. Despite the multidimensional character of automobility, regarded by Mike Featherstone (2004) as a "mode of autonomous movement", most of sociological studies focus only on car culture. Despite the current expansion of the sociology of mobility, there is no study of motorcycling as a technology of mobility.

I argue that cars and motorcycles belong to separate emotional and bodily cultures. Motorcycling involves another type of automobility, different from that of the car, begging for separate attention. There are significant differences between cars and motorcycles regarding the "body techniques" (Mauss 1979). Building on Sheller's study (2004), inspired by Merleau-Ponty's approach (1962), from a phenomenological point of view, a moving body is a form of "being in the world". Sheller focuses exclusively the "driving body", which generates the feeling of "being in a car". The motorcycling riding movement differs from the car movement. Accordingly, I hereby advance the idea of existence of a *riding body*, from a phenomenological point of view, generating the feeling of *being on a motorcycle*. In the spirit of Maurice Merleau-Ponty's phenomenology (1962), the body lying on the motorcycle is an "experimental body", characterized by agency. Following Lipovetski (2006), motorcycle may be analyzed as a form of technology based body mastery.

Taking all these studies into account, I ask the following research question: *How does someone become a biker*? In order to answer this question, I used the sociological concept of *social career*, which helped me analyze motorcycling as a social learning process. Although there are some other authors who have analyzed the social career concept, I focused on the studies of Goffman (1961) and David Matza (1969). Matza developed a social career theory applicable to the deviant behavior analysis, which may, however, be used for non-deviant behavior analyses as well (Vail 1999). Next, I will turn to the data analysis and analyze the bodily and emotional conventions that individuals acquire upon entrance into the motorcycling world.

Becoming a Motorcyclist

The motorcyclist social career "can no more be a success than a failure"; it deals with "any social strand of any person's course through life" and with "changes over time as are basic and common to the members of a social category" (Goffman 1961, 127). Therefore, *I will describe and analyze the shared transformations for most motorcyclists, regarding their social identity.*

According to Matza (1969), the deviant career requires to pass through three stages: affiliation, affinity and signification. Similar to deviant behavior, motorcycling involved a social learning process, through which individuals learn the social conventions of this world. As I will show below, many of these conventions deal with bodily practices and the social expressions of emotions. To a certain extent, the stages described by David Matza overlap with Goffman's career stages (1961) (the prepacient phase, the inpatient phase and the ex-patient phase), but Matza's model is more suited to the motorcyclists' experience, because it does not assume that entering a certain world or institution is based on constraint. For motorcyclists' social career, one deals with the transformations of the individuals' self upon entering the symbolic universe of the motorcycling.

Learning to Become a Motorcyclist – *Affinity*

Affinity formation is the first stage of the dynamic development process of the motorcyclists' social career. For Matza (1969, 90-91), "Persons, either individuality or in aggregates, develop predispositions to certain phenomena [...] as a result of their circumstances". Thus, *affinity* "may be regarded as a natural biographical tendency borne of personal and social circumstance that suggest but hardly compels a direction of movement" (93). The discussions that I had with motorcyclists indicated their tendency to consider they were born to be motorcyclists. Similarly, they claim that they have been attracted by motorcycles since their birth. Following Matza's theory, the development of affinity for motorcycles implies a self-objectivation process, i.e. individuals must become conscious of the social category they are going to be part of. This represents the condition to turn affinity from the latent stage into the manifest stage. For Vail (1999), *affinity* equals the birth of desire and in the case of motorcycling it signals the transformation of possibility into desire. One may interpret Matza's idea of "circumstances" in a broad sense as socio-historical context or, in the narrow sense, as interactional context.

Assuming the first meaning, that of socio-historical context, one may identify certain structural factors which "push" individuals to use the motorcycle. A factor mentioned by McDonald-Walker (2000) is the post-war Western climate, where motorcycling become a possibility. Similarly, in postwar Romania, motorcycles left by the various armies and, later on, those manufactured in and imported from the ex-communist countries represented an automotive option, in a context where personal car ownership was extremely rare (Romania started to manufacture *Dacia* cars under Renault license at the end of the '60s). By the 1960s, Romania also began to produce Carpati and Mobra motor bicycles, which enjoyed tremendous success. Their mass production democratized of mobility during those decades. The testimonies of two old bikers are quite relevant in that respect:

> N_ My first bike...well, my first bike was a one-piston BMW 350 that I bought from a guy that sold World Word II German motorcycles parts and motorcycles. It was a light motorcycle, very reliable, that helped me a lot with my field work as a geologist. The guy I have bought it from lived on the same street as me and had a lot of scarp iron in his yard, as well as wheels, motors, motor bicycles… I don't know how he got them, but they were made during the war, that's sure. (N., 84 year old, ex-biker).

> S_ I bought my first bike when I was in the vocational school. There were some boys in my town that had motorcycles and who met weekly at the downtown pastry shop. My father was against it, but my mom helped me with some money and I bought it. Not everyone had a car then, as it happens now, I used to ride it everyday to get to the factory. I changed a lot of motorbikes, but I gave it up when I joined the Police forces, it didn't fit well with my uniform. (S., 52 year old).

The motorcycle also represented a pre-marital mobility option. My informant *N.* told me that he had bought his first motorbike in 1979 and sold it 5 years later. Two events were decisive for the decisions: the fact he attended the police sub-officers academy, eventually becoming a policeman, and that fact that he got married. In this social context, all these suggest that motorcycle is also connected to the concept of pre-marital masculinity. Moreover, that premarital mobility did fit the "weight" of police sub-officer status.

In other cases, *affinity* for motorcycles developed inside the family, as illustrated by testimony of another informant: *Oh, my first bike was my grandfather's* IJ*! I used to ride it when I was a little boy... I was crazy about playing on it... he did not let me play on it a lot 'cause he was afraid it could fall on me... I used to help him repair it; I stuck like a bur when someone repaired anything.... (D_42 year old, engineer).* The automobility climate during the postwar and the communist period and family influence are important factors for motorcycle affinity development in the case of old bikers (I conventionally call *old bikers* the individuals that also rode motorcycles before 1989, i.e. the fall of communism in Romania). A subsequent structural factor identified in the US (Lyng 1990), but valid for Romania as well mainly after 1990 (Ivan and Frunzaru 2010) is the tendency of individuals to assume voluntary risks, such as hang gliding, skydiving, scuba diving or rock climbing.

Currently, the riding body may be approached an "iconic body" (Varga 2010) because the motorcycle communicates meanings. In other words, the technology people use to move communicates certain information about them, becoming a form of self-expression. Thus, the contact with the symbolic world created by motorcycles participates in the formation of *affinity* for this type of body technology. The internet as a specific technology provided easy access to the *iconic body of a motorcyclist.* In my interviews, there was the recurrent idea that, before they decided to buy a motorbike, individuals first accessed motorcycles internet images. Today, one may say that this type of technology provides an anticipating socialization in the motorcycle usage for individuals attracted this practice. Therefore, *affinity* for motorcycles is no longer entirely dependent on space and time circumstances because, as Pirani and Varga (2010, xiii) argue, "while corporeal existence of human being is linked to space and time, electronic technologies overcome the limitation caused by the spatial and temporary existence and enable the users to expand the horizon of space and time, and thus influence the relationship between mind, body and technology". Following Goffman's theory, for the new bikers, the internet technology become an important *agent* of penetrating *motorcyclists' world*, helping the development of the first stage of motorcycling, that of *pre-biker*:

> F_ I had a friend from high school that used to tell me all the time that she attended the motor school with her boyfriend. That's how it occurred to me that I could buy myself a motorcycle. I had a strong wish for that in high school, but my parents wouldn't let me do it. The day she told me having got the license, I got stuck in the Internet. In fact, when I reached the office in the morning, before starting any work, I used to give the Internet the up-and-down looking for motorcycles. [...] I loved high-speed motorcycles and I was making all sorts of calculations to see what type of motorbike I could afford... [...] I still do so, I check it everyday to see what guys have posted for me (F. 33 year old).

Among interactional context elements, direct contact with other bikers and body experiences on the bike seem to be extremely important for the transformation of the motorcycle from possibility into desire. The first experience as a rider, either as a passenger or as a driver, seems to play the role of a founding act. For new bikers, it functions as a kind of revelation, since it is a body experience with heavy emotional load. It is an experience they could hardly describe, maybe because it generates a ultimate experience, a mixture of fear and pleasure. For C. (26 year old), the motorcycle discovery comed out in strong positive body experiences. This subject had previously told me that he had often seen motorcycles passing by in the street, but he had not had been attracted by them:

> "C_ When I first mounted on a motorcycle, it was quite unreal, a cousin of mine gave me a ride. The feeling was extraordinary, it was summer and hot and I felt the wind caressing my face. And that sense of motion you have when you take a curve, it's hard to describe. I held him tight – at first, I felt embarrassed, but fear made me hang on him tighter."

Another subject, E., fits the same experience category. He first rode a motorcycle when he started to attend the motor school together with a friend:

> E_:I couldn't imagine it was so cool, if you know what I mean, I could hardly believe that I could ride without falling. I had that special sensation, when you touch a motorbike, of something heavy, how can I put it, how could I ride without falling, how to coordinate my hands on the controls? And the instructor stayed off and shouted at me: squeeze the bike with your legs! [...] And the helmet, the gloves.... put me under high pressure, do you know the feeling when you go underwater and try to hear anything? (E_, 24 years old).

Such descriptions of first experiences on a motorcycle highlight the idea that mounting on a bike is the phenomenological equivalent of a break in the natural course of the daily body experience causing the feeling of "lived body", a form of "being in the world" (Merleau-Ponty 1962). In such a situation, if the individuals lack certain "body techniques" (Mauss 1979), the body becomes a "subject body" and mediate the relationship this individual has with the world, and riding the motorcycle turns into an embodied experience. In this example, the helmet, the gloves, the legs squeezing the motorbike "suggest the conjoining of human and machinic bodies" (Sheller 2004, 225, 227): "Motion and emotion, we could say, are kinesthetically intertwined and produced together through a conjunction of bodies, technologies and cultural practices (that are always historically and geographically located". In this stage, they have not learnt yet how to use their bodies.

The decision to use the motorcycle is embodied. In general, the depictions of affinity process ignore the embodied nature of the decision to choose a certain social practice. Sociological studies on bikers also lack the embodiment dimension. Many interviewed subjects declared that their decision to become a biker was not welcomed by their non-bike users close relatives, parents, friends or acquaintances. The nature of this expressed reservation regarding motorcycles comes from their association with risk.

Learning to Become a Motorcyclist – *Affiliation*

Once the motorcycle riding desire had been awakened, the aspirant motorcyclists pass through another range of transformations during the *affiliation* stage. This stage consists of the social learning of certain specific behaviors from the members of the group to whom the individuals wants to belong (Matza 1969). Formally, this stage begins with obtaining the driving license. In Romania, all motorbikes of 50 cm^3 or higher require a driving license. However, as most of the interviews and forum discussions indicate, it is not enough to graduate the driving school and get a motorcycle driving license in order to become a biker.

As far as individual performance is concerned, this is a problematical stage for the newcomer into the motor world. He/she has joined to a certain social category, but he/she has not yet appropriated well the role prescriptions implied by the respective category. Moreover, the motorcycling world is highly heterogeneous, being populated by different types and styles of motorcycles (speed motorcycles, touring motorcycles, chopper/custom motorcycles, off-road motorcycles, super-motorcycles, "communist" motorcycles). The choice of the motorcycle type becomes a form of self-presentation (Goffman 1959).

For many bikers, *affiliation* begins with the motor school registration in order to get a driving license. During one afternoon that I spent at the motorcyclist training ground, I overheard a discussion that the instructor had with his students. He was explaining the importance of learning the motorbike riding techniques: "*If you don't know how to drive a car, you cause an accident and you can kill somebody. If you don't know how to ride a motorcycle, you kill yourself*", he said. This piece of advice underlines the biker's body fragility compared to a car driver.

During this phase, newcomers into the motorcyclist social category feel the pressure to demonstrate their affiliation at the performance level. Purchasing biker outfit is a means of affiliation. I noticed many motor school students and beginners wearing expensive biker cloths (helmets, jackets, boots, protective pants, sunglasses etc). Various conversations indicated the fact that some of them had bought such objects long before obtaining the motorcycle driving license. There are two main reasons for that: the idea of biker's body fragility instilled by the motor school and demonstrating their affiliation to the rest of the community. I tend to believe that, for some beginners, wearing the protection equipment is more a sign of performative affiliation than an expression of assimilation of the safety rules. For them, it is important to look like motorcyclists.

For Matza (1969, 101), affiliation "refers to adoption or receiving of a son into a family", the process of "converting" somebody to motorcycling, i.e. he/she "begins to do something new for him/herself, but already familiar to the others". In other words, it is the process during which an individual starting to use the motorcycle learns to behave as a motorcyclist from other bikers. During the affiliation process, the individual learns the behavioral conventions related to the motorcyclists social career. As I will argue below, among these behavioral conventions, bodily and emotional conventions play a significant role in the development of such a social career. Individuals also learn to relate to the motorcycle as an object. This is the stage of the expansion of the biker self, when the motorcycle becomes an extension of the body and of the social self. The new biker adheres to the ideological field having the motorcycle at the center, to which he/she starts to bring his/her contribution to it. Lying "at the beginning" of his/her career, he/she is compelled to find mechanisms enabling the management of his/her relationship with the motorcycle. He/she has not learnt well enough the necessary body techniques in order to get a high degree of confidence, required for a motorcyclist to ride safely. Therefore, he/she has to learn how to use his/her body when taking curves, how to keep a still motorcycle steady, how to slow down using both wheels, how to use mirrors when riding etc. All these are practiced at the motor school, but become a real problem in traffic.

One of the behavioral conventions learnt during that phase is that he/she has to greet other bikers met in traffic. Greetings are usually made using the left hand or rising the left leg sidewise when traffic conditions do not allow loosing the clutch. One can also greet others by an up-down movement of the head or using the headlight. The conversations I had with beginners revealed that this social convention is not easy to observe if individuals do not master the motorcycle riding techniques very well. There is a special topic on the motor forum, where bikers narrate how they greet in traffic:

> Mjollnir_ I tell you, it's really uncool when you are a beginner and somebody greets you and you are just too afraid to take your hand off the handle bar. Eventually, I asked my passenger to salute in my place. During the 350 km I rode yesterday, I was greeted in 3 different ways: a guy even took both his hands off to say hi (omg!), most of them rising one hand over the handle bar and there was a guy on a superb chopper who saluted me bringing his hand under the handle bar. (Aug 20 2007, 09:15 AM).

Fear of the heavy traffic in the city is one emotion that one has to manage at this stage. The most common way of daring to ride is to ride at night, when traffic is low; they just follow more experienced bikers or friends who drive cars. Also, many of them alter the exhaust system, so that the motor sounds' are stronger, motivating that is safer to be heard in traffic. (Alternatively, one may interpret this as the search of motorcycling authenticity). Emotion is sometimes stressed by unfortunate events experienced by almost all bikers: minor accidents, falls, and stories about other bikers' accidents. In this case, fear has a psycho-social nature, implying the re-signification of biological fear through interpersonal communication and influence of mass-media. This emotion is found at the base of the "social fears pyramid" (Chelcea 2010, 7-10). In other words, individual experience of fear is mediated in the motorcycle emotional culture where the idea of body safety is central. Later on, when the biker becomes self-confident, fear turns into a background emotion.

Aversion against car drivers (mocked as *tin can individuals*) is a feeling to be learnt during this period. Strong negative stereotypes develop between motor bikers and car drivers. This feeling is part of the emotional culture of motorcycling and is expressed in body-related terms: *cans* are protected bodies, while *organ donors* (bikers) are exposed bodies. Here is such a comment: dan_1979b_: *I was riding smoothly at about 50km/h* [in

Bucharest]. [*In one intersection*], *when the stop sign turned green for me, a jerk with his tin can popped up from my right hand side. What was I supposed to do other than step on the brakes as much as I could...The rear end of my bike shifted many time to the right side to the left. Thank God that the jerk did not stop in the intersection, because otherwise I would have crushed straight into him (April 13, 2010, 03:31 PM).*

Moreover, the management of the embarrassment emotion is an important element for the construction of the biker's social career. This emotion can be linked to the performative character of this affiliation stage. It is the result of the role distance (Goffman 1956). For beginners, the embarrassment emotion is caused by the performance of a role for which they are not yet ready. Here goes "T_", who wrote:*When I first went to the Arch, I passed by it three times until I decided to enter the park. I was afraid not to fall down in front of everybody and to make a fool of myself" (T., 26 year old).*The feeling of embarrassment may also come from one's self-perception of his/her body riding the motorcycle, marking the passage to another a better motorcycle:

> D_ "My first motorbike was a Mobra [the Communist era mass produced Romanian motorcycle] and I was really pleased with it. Until I saw myself in the windows of the buildings erected in Victoria Street just after the '90s. When I saw how pitiful I was, a big guy on a tiny Mobra, I said I had to get rid of it. I changed the second one, an Ij, when my wife, a high school student by then, told me: <you don't come again to me with dirt under your nails!>" (D_., 42 years old).

The precarious statute of a riding body confers to the individual converted to motorcycling a kind of *contemporary heroism.* I noticed that many discussions among the motorcyclists meeting at the Arch of Triumph in Bucharest are about accidents or possible accidents in which they were or were about to be involved, but escaped. The online motorcycling forum that I analyzed also includes many discussions of that nature. Bikers participate in the community through this type of stories in which the negative characters are the car driver. This is the expression of the affiliation need making bikers ascribe positive qualities to in-group members and negative qualities to out-group individuals, i.e. car drivers.

The feeling to be in the car, completed by a wide range of emotions (driving joy, anxiety), generates a feeling of safety (Sheller 2004). While the car creates "the feeling of being in a car", i.e. inside, the motorcycle creates *the feeling of being on a motorcycle*, i.e. outside. *In* the car, one's body is protected, during the journey the car is a kind a house sheltering the body and one's sight is limited to the view offered by the windows of the house; *on* the motorcycle, one's body is exposed, but the sight horizon is much larger, the contact with the environment is direct. Also, the proprioceptivity activates other sensations such as smells, the air touching one's body, curve swinging etc. On the bike, the self "colonizes" the body and struggles to dominate it. Inside the car, the individual's statute is that of a spectator-body; on the motorbike, the individual's statute is that of actor-body. The former contemplates reality, while the second actively participates in reality. In the car, the body is taken for granted; on the motorcycle, the body is problematic. McDonald-Walker (2000) reported a similar experience of her research subjects (see also Holmstrom 2002, xix - for a similar description).

Learning to Become a Motorcyclist – *Signification*

According to Matza's theory (1969), during the *signification* stage, individuals come to internalize their new statute regarding this social practice, re-signifying their whole identity. During the affinity stage individuals experience the awakening of the desire to become a biker. Then, during the affiliation stage, they start attributing themselves signs of their belonging. In the last state they focus on the transformations of the motorcyclist's self. It is the biker's social career stage that best express what it means to be keen on motorcycling. Once having reached this stage, the motorcycle becomes takes the qualities of *axis mundi*.

The prototypical biker of this stage is the *veteran*, i.e. the individual activating as a motorcyclist for several years. In 2009, the theme of veterans' meetings was first launched as a discussion forum topic. The analysis of those discussions underline that experienced bikers tend to create certain forms of association in order to distinguish themselves from other motorcyclists. They debate the social definition of the experienced motorcyclists. Thus, one of the most passionate debates under this topic is about the number of experience years one should have in order to join the veterans' category. Some argue that one needs to have 10 years of experience, others claim that a smaller number of years will do. Some suggested that experience should be evaluated and acknowledged by means of yellow laces that bikers visibly could display on their jackets. One participant suggested that one lace should count for every 10 years, although not all participants in the discussion agreed with this solution. In this context, following Bourdieu's conception (1984) regarding the stratification of social and cultural practices, the lace may represents a form of objectivation of social distinction: a high number of traces correspond to a higher distinction, while the bearers of these signs could be regarded as a *cultural aristocracy,* relative to the other members of the motor community in Romania. The veterans' tendency to distinguish themselves from the others may be found in the language used. I found out that, in many cases, in their daily conversations, veteran bikers call both non-bikers and non-experienced bikers "civilian", developing a sense of pride when comparing to them.

Another type of social identity re-signification during this stage is illustrated by the owners of motorcycles produced in ex-communist countries. Due to the existing socio-historical context, there are many such owners in Romania. In their case, the distinction criteria are represented by technical skills and mechanical know-how. These individuals develop another version of what passionate bikers means, focusing on their ability to repair their motorcycles by themselves. A specific feature of these motorcycle users is the tendency to assign *agency* to motorcycles. *Agency* refers to the capacity of objects to have intentions and their own personality (Dant 2005, 60). These owners talk about the respective objects as if they were alive, having their own individuality and will. These types of motorcycles are "stubborn", "devilish", they get break up frequently, turning the motorcyclist into a permanently fighter with the object of his/her passion. The relationship they have with these motorcycles is highly antagonistic at emotional level. They are hated since they break down very often, but they are loved because they can be dominated by socially accumulated technical knowledge. The internet forum section dedicated to the communist motorcycles is almost exclusively focused on mechanical knowledge. This topic becomes a collective memory support helping bike owners to manage their relationship with the object.

As for the bodily concerns, safety is the key world during this stage. Old age bikers live under the specter of the "defeated body" (LeBreton 1990), i.e. the body that listens neither to the soul, nor to the mind. During a communist motorcyclists' meeting, I heard a discussion among individuals who tried to find solutions for the moment when, due to their old age, they would not be able to ride a motorcycle anymore. The sidecar was considered a viable solution. Unlike the car driving, for motorcycling, incompatibility with an aged body is felt more intensely.

Conclusions

To become a motorcyclist is to experience change. This is constructed at the intersection of the structural transformations of the society, the general automobility climate (McDonald-Walker, 2000), the tendency to voluntarily assume risks (Lyng 1990; Ivan and Frunzaru 2010), the iconic statute of the body (Varga 2010), the development of online technology etc. and individual experiences. The social career theories developed by Goffman (1961) and Matza (1969) constitute a rich conceptual framework, useful for understanding how one becomes a motorcycle user. They allow for a dynamic vision of the way social practices are acquired, developed and socially re-signified.

I also stressed that all body and emotional experiences of the individual becoming a biker form the core of the biker's social career, i.e. a range of common experiences for all motorcyclists. Moreover, becoming a motorcyclist involves social learning. It starts with the formation of the *affinity* to ride a motorcycle and continues with *affiliation* – the stage of the social career in which individuals who wish to become motorcyclists learn from other bikers how to use their body to that end and how to manage the emotional experiences associated with the motorcycle use. It culminates with the *signification* stage, mainly characterized by the transformations of the structure of the self, with the motorcycle becoming a kind of *axis mundi* for the motor rider.

In contemporary societies, electronic technology plays a major role in the course of one's motorcycling career. It mediates all the stages of a bikers' social career. The sites dedicated to motorcycle users may be viewed both as virtual social spaces where bikers may interact and as a means to preserve, structure and transform the social memory of the motorcyclist community.

References

Alford, Steven and Suzanne Ferriss (2006), Classifying Motorcycle Writing: A Preliminary Consideration. *International Journal of Motorcycle Studies*. http://ijms.nova.edu/July2006/IJMS_Resources.AlfordFerriss.html.

Bourdieu, Pierre. 1984. *Distinction. A Social Critique of the Judgement of Taste*. Cambridge: Harvard University Press.

Bourne, Craig (2007), *Philosophical Ridings. Motorcycles and the Meaning of Life*. Oxford: Oneworld Publication.

Carroll, Hamilton. (2008), Men's Soap: Automotive Television Programming and Contemporary Working-Class Masculinities. *Television New Media*, 9: 263-283.

Chelcea, Septimiu (2010), Piramida fricilor sociale. Fricile sociale în Romania – o schiță psihosociologică. *Sociologie Românească*, vii: 3-18.

Dant, Tim. (2005), *Materiality and Society*. New York: Open University Press.

Featherstone, M,.(2004), Automobilities: An Introduction. *Theory, Culture and Society* 21: 1-24.

Giddens, A. (1990), *Consequences of Modernity*. Stanford: Stanford University Press.

Goffman, E., (1956), Embarrassment and Social Organization. *American Journal of Sociology* 62: 264-274.

—. (1959),. *Presentation of Self in Everyday Life*. Garden City, New York: Doubleday.

—. (1961), *Asylums. Essays on the Social Situation of Mental Patients and Other Inmates*. New York: Anchor Books.

Halnon, K., B. and Cohen S. (2006), Muscles, Motorcycles and Tattoos: Gentrification in a new frontier. *Journal Consumer Culture* 6: 33-56.

Holmstrom, Darwin. (2002), Introduction to *The Complete Idiot's Guide to Motorcycle,* by The Editors of Motorcyclist Magazine with Darwin Holmstrom, xix-xxi. Alpha: A Pearson Education Company.

Le Breton, D.(1990),. *Anthropologie du Corps et Modernite.* Paris: Press. Univ. Fr.

Librett, M.. (2008). Wild pigs and outlaw: The kindred world of policing and outlaw bikers. *Crime, Media, Culture* 4: 257-269.

Lipovetsky, G.. (2006).. *Le bonheur paradoxal. Essai sur la societe* d'hyperconsommation. Paris: Editions Gallimard.

Lyng, S.(1990). "A social psichoologycal analysis of voluntary risk taking". *American Journal of Sociology*, 95 : 851-886.

Ivan, Loredana and Valeriu Frunzaru. (2010),. Being in control by taking voluntary risk in contemporary society. *Sociologie românească* 8: 80-86.

Matza, D. (1969), *Becoming deviant*. Englewood Cliffs: Prentice-Hall.

Mauss, Marcel. (1979), "The notion of body techniques". In *Sociology and Psychology*. Essays. London: Routledge and Kegan Paul.

McDonald-Walker, S.. (2000),. *Bikers Culture, Politics and Power.* New York: Berg.

Mellstrom, U.. (2002),. Patriarchal Machines and Masculine Embodiment. *Science Technology Human Values* 27: 460-478.

—. (2004). „Machines and Masculin Subiectivity: Technology as an Integral Part of Men's Life Experiences". *Men and Masculinities* 6: 368-382.

Merleau-Ponty, M. (1962),. *The Phenomenology of Perception.* London: Routledge and Kegan Paul.

Natalier, K.. (2001),. „Motorcyclists' Interpretations of Risk and Hazard". *Journal of Sociology* 37: 65-80.

Nyanzi, S. (2008), Male Promiscuity. The Negotiantion of Masculinity by Motorbike Taxi-Riders in Masaka, Uganda. *Men and Masculinities* 20: 1-17.

Pirani, B.M. and Varga, I.,. 2010. Introduction to *Acting Bodies and Social Networks. A Bridge between Technology and Working Memory,* eds. Pirani, B.M. and Varga, I., xiii-xxxix. Lanham: University Press of America.

Sheller, M.. (2004),. Automotive emotions: Feeling the Car. *The Theory, Culture and Society* 21: 221-242.

Truitt, A..(2008),. „On the back of a motorbike: Middle-class Mobility in Ho Chi Minh City", Vietnam. *American Ethnologist* 35: 3-19.

Urry, J.. (2001),. *Sociology beyond Societies. Mobilities for Twenty-First Century*. London: Routledge.

Vail, A.. (1999).." The Outside of a Thinh Is Half a Back: Negotiating the Canvas among Fine Art Tattoo Collectors". *Journal of Art Management*, 28: 261-277.

Varga, I.. 2010. "Icons, symbols, cultures". In *Acting Bodies and Social Networks. A Bridge between Technology and Working Memory,* eds. Pirani B.M. and Varga, I., 53-60. Lanham: University Press of America.

Veno, A. (ed.). (2007),. *The Mammoth Book of Bikers.* Philadelphia: Press.

Wolf, D. R. (2000),. *The Rebels. A Brotherhood of Outlaw Bikers.* Toronto: University of Toronto.

CHAPTER SIX

MEMORY AND SOCIAL MEDIA: NEW FORMS OF REMEMBERING AND FORGETTING

ROBERTA BARTOLETTI

Abstract

The focus of my contribute is on the relation between individual and collective memory and social media. Social media are taken into account not only as communication realm but also as memory places, where individuals construct their memories and select what will be remembered or erased. Secondly, we can suppose that in social media individual and collective memories can be constructed in a more participated way, thank to connection and contents sharing technologies. Starting from the hypothesis of a mutual shaping of memory and media and that the users are able to shape technologies of memory in order to construct a meaningful world, the paper intend reflect on potential and actual capacities of social technologies to generate new pattern of remembrance and forgetting.I attempt to inquire about this question through the analysis of some case studies. We can indeed observe many new technologies of memory at work and many different kind of memory practices in the WWW: while Wikipedia can be observed as an objectification of a participate cultural memory, the web logs emerge as places of construction of individual memory and identity and Flickr as a shared archive of personal visual memories. Concurrently, new forms of participate social archive about historical events are rising and so on. The aim is to design a first phenomenology of grassroots memories in connection to social media and to reflect about the "social media memory", that is characterized by new problems and new possibilities related both to remembrance and to oblivion.

Introduction

My contribution will regard changes involving memory in the era of the social web, at a time in which we are witnessing a clear novelty in the panorama of media: the exponential growth in the participation of users in the production of communication and culture. The hypothesis from which I intend to begin is that on the Web users construct not only communication and sociality[1], but also memory, an aspect which has to a large extent been ignored up to now. This creation of memory occurs on blogs and social networks and in general in every platform open to grassroots participation and sharing of contents that are generated by users, or at least selected because they view them as significant. My focus will therefore be on the networked production of memory online, which takes place both at an individual level and at a collective level. I believe that empirical proof of this supports the hypothesis that the web can constitute not only a place for communication, as is self-evident, but also a place for remembering and for the narration of life experiences – which can be individual and collective, textual and visual – through which the work of memory becomes visible. As a consequence, the social Web has become a place in which identity and meaning are constructed, and to which memory is closely linked through a process in which the body (and its memory) certainly cannot be excluded. While reflections on the production of cultural content, communication and sociality online are already fairly abundant in the literature – and others regarding identity are not lacking – reflections on memory in social media and on social networks sites seem to me to still be taking their first tentative – albeit significant – steps. Accordingly, this sets my work in a field that is still relatively unexplored.

My observations will focus on digital arenas where memory is at work and on the memory practices defined in social media – in some cases, as we shall see, these practices are real rituals. While this will be the empirical focus of this paper, the question which I am attempting to answer through the observation of places and practices online is whether and how the coupling of individual memories (consciences) and social media produces changes in our forms of remembering and of forgetting, which themselves have implications at the individual and collective level and at the level of culture at large. In order to attempt this, I have identified several preliminary questions which I feel it will be crucial to address in future research on networked memory, which must necessarily

[1] See Mazzoli (Ed.) 2009 regarding aspects related to communication and sociality on social network sites.

be interdisciplinary in nature. I should also state beforehand that it is not the aim of this paper to supply answers to all of the issues identified, nor will I be able to allocate to each of them the in-depth examination they actually require; nonetheless, I think it is important in any case to clearly identify them as, together, they form a fairly complete and complex picture of the questions regarding memory in the era of networked digital media. If this is so, even simply enunciating them will allow me to then clarify the framework within which I wish to move, even though my analysis will be concentrated entirely on just a few aspects.

My starting point is a constructivist conception of memory as a selective structure operating in relation to the distinction between remembering and forgetting. Within this framework, I also start from the hypothesis that the encounter between media and individual memory takes the form of a mutual shaping the outcome of which is impossible to foresee beforehand – thanks also to the creativity that characterizes the uses of users – the drifting of which in unforeseeable directions cannot be understood without an in-depth analysis of the structure, rules and affordances of the different media – in our case of the various social media – and of their specific way of shaping the human mind. This hypothesis of mutual shaping of memory and (social) media is based on the one hand upon consistent conclusions in the literature on communication and media (Silverstone, Hirsch, Morley 1992) and, on the other hand, on the literature regarding consumption and uses (first and foremost the work of Mary Douglas and Michel de Certeau), which has found confirmation in the specific although not yet vast literature on the relationship between media and memory (a good starting point for which would be the work of van Dijck 2007 and several recently published articles in the journal *Memory studies*).

Within this framework of mutual shaping, the analysis of the relationship between social media and memory is played out on three distinct levels, each leading to numerous issues and questions. At the level of culture and production of meaning, one question that I intend to examine is whether the recent evolution in social media is offering new possibilities for the production and expression of "grassroots memories", and if this is the case, under what form. The emergence of new forms of participated memory can be singled out as one of the clearest outcomes of this new era on the web, and perhaps one of the most foreseeable (and as a result perhaps it should even be considered as problematic, so as to avoid overly simplistic analyses), that involves the current statute of cultural memory. Indeed, these are collective memories that can count on new forms of exteriorization and objectification – both textual and visual forms. The second question is instead related to the changes that are involving embodied memories: that is, the effects of the relationship between social media and memories with regard to the remembering of individuals and their way of constructing the present, past and future. Finally, shifting from the issue of embodied memories and culture to the level of the social system and media system, we should ask ourselves if we can really speak of a "social media memory", and of a specific way of managing the distinctions between remembering and forgetting, between redundance and variety, that supposedly characterizes the new social media in a specific way with respect to social memory *tout court* and to the memory of (mass) media, each as delineated in Luhmann's systems theory. It will be necessary to reflect on the primacy of forgetting that in this theoretical framework characterizes (modern) social memory and particularly the memory of media, and which performs an essential coordinating function for the memory of society as a whole: we need to verify if and how this primacy is being re-defined and re-shaped, or even contested and challenged by the social media[2].

In order to respond to these questions, we need to return to several key categories of reflections on memory that can be found described in both classic and contemporary scientific literature (*e.g.*, Maurice Halbwachs, Jan and Aleida Assmann, Niklas Luhmann[3]) and on the relationship between the media and memory, on which to base an exploration of the current phenomena of memory in/of social media through diverse digital arenas inhabited by users: blogs, social network sites, platforms for content-sharing such as YouTube, Flickr and Wikipedia itself, various forms of social archives in historical events and, lastly, online virtual worlds.

1. Plural and mediated forms of memory

In everyday language and in the awareness we take for granted of our world, we tend to equate memory with the remembering, just as we normally tend to judge remembering positively and forgetting negatively. In reality, our daily aproblematic observation of memory is afflicted by an

---

[2] The problem is particularly vast, and would require an in-depth analysis of the relationship between the traditional mass media and social media, for which there is no space for discussion in these pages. I refer, for example, to the classic issues of 'agenda setting' by the mass media in relation to possible interference in the topics dealt with online, for example in the blogosphere.

[3] Obviously this is a selection in the ambit of the vast literature available on this topic, which could be critiqued for its partiality but for which I take responsibility.

enormous blind spot – oblivion – which veils the other indelible side of memory as a selective structure. We should start then, by reiterating that the term 'memory' is used to describe a selective process that allows us to distinguish between things to be remembered and things to be forgotten; memory is not then simply a recording of the past, but rather a continual reconstruction of the past in the present by means of this dual operation[4]. In this ambit, forgetting plays a role no less crucial than that of remembering, and thus cannot be understood as a simple loss or failure of memory (Connerton 2008). Forgetting has an essential function in that it eliminates aspects which if remembered could be damaging or useless, both for individuals and for society; forgetting has a function of ensuring internal coherence of memory and, by means of its cancellation, opening towards change and the future.

What's more, advanced studies on memory have led to a surpassing of the metaphor of cognitive psychology – which, starting from an analogy between the mind and computers, defines memory as the processing, storage and retrieval of information – integrating the previously overlooked role of subjective experience of remembering. Today, memory is conceived as a complex and diffuse mental faculty which does not reconstruct the past faithfully, but instead is responsible for a continuing process producing individual memories that depends on the meaning ascribed and the emotions linked to the embodied experiences of the individual. This is true above all when the events are important for one's sense of self (autobiographical memory), in conferring uniqueness to one's own life (Schacter 2001). Thus, we can state that the approach that is currently dominant in memory studies is constructivist in nature (A. Assmann), and that it has been supported by the evolution that has taken place in the neurosciences themselves, although substantial differences can be found when comparing the subjective and social levels of memory.

Indeed, memory is a plural idea if we consider the different planes on which this continuous work of selection takes place: at the level of single individuals, concrete groups (families, friends, social and political groups, nations, etc., as analysed in particular by Halbwachs in relation to his conception of collective memory) and society as a social system. In these three different arenas, a distinction is made between individual memory, collective memory and social memory. At the first two levels, the work of memory is linked to meaning and identity: individual memory is linked to the construction of reality, of sense of self (autobiographical memory) and

---

[4] Here I refer to the discussion already published in Bartoletti 2007, to which readers can refer for further discussion.

of a connection to others, while collective memory is the memory of a concrete group that roots its identity also in its memories of a shared past to which the group ascribes significance, not only giving sense to the present and open to the future, but also allowing the construction of differences between us/them (Halbwachs, Assmann). Collective memory is naturally not simply a banal sum of diverse individual memories, and as it is not an embodied process it must necessarily count on forms of objectification outside the human mind and body, starting from rituals and language. In this sense, collective memory is inherently mediated.

Finally, the idea of social memory refers to a third level, to a function of society understood as a social system in its most abstract definition. "Social memory" as defined by Niklas Luhmann is the memory of modern society, which has taken on unheard of levels of complexity and abstraction and above all a radical autonomy from the memory of individuals and from the embodied need to remember and to forget, as well as from any need for cultural identity, for grounding in myth or the search for one's origins, both at an individual and a collective level – in other words, a radical autonomy from any need for the meaning from which embodied memories instead cannot be avoided. It is not by chance that this idea of social memory has led to pronunciations regarding the uselessness of the idea itself of collective memory as a conceptual tool for understanding society today (Luhmann 1997), at least as regards the specific field of sociology. In this theoretical framework we can instead observe an immediate and direct relationship between social systems and consciences (in the form of structural coupling), which has important implications for our discussion of memory. Indeed, this perspective clarifies the direct relationship between embodied memory and media memory, which on the one hand calls into play the existence and the cogency of the specific criteria of memory of the mass media system (for ex. the primacy of novelty) and on the other hand the possibilities for and restrictions to embodied memory as permitted by the single media.

On all these different levels of memory, we can individuate different forms of mediation. At the level of individual memory, the embodying of the ability to remember and forget does not exclude the possibility of reverting to external forms that can be tools for memories or for oblivion, among which we can include rituals, objects, texts and technologies useful in recording and storing memories – ranging from diaries to family photos, to videos taken with digital cameras. All these forms of mediation concur in the processes of remembering and of forgetting, shaping them, and are themselves shaped by the work of embodied human memory (van Dijck 2007). This mediation is obviously a fundamental aspect of collective memory, and in particular in relation to its most institutionalised form, that is, cultural memory as described by Aleida and Jan Assmann. In the late Nineteen Eighties, the Assmanns in fact re-examined and re-worked Halbwachs' idea of collective memory (which is universally cited, but often with the intent to criticize it) distinguishing between communicative memory and cultural memory in relation to the type of past the memory refers to and thus to the functions of remembering. Communicative memory regards a "recent past", which reaches back in time through the life experiences of an individual and his contemporaries and which is accessible to individuals through direct experience or the tales of those who witnessed or experienced it, while cultural memory refers instead to a mythical time of one's origins. While communicative memory is preserved and passed on in interactions between contemporaries, and indeed depends on the fragility of individual life and embodied memories, cultural memory is instead typically an institutional and objectified collective memory which defends itself against the lack of permanency of individual life and affirms its role as official memory. Cultural memory is accessible through various forms of externalization and objectification of memories: for example, in books, rituals, monuments, toponymy and calendars. Alongside this traditional objectification we can today also find media products, both fictional and non-fiction, that give a 'written' form to memories and contribute to the recursive reconstruction of the past in the present. Cultural memory refers to something that is a true policy of memory, in that it regards the time of origins which gives roots to the identity of the community that together remembers. Therefore cultural memory can be left neither to chance nor to subjective emotions.

According to Jan Assmann, it is cultural memory which, within the vaster collective memory, builds the identity of a group that together remembers, in a more aware, stable and strategic way, in order to guarantee its solidarity and permanence with the passing of time. While communicative memory is instead not highly structured or hierarchized and is widespread among all the members of a group that participate in social interaction, cultural memory is limiting and has privileged holders who may range from priests to shamans to scholars but all of which follow a specific training. It is to them that society entrusts the task of watching over what is remembered or forgotten, and it is they who are official "memory entrepreneurs", to express the idea with a modern and yet appropriate metaphor. It is they who have been given the institutional role of safeguarding and passing on to future generations the cultural sense of a collectivity: its educational institutions, its academic historians, its museums, its politics, and today the news media themselves, etc. This is clearly a traditional way of looking at the means through which societies construct the past, and by doing so make sense of the present, and indeed today it is being challenged on many fronts. Among these, it is of interest to our discussion to examine the role of the emerging social media, which have shown themselves able to strongly challenge the institutionality and legitimacy of cultural memories managed strategically by official memory agencies, which in our advanced modern societies are not only plural, but are also each autonomous in the creation of their own memory (Luhmann 2000).

Mediation is, in fact and finally, an essential and dominant characteristic of social memory which, because it is an autonomous memory that in reality came onto the scene only with the appearance of the press and established itself with the modern broadcasting system of electronic and digital media, which today guarantee a radical autonomy of social memory from both collectivities and individuals; this autonomy is a constituent characteristic of this form of systemic memory (Esposito 2008). It should also be stressed that the evolution of the media, from the press to electronic and digital forms, has corresponded with a progressive decline in an order that was hierarchical in nature, founded on a centre or vertex, and to which has corresponded the emergence of a heterarchical order, founded on a plurality of centres of modern society – and thus corresponding to a shift towards a functional social differentiation (Luhmann 1997). Even the most recent evolution of Internet has confirmed this trend; however, with the advent of social media, heterarchy seems not only to be a characteristic of the social structure (and therefore of the forms of differentiation at the level of society as a whole, which translates into a relative plurality and autonomy of the different functional systems), but to also even more importantly translate into a growing widespread diffusion of grassroots production of culture and meaning. Hence, heterarchical criteria not only regard the structure of society as a whole, but are also made concrete within each and every partial social system, casting doubt on the fundamental distinctions that give a hierarchical order to their actors: producers vs. consumers, elected representatives vs. voters, administrators vs. citizens, broadcasters vs. audiences, etc. Although the social media certainly have not led to an overcoming of these distinctions (to say so would be academic demagogy), but they certainly do strongly challenge them, as we shall see[5].

[5] In other words, also in this case we are observing a kind of prosumers at work, in relation to memory production. About prosumption see Ritzer 2010.

In conclusion, although the contribution of the Assmanns introduced elements useful for understanding the complexity of the different forms of collective memory – making the distinction between embodied, biographical remembering and cultural, institutionalised, founding remembering – it does not examine the transformations produced by modern communication media which are influencing all fronts of remembering and are obviously taking on a crucial role in the autonomization of memory from embodied experiences and from concrete interaction among contemporaries. As we shall see, they can also produce pathways of autonomization of collective memory from its most institutionalised forms. What's more, this is without taking into account the undoubted importance of the coupling of memory and different technological forms of mediation, ranging from blogs to digital cameras. One weakness which was already identifiable in the idea of collective memory as expressed by Halbwachs and that has yet to be resolved with the re-elaboration by the Assmanns is an issue related to the role of the communication media and, in particular, of the many memory technologies through which memories are being constructed and experienced. These are technologies which mediate the production of collective and individual memories: they mediate recording, storage and retrieval as well as erasing[6]. No less important, they also mediate sharing with and broadcasting to others. Memory technologies include indeed traditional forms of objectification such as souvenirs, memorials, the body itself and other material objects[7], but more and more frequently they are involving digital and audiovisual technologies: films, photographs, media products in general, and Internet itself with its sites, the crucial role of which today is undeniable[8]. Individual and collective memory is being both constantly produced through memory technologies, and mediated by these in ways of which we are more or less aware: just think, for example, of the mass diffusion of photographs, first in analogical then digital form,

---

[6] Regarding this limit, see the discussion in Bartoletti 2007, and the work published by Van Dijck in 2007, which moves in the direction of overcoming it, reflecting on reciprocal shaping that takes place between media and memory. As regards memory technologies, see the recent reflections on the topic by Sturken 2008, and Van House & Churchill 2008 in the journal *Memory Studies*.

[7] See Bartoletti 2007 and 2008 on the relationship between memory and material objectification.

[8] For an interesting study by the Istituto Parri on European television's representation of history and consequently on the role of this mass medium in the shaping of cultural and collective memory, see: *Media e cultura comunitaria. Per una storia televisiva dell'Europa*, 1° Rapporto di Ricerca, Centro Stampa della Regione Emilia Romagna, March 2009.

and of the voice and sound recording technologies, of home movie production, of the Super 8 technology used in the Sixties and today's digital video cameras and camera phones, *etc.*. Indeed, memory technologies mediate the production of memories on this dual front: both through the mass production of media contents placed alongside and in competition with the official objectifications of cultural memory and by creating a mass market for supports that have given shape to a series of memory practices for the creation of memories that are mediated at their origin[9]. Just as the communication media and memory technologies influence remembering, they also influence forgetting, by introducing changes which require us to stop and reflect, even though this aspect still remains somewhat overshadowed both at the level of individual awareness and in the awareness of scholars themselves working in the field of media and memory.

My specific intent is now to explore how memory technologies (from writing to photography to audiovisual production) are working in interaction with the development of the social Web, both at the level of remembering and of forgetting, giving new forms to both individual and collective – as well as cultural and communicative – memory. In this framework, the classic dualisms that characterize a significant portion of the studies on memory and media lose their clarity and risk being misleading, if our intent is to understand the reciprocal shaping of memory and media, which today is tending toward a gradual blurring of the distinction between public *vs.* private, natural *vs.* artificial (van Dijck 2007), or better yet, between "embodied" *vs.* "mediated".

2. A few comments regarding methodology

Before moving on in my analysis, it may be useful to examine a few issues regarding method, so as to limit and clarify the universe being examined – which for the moment I will define as the social web and/or social media – both in terms of places and users involved, as well as the main items through which I intend to orient my observation of this universe made up of digital places and of people in relation to the work of remembering.

Our reference universe is first and foremost Internet, that is, a digital environment for writing and interact, a digital environment for networking. More precisely, we can limit our observation to digital environments in which the participation of users in the production of content is not only

[9] Regarding photography, and in particular the role of amateur photographers in the early phase of development and roll-out of this medium, see Sorlin 2001.

allowed but rather constituent: this is an evolution of the web that has made explicit and concrete its implicit social potential, which has become concrete through various technological forms. The social web therefore includes both social network sites in the strictest sense and, in a broader sense, other sites in which content is first selected and produced by users, then subsequently published and shared. With the term social network sites, according to the recent but 'classic' and widely agreed upon definition given by Boyd and Ellison (2007, p. 1), we intend "web-based services that allow individuals to (1) construct a public or semi-public profile within a bounded system, (2) articulate a list of other users with whom they share a connection, and (3) view and traverse their list of connections and those made by others within the system". The connection between users who organize their own contents within their personal profiles (regardless of whether this content is self-produced or simply published), probably represents the key characteristic of these digital environments, which clearly combine both a private, intimate dimension (the profile, and thus the construction in public of an image of oneself and one's world) with a social dimension (the network). Alongside social networking sites in the strictest sense – *i.e.*, Facebook, the Ning, or MySpace – the social web also includes other sites in which the production of content by users is a constituent characteristic. These content-sharing sites are always associated with a networking function – Flickr and YouTube are among the most well-known and widely used examples for the sharing of photos and videos, respectively –, blogs and forms of microblogging (including Twitter), antiquated but still widely used forums, and more complex platforms such as virtual worlds (*e.g.*, Second Life and World of Warcraft, to cite just two of those most well known to the general public).

To start, we can place the social web within the general framework of the growing importance of Internet as an environment lived in by its users, for a vast range of purposes. The total number of Internet users worldwide as of December 2009 was just under 2 billion, which is equal to 26% of the entire world population[10]. In the 27 nations of the European Union, the number of families with an Internet connection grew from 41% in 2004 to 65% in 2009, while the percentage of individuals logging on to the

[10] As of December 2009, the precise number of total Internet users was 1,802,330,457, of whom 425 million were in the EU (53% of the population), 260 million in North America (76.2%), and 765 million were in Asia (20%); see: http://www.internetworldstats.com/stats.htm, accessed May 15, 2010.

Internet from home rose from 33% in 2004 to 58% in 2009[11]. Internet is clearly becoming an environment that is evermore entwined with all sorts of dynamics and practices of daily life.

We also know, from a survey conducted in May 2009 by comScore, that two thirds of Internet users visit social network sites[12]. The widespread use of social network sites in a broad sense is a relatively recent phenomenon on the Web, since it has exploded in the last five years[13]; at the same time, it is also a global phenomenon, although it is taking on different forms in different geographical areas of the world[14]. From the map of the most popular social network sites in different countries, we can deduce that there is a differentiation by geographic area, but it is also clear that Facebook dominates the world stage. According to data provided by the platform itself, currently there are more than 400 million active Facebook users worldwide, of whom 50% log onto the site daily, and each user produces on average 70 pieces of content each month (web links, news stories, blog posts, notes, photo albums, etc.), for a grand total of more than 25 billion pieces of content shared each month. Approximately 70% of Facebook users are in countries outside the USA[15]. Flickr, which does not supply official statistics on the number of active user profiles, is believed to have reached more than 4 billion photos

---

[11] Data relative to the population between ages 16 and 74, source Eurostat, see: http://epp.eurostat.ec.europa.eu/tgm/table.do?tab=table&init=1&language=en&pcode=tsiir040&plugin=0, accessed May 20, 2010.

[12] See: http://comscore.com/Press_Events/Press_Releases/2009/7/Russia_has_World_s_Most_Engaged_Social_Networking_Audience, where, more precisely, it is stated that: "of the 1.1 billion people age 15 and older worldwide who accessed the Internet from a home or work location in May 2009, 734.2 million visited at least one social networking site during the month, representing a penetration of 65 percent of the worldwide Internet audience", accessed May 23, 2010.

[13] Although the first social networking site, SixDegrees.com, was launched in 1997, the term social network came into use only five years later thanks to the success of Friendster, and found mass use with the boom in Facebook use in 2007: see: http://www.vincos.it/2009/06/20/la-mappa-su-nova-ipotesi-sul-futuro-dei-social-network/. YouTube has just turned five years old, while Flickr was launched in 2004.

[14] For a map of the most widely used social networking sites in different countries around the world, as of December 2009, see: http://www.vincos.it/world-map-of-social-networks/, accessed May 22, 2010.

[15] Source: http://www.facebook.com/press/info.php?statistics, May 15, 2010. In Italy, as of April 2010 there were more than 16 million Facebook users, which is equivalent to 70% of habitual Internet users, see:
http://www.vincos.it/osservatorio-facebook/, Accessed May 23, 2010.

uploaded as of October 2009[16]. The data on just how widespread and significant the social web is are still fragmentary and not always easy to interpret, but by putting together all the data available, including both data provided by site owners and those produced by researchers, we can obtain a good general picture of the situation, which confirms the importance of these social places.

Although the social web represents a more significant area of the web, both in terms of content and number of users, not necessarily all of its environments are equally important from the standpoint of the work of memory. Along the same lines, within each environment and platform, not all users exploit the full potential offered in relation to memory. The objective of the present study was to observe how the work of memory can find space on the social web, both at the level of autobiographical memory and on the levels of collective and cultural memory. In particular, we can observe how different platforms support the work of memory along the continuum of private/public and individual/collective/cultural, as well as the types of practices which take form within them, that emerge from the interaction between the uses and desires of their users on the one hand and the connatural possibilities and limits of the specific sites on the other. As a consequence, our attention will be focused on several platforms which a preliminary and qualitative survey showed to be potential sites for the work of memory: Flickr and YouTube for the production of autobiographical contents and of collective memory (*e.g.*, generational or family, or local), Second Life among the online worlds, as its users are given substantial freedom to produce contents (in contrast for example with online gaming worlds, which are poorly suited to the work of memory), blogging platforms and the experiences of some content sharing sites with a specific vocation for the work of memory, both explicit and implicit (*e.g.*, Wikipedia).

As regards the phenomena at the heart of our observations, I should underline that I have attempted to maintain my focus on the three planes of memory – individual, collective and social – each with its own autonomy and interdependence. To grasp the transformations underway in embodied memories in their coupling with the social media, I will examine, in relation to memory, several characteristics highlighted in the literature on so-called *networked publics*. In terms of both expression and sociality, the characteristics of networked publics as identified by Boyd (2008) derive above all from their digital nature, from the fact that they are made up of bits, and are: persistence, replicability, scalability (visibility) and

---

[16] http://blog.flickr.net/en/2009/10/12/4000000000/, May 15, 2010.

searchability. As regards persistence, this feature derives from the fact that online expression is automatically recorded and archived and therefore, potentially, remembered. Secondly, content can be duplicated (replicability), while with the term scalability we mean that the potential visibility of content is great, although there is no guarantee that the visibility is indeed high. Finally, searchability corresponds to the fact that content can be accessed through a search. These characteristics do not belong to networked publics alone, as Boyd underlined, but they are automatically present in networked media and present themselves in new relationships, which we will try to observe in relation to the work of memory. To these characteristics we should also add connectivity, which intrinsically characterizes memories constructed on the social web. We shall see, through the examination of concrete cases, how these characteristics may affect memory, both at the individual and collective level. We should underline in particular the problem of controlling one's contents and as a result one's memories – which is linked both to the ways in which searchability is fostered, that is, to the possibility of information retrieval, and to the property of the platforms, which is often private, as this can have consequences on the work of remembering and of forgetting.

3. Networked memories for networked publics

To better understand the values of the social web with regard to individual and collective memory, we can construct a preliminary phenomenology of what we have called 'networked memories', basing it on the observation of web sites, experiences and practices of Internet users. I have identified five distinct forms of coupling between memory work and social media: private archives shared online; online research regarding ancestors; online commemoration rituals; "social archives" regarding historical events; and, lastly, forms of construction and objectification of grassroots cultural memory. These forms of coupling all move in non-linear ways along a continuum which goes from private to public, from individual autobiographical memories to collective public memories. Indeed, in all these cases, individual memory work is able to network with the collective dimension and, in some cases, with what the traditional literature has referred to as the public dimension, where the boundaries between the two opposite poles – private vs. public, individual vs. collective – are blurring.

First, we can note that there has been a *proliferation of private archives shared online.* These archives are created and managed by the users themselves, in the ambit of the possibilities offered by the various sites available for the sharing of cultural content. The contents that are shared are very heterogeneous and there is also a variation in the degree of awareness among users that are sharing their archives not only with friends and family but also with a potentially infinite public of perfect strangers. A case in point of the first type is given by Flickr, the most widely known and used site for the sharing of photographic images that are uploaded onto personal profiles by users. Flickr constitutes a veritable data bank of amateur snapshots, including many pictures of daily life, and on the one hand allows an immediate showcasing of one's photos, while also serving as a sort of extended family photo album made immediately public, visible and shareable: just think that among the 'tags' most widely used for the photos uploaded onto the site we find 'wedding' – a category with over 12 million photos published by users – alongside millions of other images regarding typical events with biographical worth[17]. YouTube, the video-sharing platform, can itself be considered an immense multimedia archive of private videos, although a considerable portion of the content uploaded is not original material produced by users but rather fragments of mass culture which are nevertheless selected, recorded and brought to the attention of the wider public of users of the site. The same considerations can be made for other forms of visual or text archives, such as *lifelogs* dedicated to life stories (van Dijck 2007) or, although they are less explicit, even weblogs giving cooking tips and recipes. On the whole, the social web can be considered an immense archive of private memories made immediately public and automatically shared by users. These memories may be private and autobiographical, or mediated by fragments of pop culture: a song, a scene from a film, or perhaps a single moment of a historic event – such as the inauguration of the American President Barack Obama – which one has first witnessed 'live', or experienced through the mediation of television, and then uploaded to the Web to reinforce remembering, with its redundancy. With regard to these sites which function as shared private archives that are thus visible in public, there is a problem related to searchability and data retrieval and thus, beforehand, of the indexing and cataloguing content, which in the case of the two platforms cited (Flickr and YouTube) is entrusted to textual

---

[17] See http://www.flickr.com/photos/tags/wedding/, data from March 11, 2010; in May 2009 there were approximately 10 million photos uploaded with this 'tag'; in March 2010 approximately 4,850,000 photos were tagged 'birthday', approximately 8,840,000 were tagged 'party', approximately 7,700,000 were tagged 'family', approximately 5,500,000 were tagged 'friends', approximately 3,700,000 were tagged 'holiday', and 6,640,000 were tagged 'vacation', just to cite a few of the tags that can most commonly be associated with memories mediated by digital photographs and shared online.

tagging by the users themselves. A further problem, which has already been debated in the literature, regards the accessibility of one's archives, which because they are externalized are thus not always immediately available, as traditional archives kept on the shelves of the family bookcase in one's own home would instead be.

In the cases of the online archives cited, we cannot identify veritable memory policies, given the spontaneity and fragmentation of the practices used for uploading memories and content. This does not mean that the Internet cannot be a place for strategic grassroots collective memory, and a place for the building of cultural memory. The Italian example of the *Banca della memoria* (Memory Bank) bears witness to this inherent possibility of the social web. This private, non-profit project was begun with the assistance of the local government of the Italian Province of Cuneo, and today has evolved into a recognized non-profit organization which also works internationally (in France, Germany, Spain, the USA and Argentina). The *Banca della Memoria* is a site that collect recordings of people born before 1940 in the form of ten-minute videos, in part recorded by the editorial board of the site and in part submitted spontaneously by users; in short, the memories of a generation of the elderly are being entrusted to the Memory Bank and made accessible to the public online. The Memory Bank has been up and running since June 2008, and to date has registered over 1,600,000 visits to the site, with almost six years of memories recorded and published[18]. With respect to user-managed sites for the sharing of cultural content, the strategic nature of this platform should be underlined, as it is working with a focus on building a cultural memory starting from countless fragments of communicative memory handed down privately but now recorded digitally – and, as a result, made public immediately and shared with an unidentified public. One cannot say that the specific content of the interviews is strategically managed by the site, because in most cases it is chosen by the users who collaborate spontaneously on the project.

In the case of the *Banca della Memoria* there is also an explicit presence of a curator of the fragments of memories which are recorded and sent in by the single users and then organized in an online archive. The curator may edit the film footage and insert subtitles and the classification of the video, both through the collection of different videos in the section "*Percorsi*" (pathways) or by alerting users to the presence of videos on the same topic[19]. In the case of sites such as Flickr or YouTube, the role of the

---

[18] http://www.memoro.org/index.php, data updated to May 15, 2010.

[19] http://www.memoro.org/it/percorsi_speciali.php, accessed on May 15, 2010. Among the various 'pathways' set up by the curators of the site, we can find

curator is taken on not by the owner or manager of the platform, but rather by third parties, in an equally fragmented way and with a wide range of objectives, whether they are bloggers who incorporate photos or videos into their sites or mass or niche media that make use of content produced by users for their own news or entertainment programs (Gehl 2009). The ways of working as well as the aims of a formal or informal curator (where one exists) of these new shared archives is one of the delicate and novel aspects which of growing interest to memory scholars. Just as would be the case in a classic institution of cultural memory, such as a museum or exhibit, a curator of digital archives has the crucial work of selecting and organizing content (memories) and thus exercises a control function with regard to the significance of these archives in relation to memory, and to culture in general.

On a plane tied even more explicitly to autobiographical memory, we can see how the social web lends itself to becoming a place for a shared search for one's origins, mediated by both public institutions (registry offices, public archives, etc.) and by private enterprises, leading to the creation of archives that are private but at the same time shared and networked. In this ambit, there has been a proliferation of sites regarding networked genealogy, among which we can cite as an example the *Ancestry* network of sites, founded in 1997, which has multiple national versions including the Italian one, all of which offer users the opportunity – free of charge – to construct their family tree. On their own profiles, users can publish data regarding their ancestors, photographic images, videos, and stories, building a true grassroots multimedia archive with a structure based on the interests and knowledge of the users. For a fee, users can instead access other users' family trees and databanks from different sources with which to integrate the information found in their search for their roots – more or less successfully, given the relentless fragmentariness, incompleteness and loss of the official archives to which Ancestry mediates access. Users are also offered the opportunity to collaborate in the digitalizing of public paper-based archives, so as to expand the databank which they will be able to use free of charge to trace their family history[20]. Ancestry's official website claims that "currently

sections on the 'Resistance in Piedmont',' the bombarding of Turin', 'September 8, 1943', as well as 'memories from abroad', and 'love and courtship'. In all these cases, the attempt is to create collective memory pathways for single individual items. As regards 'love and courtship', see for example:
http://www.memoro.org/it/percorsi_dett.php?ID=%283%29 .

[20] This is the Ancestry World Archives Project,
http://community.ancestry.it/wap/download.aspx, accessed September 26, 2009.

there are 15 million users who have access to five billion historical documents on Ancestry's global network"[21]. We are witnessing a progressive democratization of genealogy, a typical form of cultural mnemotechnique which in the past was reserved for nobility and later entrusted to a specific category of professionals, but today is developing ever more into an amateur practice supported by the new possibilities for sharing and collaboration online and, in the case of sites like that of Ancestry or its competitors, exploited by the market.

Thirdly, the social web has become a *place for ritual commemoration*, in parallel with or in alternative to other traditional forms of commemoration offline. Online, users can give form to commemorative rituals and virtual memorials. While material memorials and commemorative rituals are typical forms of objectification of official cultural memory, and can be included among the conscious strategies of a memory policy managed by public institutions, the commemoration sites and rituals set up on the web are places of grassroots memory that can join together harmoniously with the forms of official commemoration but also express counter-narratives[22]. Here, we can take as an example a tragic event such the September 11th attack on the Twin Towers, which had already been the object of forms of both spontaneous and official commemoration outside the Internet, but which also has seen the proliferation of commemorative sites on the Web. For example, the "9/11 Memorial" built on Second Life in 2007 on the occasion of the anniversary of the terrorist attack had the stated mission of "honoring the fallen of the World Trade Center, of flights 11, 77, 175, 93 and of the Pentagon", and collected the names and photographs of the victims of the attacks[23]. Every Second Life user can visit the Memorial and leave a flower, a candle, a flag or a token of their visit and make a donation to contribute to the expenses for site maintenance[24]. The building of a memorial in a virtual space online allows

---

[21] http://www.ancestry.it/about/default.aspx?section=overview, accessed September 26, 2009. As of June 30, 2009 the Ancestry network included 11 million family trees.

[22] Regarding the topic of web memorials, see Hess 2007.

[23] The URL to reach the memorial on Second Life is: http://slurl.com/secondlife/dAlliez%20Land%20Headquarters/108/149/26; accessed September 20, 2009.

[24] This regards in particular the renting of land from owners of the SIM hosting the memorial. As of September 20, 2009, donations had reached a total of L$ 57,922, while the monthly cost would be L$ 8,000 for the month of September. Both figures are expressed in the internal currency used in Second Life, the Linden Dollar, which at the current exchange rate can be exchanged at the rate of approximately L$ 250 per US $ 1, http://secondlife.com/whatis/currency.php,

users to participate in metaphorical commemorative rituals just as they would in a ceremony at a real cemetery or in front of a material monument. This participation can take place at any moment and from any place, even by people who would not have the possibility of visiting Ground Zero[25].

The social web can also be a place of celebration, a form that is crucial in supporting founding collective remembering, where the users produce rituals adapted to the grammar of the Web. This is what has happened, for example, in the case of the *Mondine 2.0* project. On the occasion of the celebrations in remembrance of the anniversary of Italy's Liberation from Nazi-Fascism which were to be held at the Casa Cervi museum which is housed in what was once the home of a family of partisans –, on April 25 2008, a concert with the participation of the chorus of the '*mondine*' was to take place. The blog of the *Mondine 2.0* project launched a campaign amongst bloggers to remember the April 25th anniversary with a special dedicated post to honor the '*mondine'*, Italian women who were originally workers in the rice fields but who became protagonists in the Resistance efforts. The campaign hoped to reach 100 bloggers by the date of the celebrations, but had involved 355 by April 25th, many of whom also participated at the 'live' event at the Casa Cervi museum[26]. Just as the boundaries between communicative and cultural collective memory have become blurred, to the extent that the former today has many places where it can be enrolled and made visible and public, the distinction between life online and "real" life can also begin to fade, at the moment in which the memories and significant relationships present wholly embrace the embodied experience of users, be it mediated and digital or an unmediated and physical. Thus, the distinction between (presumed) pure memory and mediated memory also fades.

The social web is also a place in which institutional memories can network directly with private memories: this is the case of *'social*

---

accessed September 26, 2009.

[25] This memorial is currently the property of the 'Silicon Island' group, founded by Liam Kanno aka O. Liam Wright, a builder in SL who won the competition promoted by Winfried Ferraris and Sundra Petrov for the construction of a memorial in the online world, as reported in his blog: http://www.oliamwright.com/node/18, accessed September 20, 2009.

[26] The launching of the campaign can be found on: http://www.mondine.it/bloggers-per-il-25-aprile/, while the list of 355 bloggers who joined in with links to posts dedicated to the commemoration can be found at http://www.mondine.it/2008/03/11/25-aprile-radici-resistenti/, accessed September 20, 2009.

archives', which are timidly developing. While there are many examples that could be cited, both in and outside Europe, as regards Italy I would cite the recent project of a historical archive regarding the exodus of residents of Istria-Fiume-Dalmatia to Piedmont, set up by Istoreto (*Istituto piemontese per la storia della Resistenza e dell'età contemporanea*), an Italian institute for studies of the history of the Italian Resistance and the contemporary era. Istoreto has developed an application which allows users to consult archive documentation and to participate with comments or contributions (private histories), integrating the statistical data and the information originally collected by historians in their fieldwork[27]. It is not by chance that these are fragments of history forgotten in historiographic narrations and national politics, which leave room for the work of scholars interested in the 'lesser' history and in the testimony of the protagonists of neglected events, or of their descendents, who thanks also to the connectivity of the web can create networked knowledge. Following on the wave of social archives, in relation to the already mentioned September 11 attacks, we can cite the setting up of a "September 11th Digital Archive" by academic historians in collaboration with the American Social History Project, which foresees the participation of web users who will be free to add pictures, text, and other materials that document their experience of this event[28]. While social archives will become an important reality, the boundary between communicative memory and cultural memory is destined to become ever more permeable, and the Internet can represent a more structured place where anyone can share in the building of collective memories and in cultural memory. Social archives thus will launch important challenges for public strategies of remembering that will necessarily involve academic historiography, cultural institutions and perhaps even political institutions.

Finally, the social web has shown itself to be a place for grassroots cultural memory when its forms of writing take on a legitimacy analogous to that of official sources of collective remembering — or at least a competing, competitive legitimacy. This has been the case, for example, with Wikipedia[29] and its local and thematic variations. In these new online

---

[27] Regarding this experience, and in general Web 2.0 archives or 'social archives', see the 'Archivagando' blog and in particular the article http://deffeblog.wordpress.com/2009/02/16/il-web-tra-storia-memoria-didattica/

[28] Regarding this experience, see: http://911digitalarchive.org.

[29] http://www.wikipedia.org/, the "Free Encyclopedia" in 10 languages. A variation on the theme can be found in Armeniapedia, a site dedicated to anything and everything regarding Armenia: http://www.armeniapedia.org/index.php?title=Main_Page, where the function of

encyclopedias, the cultural sense, that which is remembered and how it is recorded, is the result of a collaborative work among users. Wikipedia positions itself as an alternative to official sources (to the chagrin of academics). In this case we are not dealing with autobiographical memories, but rather with a true shared building of knowledge which is at the same time participated and influential, where authoritativeness is guaranteed on the one hand by rules for monitoring and verifying content established by the system (although they are not universally recognized as sufficiently effective in guaranteeing the reliability and truthfulness of the content, as seen by the recursive critical articles which often appear in newspapers and on the Web itself[30]). On the other hand, a recognition is emerging from the encyclopedia's users themselves, who use its entries to construct their own images of the past and of the present (and therefore also of the future).

Although this first attempt to delineate a phenomenology of networked memories in the panorama of the social web clearly lends itself to further enrichment and refinement, it does take into account the richness and significance of the experiences already present and the new problems being posed for scholars of memory and media. From this empirical observation, I believe it should be possible first to confirm that the evolution of technological media has not corresponded to the loss or crisis of memory, by some taken for granted, an issue that up to now seems to have been predominant in studies on the relationship between media and memory starting from the invention of writing. The social media have indeed clearly emerged as places that are chosen by web users both for intimate, autobiographical remembering and for the building of public memories. Rather than in terms of loss or crisis of memory, an analysis of the relationship between memory and social media should instead lead to a redefining of the issues that regard the processes of remembering and of forgetting, in terms both of new possibility and of new risks, and in terms of the forms that individual and collective memories take on in coupling with social media. In the following section, I shall discuss some preliminary thoughts in this regard.

---

the site in building a collective memory is even more evident.

[30] See for example, regarding Italy, the article published in the Corriere della Sera newspaper – and on journalist Bardazzi's blog – in the early months of 2010. Source:
http://www.corriere.it/scienze_e_tecnologie/09_maggio_07/wikipedia_falsa_citazione_621f5f22-3b00-11de-b512-00144f02aabc.shtml and
http://marcobardazzi.com/blog8/2009/12/07/i-limiti-di-wikipedia/

4. Memory and social media

Let us consider first and foremost how the development of the social web has transformed individual and collective memory work. We have seen that, online, memory work takes place thanks to the integration of social media open to the creation of content by users with memory technologies suitable for a coupling with digital places – from writing to the most recent digital technologies. We can say, then, that these are tools for use both by spontaneous grassroots individual and communicative memory, and by cultural memories created in an artificial and strategic manner by guarantors of collective memory. They thus span the distinction introduced by the Assmanns, introducing further complexity, and this complexity explodes when these technologies are coupled with the social web. On the one hand, thanks to the evolution of the social web communicative memory, the embodied memories that risk disappearing with the passing away of older generations have new possibilities both for persistence and to be heard. The Web has become a place where users can record, store and make visible to others their own memories, objectified in writing and/or in images thanks to audiovisual and digital memory technologies. In other words, the Web offers a new possibility for both individual and collective memory to not be described only by institutional memory entrepreneurs who hold the power of writing in many forms (de Certeau 2001). The Web has become a place for the objectification of these memories, although this is not sufficient to transform them into cultural memories in the sense ascribed by the Assmanns, in that the characteristics of officialness and authoritativeness of these memories are not automatically guaranteed. Therefore, this is a new form of 'grassroots' collective memory, that is objectified and cannot be classified as private, but that lacks the strategic characteristic inherent in cultural memory as it has been traditionally understood of being directed toward building a founding past for a specific collective identity (be it national, political, etc.). The legitimacy of these memories lies only in their rooting in embodied experiences, in their being shared by users online — and therefore in their spreading and on the emotional intensity of this sharing. Last but not least, their legitimacy also depends in part on their possible resonance in places of cultural memory (for example, the mass media)[31],

[31] This was already the field of work of *Oral History*, which created a grassroots history. Today, on the Web it is as if a sort of amateur, spontaneous and widespread grassroots self-ethnography were being produced.

or in official narrations themselves (think, for example, of the attention given by academic historians to individual narrations)[32].

The memories recorded and archived on the social web, which can potentially be accessed by the vast public of Internet users, can therefore regard both an autobiographical past and the founding collective past. In both cases, memories are offered for sharing not just in terms of accessibility, but also in terms of negotiation of meaning, thanks to the different forms of interaction and conversation made possible by the different sites and developed by their users. For example, in the case of Wikipedia, as we have seen, the contents are typically related to cultural memory, and the process of negotiation is implicit in the system by which the platform itself functions, as well as in the uses of the site's users[33]. In the case of Flickr or YouTube the planes of autobiographical and cultural memory are instead more interconnected, and the shared work of building the meanings of the content uploaded online is left mainly to spontaneous interaction among users. Each of these platforms would merit specific attention to understand the variety of systems of rules regarding their functioning and user practices which contributes to creating the complex panorama which we are describing herein with the term 'social web', which is a brutal generalization.

From this first survey of memories online, it is first of all clear that we are going to need to re-think our idea of memory work in social media, at least in two directions. The first of these is the coupling between technologies and memory (prevalently in the form of domestic technologies) which makes explicit the ways in which the distinction between the public and private in the construction of memory is becoming ever less convincing, while the *immediately public nature of personal memories* (which are not necessarily autobiographical) is a fundamental aspect of memory work in social media, in new ways with respect to the old media. Therefore, the social web represents not only a place where we can observe memory at work, but also a place where memory work can be observed in connection to its meaning, in agreement with what has been suggested in psychological studies on memory and in neurosciences (Schacter, Damasio).

---

[32] A classic case that is cited frequently in cultural memory studies is that of the Shoah, the historical narration of which draws also upon private narrations: think for example of the emblematic case of the public use of the private diary of Anne Frank (van Dijck 2007).

[33] For a more in-depth discussion on the case of *Wikipedia* as a place for global memory, see Pentzold 2009.

Secondly, the distinction between "pure" memory and mediated memory is beginning to lose its importance, at least as far as the mediation in social media is the immediate form in which the memory is created and automatically recorded and stored, and automatically shared, through a digital objectification that serves as a future trigger for future remembering, both of the original author and of the indeterminate public of users (and this is in continuity with the advent of electronic media and the domestic memory technologies). One novelty introduced by the social web thus lies in the exponential growth in the "objects of mediated memory"[34] which each individual can have at his or her disposal, objects which go beyond the confines of the home and are offered to a vast public made up not only of friends but also of strangers who can transform (digital) objects created by others into triggers for their own memory work. Moreover, this multiplication of the places and objects of mediated memory corresponds to what is certainly a core aspect of modern memory: that is, the pluralization and fragmentation of the processes of individual and collective memory, which have found places for expression and enhancement in social media.

Thirdly, with the arrival of social media a new problem has appeared, regarding the control over one's own archives and thus over the memory work, be it individual or collective, that derives from the prevalently proprietary nature of the platforms in which user content (digital memory objects) is recorded and stored. Indeed, the choices of individuals regarding content retrieval, and thus the process of remembering, depend not so much on decisions made in the field of culture or of public memory policies, but rather on market strategies geared towards making a profit[35]. In other words, memory work on the social Web poses the problem of the relationship between public *vs.* private control of these digital archives, which are split between a social (widespread, shared) nature of the production of contents and the private (concentrated, exclusive) ownership of the sites. Outside the social media, an exemplary case of this kind of tension can be seen in the Corbis image archive, where privatization of a collective memory has corresponded at the same time to its being materially preserved and yet substantially unavailable, and where its conservation for hypothetical retrieval in the future corresponds to what is

---

[34] For the expression "mediated memory objects" see Van Dijck 2007.

[35] See in this regard for example Zierhold 2008 who underlines effectively that the central question in the future relationship between media and memory will be the issues of ownership of the technologies and of the cultural policies related to the long-term availability of digital information.

substantially oblivion[36]. Again, in relation to the problem of control over digital information, we also need to consider problems related to the retrieval of information on the Web, which respond to relevance criteria that are internal to each specific media system. We will return to this point in the next and final section, as well as to problems related to forgetting, which from the perspective of individual memory is linked to the construction of new identities, or else serves to remove elements of past experience that could cause suffering or conflict (Connerton 2008). As regards individuals, forgetting is important in the building of one's personal identity, but there can be obstacles to oblivion when one does not have complete control over one's contents (think, for example, of the permanence of posts erased from blogs that can be retrieved digitally, or of content placed online by others that is out of our control, etc.). Therefore, we should ask ourselves whether forgetting, in its role as a voluntary, strategic operation, can escape from the full control of individuals and groups when it is entrusted to social media, because *where one loses full control over the availability of one's own content, which becomes automatically public, memory work is obstructed*, on the individual or collective level.

In conclusion, individual memory is undoubtedly influenced by media environments in which remembering coincides on the one hand with automatic recording and digital archiving, and on the other hand with automatic exhibiting to the public and potentially unlimited sharing of memories, an aspect which will require in-depth reflection. It will be necessary, for example, to understand whether automatic online archiving can also correspond to a new frontier of forgetting. Indeed, archiving can be considered a form of oblivion which came into being with writing and is being perfected with the new information technologies. The annulment described by Connerton (2008) is tied to this topic, as is the primacy of forgetting associated with forms of exteriorization of memory as a human capacity. However, I believe that in relation to online social archives this issue should be examined as a more specific problem. What's more, we have given support for the hypothesis that in social media many sites are shaped by their users to suit the needs of memory work. At the level of collective memory, we have argued, examining concrete cases, that the coupling of social media and memories has produced the emergence of multiple grassroots memories, and multiple processes involving shared remembering that influence the traditional forms of collective and cultural

[36] See for ex. the articles by Michele Smargiassi published in the Italian newspaper *La Repubblica* (7 January 2007 and 16 May 2009).

memories. We have also seen how the experiences of users online, supported by the creators and managers of social web platforms, demonstrate how the Web truly represents a place for memories and for narration, and not only for forgetting and for disembodied information. Certainly, these may not be great narratives; rather they are a grassroots chorus that, thanks to the Internet, has found a place to shelter itself and become visible, at times in disassociation from official narratives. These grassroots memories are not structured like a true cultural memory and are not strategic, but rather fragmented and scattered, if also widespread and responding to the concrete needs for memory, for roots and for identity felt by Internet users – although experiences such as Wikipedia and social archives might also give us pause to think that there has been a leap forward toward a greater structuring and organicness of these networked memories.

5. A memory *of* social media?

I'd like to conclude with some thoughts on the memory of social media, that is on the specific way in which these media manage the distinction between remembering and forgetting. In other words, I intend to verify whether a specific memory of social media exists at all, in relative autonomy from both the social memory and of the memory of media as conceived by Luhmann (the latter being centered above all on the mass media), as I mentioned at the beginning of this paper. As regards remembering, the memory of social media is based on the selections made by users and does not seem to count on curators who are sufficiently strategic in the selection of content and its organization (in the ambit of each specific site), in contrast with what happens with traditional print or broadcasting media. Accordingly, while on the one hand the social media are undoubtedly able to offer issues for social communication, keeping communication alive (taking on a function also found in traditional media), on the other hand the way in which they carry out this function is very much the spontaneous result of the production of content by users, which operate according to their own criteria regarding pertinence. The selection of users is oriented towards their need for identity and social recognition, for example, which do not necessarily always coincide with the selection criteria used by traditional media, in particular the criterion of novelty, above all given the weak role of the curators themselves (in cases where one exists). I wonder if this does not perhaps reshape the primacy of novelty, which was considered to be inherent in the media system and its memory (Luhmann), and I believe that this represents a research question that merits following up on.

What's more, social media represent an element of change also with regard to the second aspect of memory: forgetting. In particular, within the framework of Luhmann's systems theory, forgetting is indeed the most important aspect of memory, in that it opens up space so as to avoid the system overload that could be caused by the accumulation of results of the past operations and to constantly allow the production of novelty. Forgetting also takes on the crucial function of ensuring coherence into the system. In the case of social media, and of the forms of forgetting involved in ensuring the functioning of the social system, we can also take into consideration the ordering and relevance criteria set up by the sites themselves which in essence set up an internal order and hierarchy online – a perfect example is the ranking system used by Google. These criteria represent a form of forgetting that can be even more effective than canceling or erasing: a piece of content that appears more than three pages down the list of search results in all likelihood will be quickly forgotten, without there being any need for the information to be removed[37]. From another standpoint, we can also say that the participatory, spontaneous, diffuse and self-organized nature of the social web is able to introduce an element of surprise or unexpectedness in the relationship between media and social memory. The primacy of forgetting in social memory is being challenged by the impertinent permanency of digital traces of the transactions that take place in the web, and by the loss of control over what is remembered and what is forgotten online. There is considerable evidence that this is a critical point which has yet to be adequately addressed with regard to both cultural and technical aspects (we have to consider the case, for example, of forced removal of posts or blogs being blacked out by sites owners). The inadequate work performed by the 'curators' of online content could furthermore create obstacles to the key function of forgetting which involves freeing oneself from information that is useless either because it is outdated (for example, in the sciences) or in general no longer necessary. The question is: online, who carries out this work aimed at ensuring coherence?

The social media have therefore modified and will continue to modify the panorama of memory, regarding aspects of both remembering and forgetting, opening up new horizons for rewarding research in the area on the frontier between media studies e memory studies.

---

[37] A similar problem is observable in the case of Facebook: user generated contents quickly disappear with the progressive publication of posts in the wall of the users personal profile.

References

Assmann, A. (2002). *Ricordare. Forme e mutamenti della memoria culturale*. Bologna: il Mulino.

—. (2006). "The Printing Press and the Internet: from a Culture of Memory to a Culture of Attention." In *Globalization, Cultural Identities and Media Representations*, edited by Natasha Gentz and Stefan Kramer, 11-23. Albany: State University of New York Press.

Assmann, J. (1997). *La memoria culturale. Scrittura, ricordo e identità politica nelle grandi civiltà antiche*. Torino: Einaudi.

Bartoletti, R. (2007). *Memoria e comunicazione*. Milano: FrancoAngeli.

—. (2008). "La fabbrica delle memorie." *Sociologia del lavoro* 108: 94-110.

—. (2009). "The Crisis of Collective Memory and the Commodification of Nostalgia: The Case History of Heidiland in the Swiss Alps." In *Sociology of Memory: Papers from the Spectrum*, edited by Noel Packard, 212-229. Newcastle: Cambridge Scholars Publishing.

Boccia Artieri, G. (2009). "SuperNetwork. Quando le vite sono connesse." In *Network Effect. Quando la rete diventa pop*, edited by Lella Mazzoli, 21-40. Genova: Codice.

Boyd, D. (2008). *"Taken out of context. American Teen Sociality in Networked Publics"*. PhD diss., University of Berkeley.

Connerton, P. (2008). "Seven Types of Forgetting." *Memory Studies* 1(1): 59-71.

Damasio A. R. (2007). *L'errore di Cartesio. Emozione, ragione e cervello umano*. Milano: Adelphi.

De Certeau, M. (2001). *L'invenzione del quotidiano*. Roma: Edizioni Lavoro.

—. (2007). *La presa di parola e altri scritti politici*. Roma: Meltemi

Erll, A. and Ansgar N. (eds). (2007). *Cultural Memory Studies: An International and Interdisciplinary Handbook (Media and Cultural Memory VII)*. Berlin and New York: de Gruyter.

Esposito, E. (2008). "Social Forgetting: A Systems-Theory Approach." In *Cultural Memory Studies: An International and Interdisciplinary Handbook (Media and Cultural Memory VII),* edited by Ansgar Nünning and Astrid Erll. Berlin, New York: de Gruyter.

—. (2001). *La memoria sociale. Mezzi per comunicare e modi di dimenticare*. Roma-Bari: Laterza.

Gehl, R. (2009). "YouTube as archive: Who will curate this digital Wunderkammer?" *International Journal of Cultural Studies* 12(1): 43-60.

Hess, A. (2007). "In digital remembrance: vernacular memory and the rhetorical construction of web memorials." *Media, Culture & Society* 29(5): 812-830.

Luhmann, N. (1995). *Gesellschaftstruktur und Semantik. Studien zur Wissenssoziologie der modernen Gesellschaft*, Band 4. Frankfurt am Main: Suhrkamp. (2000). *La realtà dei mass media*. Milano: FrancoAngeli.

—. (1996). "Zeit und Gedächtnis" *Soziale Systeme* 2: 307-330.

—. (1997). *Die Gesellschaft der Gesellschaft*, Francoforte sul Meno: Suhrkamp

Mazzoli, Lella. (2001). *L'impronta del sociale. La comunicazione fra teorie e tecnologie*. Milano: FrancoAngeli. (ed.) (2009). *Network Effect. Quando la rete diventa pop*. Genova: Codice.

Pentzold, Christian. (2009). "Fixing the floating gap: The online encyclopaedia Wikipedia as a global memory place" *Memory Studies* 2(2): 255-272.

Piazzi, G. (2009). *Julie*. Urbino: Quattroventi.

Pirani, B. M. (2005). "Body Rhythms, Social Rhythms in Digital Societies" *Current Sociology* 53(2): 237–273.

Ritzer, G. (2010). "Focusing on the Prosumer: On Correcting an Error in the History of Social Theory." *Prosumer Revisited*. VS Verlag für Sozialwissenschaften, Germany.

Schacter D. L. (2007). *Alla ricerca della memoria. Il cervello, la mente e il passato*. Torino: Einaudi.

Silverstone R., Hirsch E., Morley D. (1992),. "Information and Communication Technologies and the Moral Economy of the Household". In *Consuming Technologies: Media and Information in Domestic Spaces*, edited by Roger Silverstone and Eric Hirsch. London: Routledge.

Sorlin, P. (2001). *I figli di Nadar. Il "secolo" dell'immagine analogica*. Einaudi: Torino.

Sturken, M. (2008). "Memory, consumerism and media: Reflections on the emergence of the field" *Memory Studies* 1(1): 73-78.

van Dijck, J. (2007). *Mediated Memories in the Digital Age*. Stanford: Stanford University Press.

Van House, N. and Churchill, Elizabeth F. (2008). "Technologies of Memory: Key issues and critical perspectives" *Memory Studies* 1(3): 295-310.

Wessel I. and Moulds M. L. (2008). "How many types of forgetting? Comments on Connerton" (2008) *Memory Studies*. 1: 287-294.

Zierold, M. (2008). “Memory and Media Cultures“. In *Cultural Memory Studies: An International and Interdisciplinary Handbook (Media and Cultural Memory VII),* edited by Ansgar Nünning and Astrid Erll, 399-407. Berlin, New York: de Gruyter.

CHAPTER SEVEN

MOTION CAPTURE TECHNOLOGIES AND BODY TECHNIQUES: AN ANTHROPOLOGICAL APPROACH TO HUMAN COMPUTER INTERACTIONS

JÉRÉMY DAMIAN AND HÉDI ZAMMOURI

Abstract

The technological environment which surrounds the body today deeply affects it in its ways of being, doing, feeling and linking up. It seems important to take into account the contemporary ways body and objects fold in each other, to the extent that they make old borders between human subject and material object indistinguishable. We particularly focus on the mediation through which a technological device and « body techniques » adjust themselves in a long time of learning, customization and embodiment. This process might be producing these two entities (person and object) as « actants » (Latour), and imply by their actions on the other a deep transformation: less on a relationship of producing knowledge than in a relationship of producing existence (Dewey).

Our aim is to highlight the role of the body in this encounter and this process, not to reduce it to a mere intermediary and make it signify instead something other than its own self (an institution, a social order...). The issue is to bring onto the scene the prior role it has, first in establishing relational process prior to personalization of motion capture technologies, then in discovering the building and setting of a new sensitivity to the object, to the self and to the world. We organized and observed a choreographic creation experiment by two dance companies. We proposed two different motion capture technologies to experiment the following links: gesture/sound, gesture/light. Finally, our paper discusses the relation to the object and the way in which technology forces us to redefine notions such as « person », « identity », engaging the body in new ways of linking up to the other and to itself, with variable and situated ontologies.

Keywords: human computer interaction, body techniques, dance, personalization, embodiment.

1. Introduction

This chapter describes an experiment in the "art/science" domain, in which motion-capture technologies translating gestures into sounds and lights were placed at the disposal of a duet of dancer-improvisers, Anne Garrigues and Émilie Borgo. This experiment is part of broader doctoral research project in socio-anthropology by the authors — Hédi Zammouri on the personalization of motion-capture technologies (supervised by Pascale Trompette, PACTE laboratory, Grenoble Universities) and Jérémy Damian working towards a pragmatic socio-anthropology of sense-perception experience (supervised by Florent Gaudez, CSRPC-ROMA, Grenoble Universities). This residency was funded by a multi-partner innovation platform that focuses on the uses, reappropriation, and repurposing of microtechnologies by various types of publics.

The question at the outset of this study was the following: What relational processes govern the personalization of motion-capture technologies? The question was framed differently by the industrial partner who commissioned this experiment: What gestures make sense for home technologies? The objective was thus to identify in this field of observation a range of gestures that "make sense"[1] and that can be exploited in the domain of ambient intelligence (ubiquitous computing) within the home. The team focusing on these questions, in different capacities and from different institutions, includes Christelle Casse, an independent ergonomics expert and psycho-pedagogue of motion and perception, and Alexis Lecharpentier, art/science engineer, developer in signal processing.

The technical equipment used during the experiments so far includes five MotionPod sensors, each comprising three accelerometers and three magnetometers on three planes, x, y, and z; a laptop computer; MAX/MSP software; Live software; an amplifier; and stereo speakers. The sensors are attached with fasteners to participants' wrists, ankles, and plexus.

2. Object, subject, person as… "bodies"

For more than twenty years a number of thinkers have managed to show that the subject/object distinction is not heuristic in methodology, in the analysis of field results, or even less so in final theorizing[2]. They have proposed the idea of hybrid collective to redefine the social forces in

[1] The simple reference to gestures that "make sense" requires a critical endeavour to define what such a formulation might mean. This study will be undertaken in subsequent research.

[2] Cf. currents of thought since the 1990s related to Callon and Latour.

action in something other than the dichotomy of humans and non-humans, in order to rethink about material culture and its impact on living creatures. Our fieldwork gave us the opportunity to test the relevance of this dichotomy. Indeed, dancers fitted with sensors do not make this distinction; they deal with the technology in a symmetrical manner, and we were obliged to do so as well. Furthermore, this distinction is no more operational in the analysis phase. The dancers and their captors can thus be viewed as "entities", as "bodies" to an even greater extent, and finally as persons. Attached (Gomart & Hennion 1999) to each other, the entities engaged in new frameworks of action and perception, each making the other do what they could not have accomplished alone. It would be judicious to make a closer theoretical analysis of the specific interaction at play during this residency and of the resource shared by the two entities involved, namely, movement.

The distinction between the *subject* and the *object* of the movement is not operative when the meeting itself is precisely what causes the movement. The improvisation sequence selected (a solo — albeit not one, as will be shown below) led us to focus on a pragmatic analysis of human relations to technology, which Gell alternatively describes as internalist and externalist, and on the relational dynamics that governs the processes of personalization and recognition of others, in connection with the concept of *objectification* (Miller 1987)[3].

We chose to focus on the interaction that can develop between the dancer and the device, making no *a priori* distinction between subject and object but distinguishing between the "material body" and the "living body" (Bergson 1896) during the improvisation, in order to examine the entities taking part in the interaction (the dancer, the sensor, the engineer, the computer, etc.). Thus, we attempt to empirically "fluidify" the concepts of *person* and *ontology* by confronting them with the particularities of our study and dealing with the "bodies" present. Finally, we will formulate questions about the concepts of *distributed personhood* (Strathern) and *agency* (Gell) outside of the specific contexts in which they emerged, namely ritual objects and works of art. The originality of our approach will be to apply these concepts to motion-capture technology and to propose a fine description of *objectification* which, well before Miller, could be found in Hegel. We will thus try to describe the *personalization* of this technology in a new, dynamic manner: *situated*. We will concentrate on

[3] Although similar in several points, our approach is different from those based on situated action (Suchman) or distributed cognition (Hutchins).

the recurring processes that govern the configurations, making it possible for motion-capture technologies to be personalized.

We will reference philosophers and anthropologists of material culture so that we can deal with this subject/object dualism, paying tribute to the contemporary sociologists who have sought not only to include objects in their analysis but also to attribute them with prime importance.

2.1. Objectification and phenomenology

Miller (1987) suggests, following Hegel's philosophy, that the term *objectification* can be used to describe a dual process in which a subject is externalized in a creative act of differentiation and subsequently reappropriates this externalization by means of an act that Hegel calls sublation (*Aufhebung*). Objectification is thus a series of processes consisting of an externalization (auto-alienation) and a sublation (reabsorption) in which both the subject and the object of the process are present in dialectic or in a movement of *co-birth*. It is in Hegel's *Phenomenology of Spirit* that the process of separation and incorporation is first described. The process of co-birth begins with an undifferentiated substance that causes a separation between the subject and the object: the subject thus sees itself as non-other, acquiring an awareness of self. The particularity of our study is to examine the opposite phenomenon: our point of departure involves entities that are clearly distributed between subject and object (the subjects: the dancer, the engineer; the objects: the sensor, the computer…). However, the experiment *disrupts* this distribution. The identities — subject, object — initially distinct, lose their relevance as criteria as the technological device is appropriated or personalized. Hegel adamantly refused to acknowledge the existence of either a subject or an object without the other; he considered them within a mutually constitutive relationship. We will proceed similarly in our analysis.

The term *objectification* thus affirms the need for a particular type of relation between human development and an external form. This relation is never static; it is always a process of change which cannot be reduced to its two composite parts, subject and object. The action of externalization and sublation is always constitutive, never simply reflexive, and is thus not a process of significance. The objective of this chapter is to show how this process of personalization is articulated.

2.2. "Active" participation of the object in the interaction

It is important to see how the field of social sciences has, for the past three decades, undertaken a veritable program of rehabilitating objects in the construction of social relations, questioning precisely the traditional distribution between subjects and objects. Dick Pels, Kevin Hetherington, and Frédéric Vandenberghe (2002) in particular, discuss the return of objects in contemporary social theory, following the waning of post-structuralist and constructivist currents. They insist on the sensory immediacy of the objects through which we live, work, and converse. Social and material relations are necessarily intertwined insofar as "things" have a socializing impact or effect; the social sphere needs a certain type of existence or material presence to hold together. These reflections lead us directly to reconsider the relational configurations involving objects. From an epistemological point of view, this agnostic turn[4] leads to an "ontological relativism" that could structure a new, more comprehensive social ontology.

In the light of the authors cited, we were able to reconsider the question of mediation and the creation of subjectivity, insisting on the Actor-Network Theory (ANT) argument (Latour 2005) that the social sphere is technically constructed, as opposed to a one-dimensional view that considers only the social construction of the technique. To resume, objects perform social interactions by shaping the movements of those who deploy them, suggesting particular encounters. ANT's idea of *actant,* in which agency (Gell 1998)[5] is not viewed as a set of essentials but rather as relations, as if a physical force incited us to act (actantial causation), is thus relevant. This original perspective on the social sphere revives the old discussions opposing idealism vs. materialism, realism vs. constructivism, agency vs. structure, and essentialism vs. (ontological) fluidity.

Rather than asking what an object is, or what a thing is if essentialism is obsolete, we chose to ask: what do objects reveal about human interaction, if they are viewed as models or molds, and are consequently good analyzers for establishing the form of interaction and *personhood*? Probing further, we suggest another line of questioning that reflects the previous questions but nevertheless represents a shift: how are objects – and more broadly, all non-human entities — able to *take part in* interaction, that is, become potential, credible "active subjects"? Our

---

[4] Which we link to "methodological philistinism" (Gell 1998) required by an analytical approach towards art, an offshoot of the methodological atheism of anthropology of religions.

[5] We refer to Gell's idea of "force without will."

discussion will take advantage of the fluidity and the instability of the (multiple) ontological borders that separate things from persons — upsetting the traditional hierarchies that separate subjects and objects, cultures and natures, human and non-human.

Following William James's radical empiricism — "To be radical, an empiricism must neither admit in its constructions any element that is not directly experienced, nor exclude from them any element that is directly experienced" (James 1912:58) — means taking seriously a whole series of situations that the dancers experience, such as the connection with a motion sensor — situations that would more readily be classed in the field of "belief." We see "bodies" of different potentials in the entities present in the improvisation. This has encouraged us to move towards more sensual, more emotion-related types of experiments, referring to modes of bodily experience that emphasize greater, more immediate awareness of people and things (thus less rational). Accepting that what is social can be reached through sensation amounts to reducing the relevance of the concept of *re-presentation*, and we prefer to study the immediacy of the potentiality of "encounters" in a broader sense.

2.3. Presentation of the sequence of improvisation

First we will proceed with a description of the sequence, and then we will provide an original analytical approach inspired by elements for notating and codifying movement (Waille 2009), in order to suggest an interpretive framework that is more centered on the details of the relation (Houseman 2006) and its potentiality, as well as on the co-presence (Piette 2008) of material and living bodies (Bergson 2004).

The present analysis is based on a single 35-minute sequence, even though at the moment of publication we dispose of dozens of hours of video recordings, interviews, and descriptions of the dancers' perceptions of the interface (protocol set by the ergonomist). We take our inspiration from the principle of clinical anthropology[6] that consists of minutely deciphering one case rather than attempting to establish a rule. This premise comes from a biological metaphor in which "one sole micro-organism can reveal the mechanisms essential for understanding the entire living organism" (Severi and Houseman 1994). According to the principles of this interactional anthropology, when these relations (or networks of relations) are analyzed by means of a sufficiently thorough description of an object or a particular situation, they may be generalizable.

---

[6] Led by Carlo Severi and Michael Houseman.

In fact, the sequence we have chosen struck us immediately as a valid, forceful illustration of the manner in which the device alters human interactions and the motion sensor intervenes as a credible partner in the interaction.

Our ethnographic description cannot be considered to belong to cultural, cognitive, or communication (Bateson) anthropology, or even less social anthropology (Lévi-Strauss), nor does it correspond to the literature on technological interfaces. Our work in the field of the anthropology of human-computer interactions should be seen through the prism and the framework of linguistic and ethological anthropology, of which Stephen Levinson (2006) is the figurehead. His approach identifies the prerequisites and universal bases necessary for interaction, including shared resources; turn-taking; rites of mutual access; shared temporality; kinetic, vocal, and motor activity; visual, auditory, and haptic reception; intention attribution; cooperation; and the use of ethological facial displays, postures, eye contact, intonation, spatialization of discourse, co-verbal gesture, touch, proxemics, and axial orientation — dimensions that we found operative between the different "bodies" in the dance interaction. Comparing human-human interaction to human-machine interaction will make it possible to measure the extent to which an object acquires the competencies of a person during the interaction.

Thus, the basis of this article is an improvisational dance solo performed by Anne Garrigues on 14 October 2009 in our first session of residency, with motion sensors dealing with gesture/sound relations, even though other explorations or statements, before or after this sequence, and other comparable experiments[7], may be used to amplify or develop the ideas formulated here. The sequence took place in a *dōjō* in Grenoble (France) where we established our residency.

Alexis, the engineer, suggested to Anne that she modulate the volume in function of the orientation of the sensor on just one axis. Thus, she set the magnetometer at an angle to the two sensors used. She began by standing with one sensor on her right wrist and another on the exterior of her right hand. She looked off into space during the entire beginning of the sequence of exploration. She sometimes looked at her hand moving; other times she stared at a fixed point or at something within. Then she sat on her knees with a look of astonishment. The rest of time, her left hand looked as if it were waiting, motionless, fixed, or "dead."

[7] Cf. in particular, *Interagir avec les nouvelles technologies (Interacting with New Technologies), Nouvelles de Danse*, 2004 (52), Bruxelles: Contredanse.

Later she accidentally broke the protective cover on the MotionPod. She left, took off the broken sensor, and returned with one sensor strapped inside her right wrist and another on her right bicep. She opened her mouth and said "So…" but then went no further with her utterance. She appeared to converse silently and slowly with the sensor on her wrist. Then she accelerated her dance and looked at the sensor. She asked whether the pitch level was still set. She laughed and then made a swipe in the air with her hand. Each time that she managed to create a specific shrill sound, she nodded her head as if to confirm her "discovery" to herself. As she turned the sensor around her wrist like an oversized watchband, the sound was produced again. She looked questioningly at Alexis and then tucked the sensor into her palm. Several times, she touched her hand like a foreign body, and she touched one sensor directly. For her last attempt, she turned the two sensors toward the interior of her arm, one in her palm and the other on her wrist. Then she laid one of the sensors on the ground and pointed to the spot, saying, "There." She set her arms at an angle, one horizontal and the other vertical. Then she danced, looking at the MotionPod on the ground. Afterward, she placed her hand on the MotionPod, and the two sensors were thus at the same level. When she produced a loud sound, she looked at the sensor and appeared to try to hush it with a frown. At that point, she detached her other sensor and grasped it in her hand. Holding it thus, she gained a real mastery of the interface in just a few minutes, obtaining a greater diversity of sounds. She transferred it from one hand to the other, and smiled for the first time, gazing at it in her hand like a precious stone. She lay it on the ground, obtaining a continuous sound (fairy-like, in a dreamy, new-age style), and then began walking in a circle around the sensor. She performed a free dance with broad, round movements at varying speed. She seized the second sensor and modulated the volume with her gestures. She finally managed to stop it with a precise, rapid movement, putting it down on the floor with a lively gesture, on the "zone of silence."[8] We were intrigued. She suddenly displayed a joyful smile, like a riotous child with a new toy. Her laughter and joy were contagious for the others present. She rolled the sensor up in its band and hid it in her hands, then played at revealing it and hiding it again. In the end, Anne held her left wrist for a long while, as if it hurt slightly. The sequence lasted 35 minutes.

In this first section, it can be seen that humans were not simply using a technology; they were mediated by it. This paradigm implies critical

---

[8] This is a term that the dancers regularly used, corresponding to the posture/position of the limb with a sensor that creates silence, that is, that was set not to emit sound. Note the shift towards a spatial metaphor.

questions about otherness, objectification, and alienation, reinforcing the argument that objects require stories, and (social) spokespersons. In this experience of the encounter between a dancer and motion-sensor technology, the dancer actually took on the role of spokesperson, giving existence to the sensor in an original manner, and learning to handle it. However, the device as a whole made us aware that the sensors were present *by and for themselves* during the interaction, that is, they were offered to the dancer in a paradoxical "mode of presence" (Piette 2008), sufficient to completely *incarnate* an entity of the interaction. The following section describes their capacities of *consistency* (Doré 2010) as a social actor (actant).

3. The Device in Motion

In the sequence that we chose, a dancer and an engineer were "connected" through a motion sensor device worn by the dancer. The former controlled the parameters of the sensors by acting on the sounds that they emitted, and the latter explored these parameters using her body. Both actors therefore took an active part in this improvisation, the dancer by exploring the possibilities of the sensor she was wearing, the engineer by controlling its parameters. Yet, it is clear that in describing this interaction between the dancer and the engineer, we cannot neglect the role played by the sensor itself. First, the simple fact of its physical presence affected the dancer's movements. Second, it had a *performance* of its own, that is, the sounds it emitted in connection with her movements. However, the sensor itself is powerless if it is not connected to a computer equipped with specific software programs, MAX and Live. For the sensors to emit sound, the software that processes the information that they send need to be connected to an amplifier and to speakers. Our initial configuration – a face-to-face encounter between a dancer and an engineer – was therefore considerably complexified by the series of mediations that connected them. But can we say that these mediations were solely there to "connect" the two "human" protagonists? Or were the dancer and the engineer interacting *through* these technologies? Goffman asks, "Under what circumstances do we think things are real?" (1974, P. 2) When it comes to reality, he argues, what matters is the conviction that a thing is real. The question therefore shifts to the conditions that nourish this conviction, and we posit that sensations can fuel such a conviction.[9]

---

[9] This is what James (2007:230) calls "direct acquaintances" "All true processes wind up leading us somewhere, to experiences that can be felt and that are directly

3.1. Complexifying interaction: broadening the scale of actants

If there was indeed a dialogue between the dancer and the engineer, it passed via non-human actants that did not simply transmit the information from one protagonist to another, but altered it, modified it, and finally brought it to existence. It is the very role of the sensors to alter, through auditory sensations, the nature of the interaction between the dancer and the engineer. After a period of adaptation, after the initial encounter[10], the dancers felt connected to the sensors that they were wearing. In other words, they could play with them and be surprised by them, without the engineer having to intervene. During the first sessions, Alexis would simply adjust the sensors before the dancers began improvising and would not touch them again. The improvisation was therefore solely between the dancer and the sensor, or between the dancer and the soundscape generated by her movements. This soundscape is an intricate interaction between both the dancer's body and the sensor's activity; it cannot be reduced to either one of the two components. In order to describe and understand this sequence, it is thus fundamental that we take into account the device's role and its ability to *act* on the dancer. By doing so, we broaden the range of actants — "partners" in dance jargon — to include non-human entities (Latour 2005, Descola 2005): sensors, software, amplifiers, etc. We call a "partner" anything that *actively prompts* the dancer and/or the engineer to *act*. Broadening the range of actants to non-humans implies that we acknowledge that they make the protagonists act (dance, program) in a different manner. It is obvious for dancers, much more than for sociologists or for anthropologists, that the ground, the air, the music are "partners" in this sense.

A classic description of this device, one that would cautiously divide the roles between those who act and that which is acted upon, one that would describe this social interaction strictly as the relation between two individuals, would not be able to take into account how the sensor itself acts on the dancer. The technological interface plays a curious role that consists in "distributing" the various actants into the "bodies." The sensor worn by the dancer comprised, in part, the engineer's action; a part of each movement was thus performed by what the sensor made the dancer do, and by how the engineer acted on the sensor's parameters. In turn, part of what the engineer saw and worked on contained the dancer's movements, translated into gyrometers, potentiometers, etc. Furthermore, motion could

---

verified."

[10] Dewey calls an "encounter" the recognition of alterity and the creation of a common space for the various points of view. Cf. Zask (2004).

be heard in sound, and sound appeared in the dancer's movements. Everything seemed to interact together to the point that it seemed almost impossible to define the source of any movement and inspiration or to assign each component a predefined, static, stable role. Each actant took part in the creative process and played a role in a workspace usually reserved to humans.

3.2. Describing the device through movement

From this double premise of a broadened range of actants and a fundamental indetermination as to the source of each movement and sound — or to the identity of what is acting — we have no choice but to abandon a traditional description of this device. Instead, we choose to describe it from the point of view of the only constant[11] continuously engaged "entity": movement. By doing so, we are focusing on the common denominator of each component of the device. Moreover, it means that we consider the relations between these entities in terms of the circulation of a single motion uniting them. We therefore do not only consider each separate entity, but also their relation to one another. Because we experience not only things but also the relations between things (James 1912), let us therefore consider that anthropology is able to render this experience of relations. For once, this relation can be described as a flow that alters and transforms itself, and we can trace its journey and its evolution. For once, the connection between two beings appears as a series of tangible elements (Chateauraynaud 2004). For once, this connection is offered through its dynamic nature. We will try to make the most of it.

Taking the device as a whole frees us from the quest for a founding principle and incites us to embrace this fundamental uncertainty. The impossibility of pinpointing "who is acting" enables us to *activate* a much greater number of actants and to see how each affects the movement (Despret 2004). In fact, the circulation of motion becomes a constitutive element of each entity engaged in the device, rather than a simple link between predefined distinct constituents. The relation, in this case the movement, creates and stabilizes the entities' identities *from the inside* in what Stengers (1997, P. 64) calls a "mutual capture."

Although we are working on movement, we do not maintain that movement is always the same everywhere. Movement affects each entity in a different way, and in return, each entity itself produces a certain

[11] The two dancers involved in this creative process called their work "constant without gravity."

quantity and quality of motion. The relations between each entity and the others are in no way identical. For instance, the way in which the dancer interacted with the sound emitted by the speakers had very little to do with how she was connected to the MotionPod, or to the software. By describing the device from the point of view of movement, we understand how its circulation associated, united, and held together a heterogeneous ensemble of entities, in what James (1911) calls a "concatenated union":

> Two parts, themselves disjoined, may nevertheless hang together by intermediaries with which they are severally connected, and the whole world eventually may hang together similarly, inasmuch as some **path of conjunctive transition** by which to pass from one of its parts to another may always be discernible. (P. 108; emphasis added)

The various terms (entities) that constitute the device are all interconnected. Nevertheless, they are not necessarily *directly* connected to one another. The engineer can only "reach" the dancer through his keyboard, his software (MAX and Live), and the speakers; the sensors have very little to do with this equipment, yet they are connected, they belong to a common device, they are traversed by the same movement. James, in his concept of concatenated union, uses the image of a "partial conflux" that could characterize the way in which the sensors and the speakers, for instance, were contiguous without actually interacting:

> In a concatenated world a partial conflux often is experienced. Our concepts and our sensations are confluent, successive states of the same ego, and feelings of the same body are confluent. Where the experience is not of conflux, it may be of conterminousness (things with but one thing between); or of contiguousness (nothing between); or of likeness; or of nearness; or of simultaneousness; or of inness; or of onness; or of forness; or of simple withness; or even of mere andness, which last relation would make of however disjointed a world otherwise, at any rate for that occasion a universe of discourse. (P. 98)

Viewing the activity through movement enables us to trace "paths of conjunctive transition," uniting by concatenation even the entities that seem least connected to one another. This flow of movement can be described by analyzing each entity's — we could also say each "body's" — relation to the flow. In other words, we will analyze how this movement meets each body and how each body both affects and is affected by the movement. We will distinguish between two modes, affection and restitution, or between two types of entities. This will help us describe even more precisely the true "mediators" (Latour 2005) at work in the device, the credible and possible interactants besides the dancer and the engineer.

3.3. Which entities at play? From "entities" to "bodies"

These are the bodies, very schematically, through which movement circulates: a dancer, a motion sensor, two software programs, an engineer, an amplifier, two speakers, and sound. This circulation of movement is represented in Figure 1, which presents the bodies that take part in this device, rather than an actual sequence. For the purpose of our description, we will begin it with the dancer, but we could also have started with, say, a sound — which would be nothing if it were not for the engineer who programs it, etc.

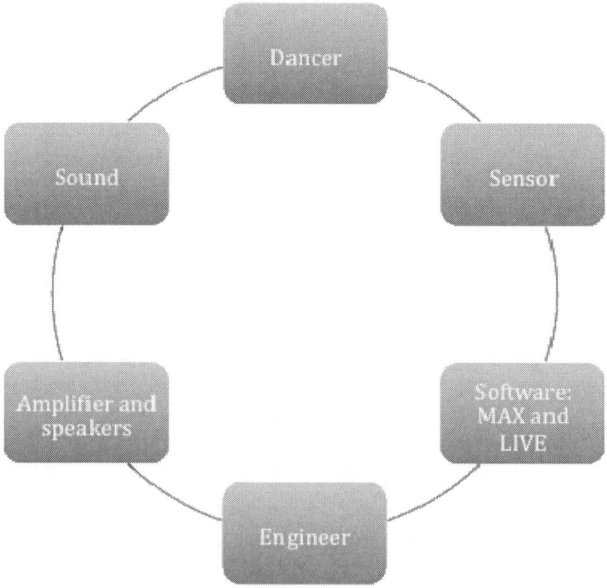

Figure 1. The circulation of movement through the device's entities

The dancer (1) makes a movement captured by the sensor (2), which then sends it through waves of information to software (3) that translates it into potentiometers that measure and display the intensity and the orientation of this movement. This, in turn, sets the engineer (4) in motion. Through his keyboard, he produces or alters the information that enters the amplifier (5). This information is then converted into electric impulses that activate the speakers (6). Finally, this sets their membrane in motion to create sound waves (which are also in movement) that meet both the dancer's body and the engineer's at the same time. The more this movement goes around, the more its trajectory becomes complex and the more its network becomes dense. There is no predefined order; there is only the constancy of movement. In what follows, we propose to unite the sensor and the software, since they cannot exist without one another, and call this new entity the "technological interface."

In order to define more precisely, from the point of view of movement, the various entities that constitute the device and to try to answer our initial question as to what enables human beings to feel an interaction between themselves and non-human entities, the distinction Bergson (1911) draws between the *material body* and the *living body*, and their distinct ways of rendering movement is useful. Material bodies (e.g., a rock) integrally restitute the movement they receive. Living bodies keep from a given movement only what is useful for their own action. When a rock rolls towards another rock and hits it, it transfers its energy and velocity according to the laws of physics. If the same rock rolls towards a man, he will try to dodge it if he perceives it. From his perception of the rock, the man processes its speed and the potential danger that it represents more so than its qualities (e.g., roundness, color, etc.). This perception entails a reaction that will most probably consist in dodging the rolling rock or trying to protect himself in any way possible. But in no way (except if there is no perception of the rolling rock) will he stay put and wait for the rock to transfer its energy to him.

What distinguishes material bodies from living bodies is the latter's capacity to perceive movement. This implies that they do not simply suffer the effects of movement, but that they analyze it, classify it, and then respond to it. While a rock directly transmits and absorbs a movement according to the laws of physics (action = mechanical and determined reaction), the same action on a living body is separated from its physical reaction by an interval, by a gap. This gap is the distance between the center of action and the *center of indetermination* that is the brain. This *center of indetermination* is capital in Bergson's demonstration, as well as in ours. We cannot legitimately deduce, from any given input of information that it receives, what its reaction will be. *It does not mechanically transmit data*. Because living bodies perceive, they eliminate what they do not need; they select what is useful for their purpose. Perception is not contemplation; it is action.

Why should such a distinction interest us? Using this distinction, we will now call the device's entities "bodies." We will now be able to identify the bodies that restitute movement in its integrality (material bodies) and those that introduce an interval that alters this movement (living bodies). This distinction is analogous to another one in the field of sociology, between "intermediaries" and "mediators" (Latour 2005). The main difference is that Bergson blurs the classic distinction between subject and object, thus rendering it impossible to assign definitive roles to entities. Both material bodies and living bodies can share certain properties and capacities. Instead of subjects and objects, we therefore find ourselves with various "bodies" that possess distinct capacities to perceive movement, to be affected by it, to react to it, and to express it.

We therefore find ourselves with three "living" bodies, three bodies that are able to restitute a movement by adding an element of indetermination: the dancer, the engineer, and the technological interface. As S. Rubidge (2001) argues, such interfaces become actual partners in the creative process:

> In using complex interactive systems, interactants have the (subjective) impression that the system, or more precisely the environment that generates this system, is interacting with them, rather than producing a standard reaction to a standard behavior. The system and the interactant are thus engaged in an actual dialogue.

We invite our readers to suspend, for the length of an improvisation, or of a chapter, their old certainties, and to feel what experiences have to offer.

3.4. From a "material" body to a "living" body

By themselves, the sensors are merely material bodies that measure and record quantities of acceleration (accelerometers) and changes in direction (magnetometers). But combined with the two software programs, MAX and Live, they form an interface that can be programmed to produce what Bergson attributed solely to living bodies: indetermination. For instance, Alexis can program the sensors and the software to produce different sounds depending on the dancer's location on the dance floor, but rather than assigning a note to each location, he can program the sensors to produce random notes. It is therefore impossible to predict what sound a movement will produce. If we pursue Bergson's reasoning, we can say that the selection made by the center of indetermination that is the technological interface can be understood as a perception: movement is perceived as the interface selects recognizable information relayed by the sensor[12]. But Bergson quickly moves from perception to consciousness and from perception to life, and to the fundamental indetermination that defines it.[13]

By operating a series of random choices, the technological interface *acts* as a living body would during an improvisation session. It becomes a potential partner. The difference between material bodies and living bodies lies in the latter's ability to introduce indetermination, surprise, freedom. George Lewis (Casserly and Lewis, 1997, 90), a pioneer in interactive music, said about *Voyager*, his "improvising" software: "Many people imagine that it always reacts the same way to a given sequence, but it is precisely built to do the opposite. It was conceived to give different answers, realistic answers, and to make choices by itself, even when we are not playing". During a recent improvisation session, Émilie — the second dancer — wondered what an interaction was for her: "I think I felt an interaction. I felt at play, I felt surprised." When we asked her what caused that sensation, she answered, "Feeling that something in front of me is making choices. I think I felt (talking to the engineer) its freedom." We define a dance partner as someone or something that can make choices and provoke surprise, and that can react indeterminately to exterior stimulations. We are in the same situation as the one described by Lewis : "When I play on *Voyager*, the computer is only another way for me to connect to those around me." As with our device, the goal is interaction and the fact of passing through a machine *does* alter that relation. Our hypothesis is that this detour is an *actant* that possesses, as our dancer and George Lewis experienced, a series of skills that make it credible as an interactant, that make it a real partner in improvisation.

3.5. Ontological confusion: who are we dealing with?

In the moment of improvisation, in this circulation of movement, the technological interface might have acted in response to what the dancers expected from a dance partner, due in particular to its competence for indetermination. We will next examine, in what we will call a *time of "recognition"* (Ricœur, 1986), the way in which the dancers were prepared to recognize the technological interface, embodied by the sensor that they

[12] The same way in which we only perceive that to which we have become sensitive. See the sociology of taste developed by A. Hennion and G. Teil (2004).

[13] "From this indetermination, accepted as a fact, we have been able to infer the necessity of a perception, that is to say, of a variable relation between the living being and the more or less distant influence of the objects which interest it. How is it that this perception is consciousness…?" (Bergson 2004:24)

wore on themselves, like a body with which to improvise, proceeding by analogy between the body of the sensor and their own bodies.

During one session, the dancers were asked to compare the anatomy of part of the inner ear, the vestibular system, to that of the sensor. The inner ear can be divided into two systems: the auditory system deals with hearing and the vestibular system manages information concerning the orientation of our bodies in space, their speed, and their balance. The MotionPod sensors function like the inner ear, sharing the two capacities of the vestibular system: recording speed and orientation. The comparison surprised the dancers. And yet the two "bodies" present share the same receiver which again bears upon the "identities" of the "bodies present." A whole series of remarks on the *ontology* of the sensors reveals this confusion:

> "The sensor is influenc… is affected by that."
> "It (the sensor) could capture gravity if you opened its capacity of perception, uh… of recording… I was going to say perception."
> "Does it know how much it weighs? Its weight? Its mass, sorry. Because it really is like us."

This "It really is like us," was again expressed by Anne, commenting on the sensitivity of the MotionPod on the acceleration mode: "Oh, well, it is exactly like us!! It works exactly like the inner ear. The variation is what is registered, not the constant."

The confusion that was expressed by our dancer — shared, in fact, by our whole team —gives the impression that she was trying to determine "who" she was dealing with. The boundary between the questions "What is Anne dancing with?" and "Who is Anne dancing with?" becomes indistinct. What was sought in a whole set of discussions peripheral to the dance explorations was quite simple a characterization of the *ontology* of the sensors — a qualifying test — of what they can or cannot do according to *what they are* and *what they know how to do*. Their competencies are tested and rated, not so much according to a fundamental identity, which Descola calls "interiority" (2005), but rather according to their "externality", i.e., their bodies and their common capacity to be affected by the same set of phenomena (acceleration, orientation). For example, at one time Alexis wanted to show Anne that the sensor was sensitive to gravity:

> "If you put it on its side, it knows it's like that." "Oh gosh, it does know a lot! That's reassuring, because dialoguing with something that doesn't know gravity would have been complicated."

The dancer is part of a process of "recognition" in which she recognizes and attributes to the MotionPod a set of competencies[14] that make possible a "dialogue" with it, that is, an interactivity between her body and that of a machine, as a necessary precondition for comprehension to be possible. This comprehension is grounded in the shared sensitivity of the two bodies to the same effects, concretely resulting in what Ricœur calls "common categories of action" (1986:251), between the dancer and the machine: the generation, modulation, and organization of matter made of sound and light. Ricœur specifies: "The categories of common action make specific relations between contemporaries possible." Furthermore, he matches this possible connection, based on common categories of actions with another principle, that of *analogy* or *recognition* that attributes to the other the power to be different or, more simply, someone to whom one can relate[15]. In Ricœur's language, Anne proceeded by analogy by recognizing the MotionPod — this "other" — not merely an object but a full-fledged subject. Analogy in fact depends precisely on this suspected and attributed capacity of all subjects to say "I."[16] By a "transfer in imagination," she discovered in it qualities that established a *continuity* between her body and that of the sensor.

And finally, this confusion created curious scenes, such as the end of the sequence that we have studied. In her improvisation, Anne and the sensor produced a strident sound. She looked at it as if to tell it to be quiet, scowling at the sound that they co-produced.

The relation to the sensor is not simply rational. It involves a very particular type of expressiveness (the role of the face in the interaction) — not that the dancer has a real dialog with the device, and the sensor does not see the expression on her face. Nevertheless, the interaction that she has with the sensor is sufficiently elaborate and credible for her to indulge in such mimicry as a mode of exchange with the object.

---

[14] Competency that the dancers were ready to recognize in the sensors insofar as they continually formulated analogies between their functions and the functions of the human body: the sensors were identified variously as "muscles," "nervous system," and "inner ear." Computer programmers in fact continue to draw inspiration from living bodies, human physiology in particular, to develop new software, such as from "neuronal networks." (Bret & Tramus 2004)

[15] By "transcendental" mechanisms that Ricœur calls "transfer in imagination" or "intropathy." "Transcendental" is worrying and cumbersome for socio-anthropologists. We try to give these mechanisms a place in the actual experience itself. The process described here is much like the Amazonian perspectivism studied by Viveiros de Castro (1998).

[16] Cf. the whole demonstration of the "analogy of the Ego" (Ricœur 1986:252-253).

Figure 2. Anne ordering the sensor to be quiet

3.6. The three types of recognition

In the interactions above, in view of the sensor's anatomy and potential indetermination (and thus interaction) we can distinguish three possible types of "recognition," via contact, similar constitution, or a potential for interaction. Two of these types were suggested by Frazer's model for "magic action": magic of contact, in which influence passes from one object to another "by contact," and empathetic magic, in which two entities influence each other because they share common properties. The first implies a sort of "contamination by direct contact" — by definition, the capture of personal movement. The dancer's movement contaminates the sensor — and vice versa — by means of the contact of their two "bodies." The second form of recognition is based on the dancer's and the sensor's common physical constitution, which Gell calls *externalist animation*. This form of recognition was seen to be primordial for dialogue and comprehension between dancer and sensor, as shown above.[17]

Gell offers the third form of recognition, which corresponds to what he calls *internalist animation* or the attribution of an intentional psychology or "agency" (potential interaction) to a non-human entity whose physical characteristics and capacities justify such a recognition. The projection of agency on the thing perceived is, cognitively speaking, the source of its animation. This allows us to postulate the possibility of a situated ontology, which is relational rather than taxonomic, as stated above. Moreover, the distinction between *animation* and *inanimation* avoids the distinction between living and non-living as "natural species" (Boyer 1996:86), just as previously the distinction between living body and

[17] Anne was, for example, relieved to know that the sensor "understood" and "felt" gravity.

material body avoided having to differentiate between human and non-human. If an object is animate, it is not because it has been attributed biological life but rather because its capability of being a potential "subject" has been recognized (Viveiros de Castro 1998). The nuance is important and, as Boyer remarks, "The projection of human physical functions [onto gods, spirits, objects, etc.] generally results in a projection, tacit or otherwise, of intentional psychology, but a projection of intentional psychology generally does not suppose the projection of any other human quality."[18] (P. 92)

Consequently, to attribute the status of "social agent" to a thing or a person, it is less important to define what they *are* in themselves than to describe the place that they occupy in a network of social relations. The only proviso necessary is that there actually be people/human agents "in the neighborhood" of these objects in order to activate their agency. Whereas Western philosophy has a great deal of difficulty identifying exactly what the difference is between human persons engaged in "actions" and simple objects that obey causal laws, anthropology, on the other hand, seems to be able to grasp why certain persons remain (in a contextual way) indifferent to this distinction.

We will now focus on the fine decoding of the network of relations that links the "bodies" present in the dance improvisation.

4. Potential affections of bodies: personalization

This work on the qualification of a technological interface as a credible partner of interaction now brings us to focus on the concepts of personhood and personalization. We started by theoretically rejecting the concept of personhood based on a subject-object dualism. Then, in our study, we had to characterize the presence of "bodies" equipped with specific capabilities of interaction, in particular those of a non-human. We set out a description of the device in terms of movement by distinguishing material bodies and living bodies, in order to include only the latter as real partners of the interaction. We finished the preceding part by asking what might provide the sensation of recognizing an object as a full-fledged interactant, distinguishing the diverse forms of animation that governed this recognition, that is, everything that made it possible for the sensor to "act," to acquire agency, or to be deemed a potential interactant. The last part of this chapter will be devoted to the concept of *personhood*[19], with

[18] For example, ghosts do not have human bodies.

[19] Understood in a minimalistic, pragmatic sense: someone or something capable

the process of objectification described above as the first step toward the dynamic personalization of this motion-capture technology. Examining the type of connection that unites the bodies described above, we will look closely at the specific ways in which the device generates a connection between bodies and also from a body to itself.

The device would indeed allow the dancer to "externalize," by means of the mediation that it offers and the circulation of motion, and return to self. But how are the bodies involved mutually affected in the improvisation experience, in the interaction? By analyzing this in light of Stephen Levinson's ethologic and linguistic anthropology, we will attempt to understand the specificities of the relation between the dancer, the engineer, and the sensor in order to decipher a return to self that we call personalization. Lastly, we will compare Marilyn Strathern's concept of *distributed personhood* to personalization, which we prefer.

4.1. Co-affection

Above we discussed the confusion that made it difficult to categorize a technological interface envisioned as a "body" equipped with competencies of interaction. This confusion contaminates the body of the dancer, in what interests us less for what it *is* than for its capacities to be affected by potential movements, whether they are human or non-human. And yet, non-human objects (or invisible ones) can become a "body" with which to interact and dance, as shown in the following quotation:

> I previously described how in certain situations I could feel an impulse given by the sound that penetrated me, coming into my body and generating a particular set of gestures. Sound intervenes in dance, as does a dancer. It itself generates gesture… sound thus constitutes a body. On several occasions, I was manipulated by the sound as I might have been by another dancer, in the sense that I felt a corporeal, muscular feeling in its impact.… The shock of one body against another is produced.… The amplified sound clearly acts on the body of the dancer like an external body. (Buisson 2009)

During the improvisation sequence that we analyzed, Alexis proposed that Anne should try wearing two sensors in relation, regulating each other. She attached them both to her right arm, one on the back of her hand and the other on her wrist. The sensor attached to her hand immediately felt uncomfortable: "Yeah, I'm not wild about that spot. I'm going to try another position that seems better for expression.… The sensor there really

---

of being in a relation.

hinders me — more than on other parts of the body." She consequently decided to move it to her biceps. The placement of the sensor makes a difference in the body's potential, as Anne pointed out: "The interest of these technologies is to mobilize certain parts of the body and to reveal possibilities."

However, Anne potentialized the sensor's "body" by reading a sort of poetry into it that was not inherent, but which appeared in her presence, depending on where she placed it on her body. She associated her own autobiographical memory with the sensor, bringing in a previous creation called "Holding Hand," centered on the expressiveness of her hands. Finally, the dancer and the machine ended up less expressing themselves one *through* the other than composing together an expressive voice mixing movement and sound. What emerged was a potential of feeling, affection, and expression that came from a sort of "co-affection."

The device brought together three "articulated living bodies" through movement. Each body gradually became aware of the others and was able to be affected by them and to affect the others in return. They became mutually *interested* by means of the specific mode of intervention and expression that characterized them in the improvisation. Each body, however, had its own a potential of expression, confirmed by the way in which it "interested" the others, or by its capacity of being perceived by them. Expression here is not a linguistic act but rather *embodied action* (Leenhardt 1947): no longer an act of signification but of creation — in particular, the creation of a connection among the three bodies engaged in the improvisation.

What connected these three bodies (dancer, sensor, engineer) must consequently be considered in its specificity. It was not a traditional interaction but one involving a specific mode of presence (Piette 2008) that connected them in a singular way, either in a soundscape or in a play of light. The true "meeting place" of the three bodies was not the dance floor but the sound and light that was produced and that embodied the relation at work among them. Each body, by its mode of presence, had a mode of expression that "interested" the other two, embodied in sound but unique to itself: the engineer did not dance nor did he represent accelerations and orientations in information to produce sounds; the dancer could not express herself like the sensors either, nor could she program the MAX and Live software to intervene in the improvisation; and the sensor could not dance and could not program.

We use the term *concretion* to designate the phenomenon by which sound or light embodied and manifested the presence of the bodies in their relation, and in which the sound was both the indicator of the presence and also a whole that could not be separated from the action of the entities taken individually. It may also be viewed as an intertwined association between the movements of the body and of the music that reproduces movement in sound and the sound that emerges and/or is modulated by movement. The device comprises a union of things that are normally separate and isolated; this is witnessed by the logically absurd formulations of the dancers: "You hear the vision," "You hear the movement!" or "I hear your position." That is what we call *concretion*. In geology, a concretion is an aggregation of particles, a phenomenon of thickening or solidifying. Here, the term is applied to the meeting of several heterogeneous entities (a wrist, a sensor that resembles a wristwatch) and particles (articulated movement, modulation of the sound volume, light intensity) coming from outside sources that are solidified together in one body. Conceiving of this interface as an "aggregator" makes it easier to understand why and how it blends so well with danced improvisation, where gestures are usually abstract, elliptic, and nonfigurative.

This concretion exists on at least two levels. It appears first, pragmatically, in the material attachment of the dancer's body and the body of the sensor, tightly bound or connected, because the gestures produce or modulate sounds or light. The music, for example, becomes "graspable," "palpable" or "tangible," like a solid object that can be held. Second, when a gesture is directly and immediately "distilled" to a sound/light output, that reduces its polysemy; it becomes no more than modulation of sound or light. This reduction implies that the acceleration or orientation of the gesture "is reduced" to an increase in volume, and that immobility "doubles" silence, by concreting gesture and sound. The device thus creates performative bonds between entities of different natures (human gesture, sound, light, computer competency).

4.2. The concretion of presences: affection, reflexivity, and interaction

As a whole, the device thus demonstrates the co-presence (Piette 2008) of bodies in a relation, but it is also the means or the tool by which each body can reach the others. All in all, it is both a translation of the working relationship and also an instrument that intervenes in the relation. It "translates" but also "acts"; it is the performance of the action "in concretion" of the three bodies, and it also performs the attitude of each one of these entities. This specific type of relation, unlike more classic interactions, conforms nevertheless to the structure of elements governing all interactions, proposed by Stephen Levinson (2006). The three bodies share the same *resource* — movement — in a specific *temporality* that attests that the three bodies are all simultaneously, continually connected. Each one *waits* for the others to express themselves in their own mode of expression and manifest their presence in *rites of mutual access*: the sensor flickers to announce its activation, the body of the dancer wildly sets off the needles of the MAX software to announce the movement, and the engineer is triggered into action by the sound. Each signal of presence makes the others do something, thus confirming mutual access. The example of the dancer ordering the sensor to be quiet clearly showed the *intentional and emotional dispositions* that this connection caused. *Cooperation* is illustrated by the joyful play that the device provoked, as witnessed by the length of time that the dancers spent in relations with the device. By fulfilling all the prerequisites of what Levinson calls a "human interaction engine" this confirms the intimate sensation of the relation of these bodies involved, in spite of the fact that non-humans are involved.

However, the device not only intensifies the capacities of a body to be affected and potentialized by other bodies, it also allows the dancer's body to be affected by itself, thanks to the mediation of the technological interface. At the end of one exploration, the dancer declared, "The sound came and activated my ears as points of awareness." This was after a rather complex sequence: (1) The dancer moved with the sensor placed on part of her body. (2) The sensor focused a particular awareness on this part. (3) The movements of this part of the body started a sound. (4) This sound activated the awareness of another part of the body, the ears. (5) The sound that the ears heard returned to inform the part of the body with the sensor of its movement.

The dancer's body, the sensor, and the engineer create a reflexive loop (feedback) inside which actions (movements) and reflections (conscience) circulate so rapidly that it is not possible to view them in relations of precedence or causality. A movement sparks awareness and awareness sparks movement, thanks to the mediation of the sensor and the consciousness of the dancer. The dancer is affected (and potentialized) by her own movement that is retranslated and given form in sound and light (concretion). This relation of the dancer to herself passes through the sensor.

Objectification, we stated above, is the process in which a subject is externalized in material form. The recognition of this object as source of affection and potentialization constitutes a return to oneself, which we could call personalization. The production of feedback generated by a motion sensor seems to exemplify such a process. The mediation of the device gives the dancer information about the parameters of her movement and the qualities of her sensations, two domains that in normal life most often go "unperceived" (Bois 2001). It provides, in other words, a perception of oneself mediated by the sensor. Anne, in fact, saw this as the interest of the device for her work in improvisation: "I believe that is the relevance of the device: simultaneity. It's a system that gives us an outside view of what we're doing."

But what is the theoretical route, which starts with the connection of these bodies and ends back at the process of recognition (from the point of view of the dancer in the following demonstration), that is called personalization? In other words, what is accomplished by this pragmatic complexity that mixes material and living bodies in the person of Anne? This device enabled us to view the body as a set of relations that are connected with other bodies — human or not — and also to itself. As Kaplan (2008) shows, each body is actually composed of "multiple corporal spaces," or arrangements that allow for different forms of body schemata. Objectification is one particular form of schema, the crystallization of an immaterial externalization (a solo) in a material form. Personalization would thus, as will be seen below, consist of an ultimate stage in which the connection of oneself to oneself via the "body" of the sensor creates a sort of mirror effect, passing through an artifact whose reflection redefines the person "captured." The relational configurations, the relative positions of the participants, are modified by this interactive co-construction: Anne has changed her view of technology, according it a poetic quality that she thought that only she possessed before the encounter. The sensor reveals the potentialities of Anne's dance, offering her its dimension of pure technicality. All this reinforces our view of the sequence as an undertaking in *qualification,* in which a body becomes a credible partner in interaction, capable of affecting or potentializing the other bodies present, with a trajectory of establishment and complexity made possible by a device that is defined by the relational configurations that here brought together an engineer, a dancer, a motion sensor and a battery of computer and electro-acoustical equipment.

4.3. Prehensibility of the self

If we try to connect what we said about concretion with what we have just established concerning relational configurations, two questions arise. What is uttered by this "combined voice" that emanates from the concretion of movement and of sound? What does it say about situated processes by means of which representations, categories, ontologies, and cosmologies are born in the course of the action?

Above, when Anne was wearing two sensors on her right arm, feeling constrained, she suddenly decided to remove one of them; she lay it on the floor, took hold of it in her hand in order to handle it. At the very moment in which she did so, she demonstrated the prehensibility of the device, or its potential of grasp, as she explained afterwards:

> What's more, it's true that it's the hands, so there is poetry. That is, more poetry because there is more human presence; that is, the hand has an immediate meaning, so it is different from the movement of the whole body. And besides, there are the objects. That says something very human. Well, I mean [taking the second sensor with the other hand, standing] aaaaah. You can play with all that, right? [She observes them and seems to be weighing them to compare them.] That's interesting — I think that's a lead that I'd like to follow. [She hides the MotionPod in her hand.]... We are working with a device doing research. The poetry isn't necessarily there; it's not on a set theme; it will come out of what is there. So can we make poetry with this [holding the sensor up directly to the camera in her palm]? That's what's going to interest me most.

Her technical and pragmatic grasp on the sensor transformed the device into a tangible interface. But it is also a reciprocal grasp, in terms of "self-insertion" (objectification), that Anne was handling, showing and hiding the sensor, caressing it, observing it in her hand, enhancing it, by interjecting into it her solo "Holding Hand," reactivating and incarnating this layer of her person in the object. By discovering this new potential of the sensor, she re-accessed her person, which should be viewed as a set of space-time coordinates, incarnated in a body, a biological organism, but also as a succession of biographical events, memories, and traces. By experiencing this encounter with a new facet of technology, she experienced an encounter with herself altered, transformed, renewed, mixed:

> I'm bringing in things from my solo, "Holding Hand," that I created with my hands in May. So here are all the things that I went through in six months to produce that. They're... They provide me with "matter" and I'm developing them there today.

The poetry that came out of this illustrates the double affection mentioned above. We witnessed a technical construction of poetry in which Anne used her technical repertory of gestures from "Holding Hand" and the sensor produced a poetry unsuspected until then. This was how Anne externalized her solo into the object and objectified part of her person, her vision of art, soft and fairy-like, opposing the fears of a cold technicality that she had had before the residency. This opposition exemplifies the relational reconfiguration that we observed.

This process of personalizing the motion-capture technology did not take place effortlessly; we observed a preliminary phase of depersonalization and anonymization in which the sensor was detached, set on the ground, and was observed with distance and detachment, allowing Anne to enter her own dance and to reconnect with herself before taking up her dance with the sensor again. This connection was not broken when the case came off the sensor and the sensor detached itself from its former objectification in a process that recalls how ritual objects are emptied, purified, freed of their *genius loci* before they can be animated (Zammouri 2008). These observations allow us to affirm, following Gell and the art objects that he studied, that technology has a physiognomy (Gell) and a personality that can be grasped only during interaction, if what they cause, suggest, allow, enclose, and force are studied ethnographically. These additional remarks allow us to go beyond an anthropomorphization (Guthrie 1993) and an animation of the sensor, in order to reach the relational processes that govern the personalization of this technology[20].

4.4. The personalization of the device

In order to grasp person(hood), it is necessary to think about network and distribution. Anne and Alexis were everywhere there was movement being captured; the sensor was everywhere that there was movement. How does the concept of *distributed personhood* enable us to define personalization as inherent in a network of relations and distributed entities? Marilyn Strathern, specialist in Melanesian cultures, took her inspiration from idealist philosophy[21] in order to posit a view of a social world composed of relations rather than things or persons. The key functions that she identifies are the relations of relations, in which the social order is enriched by images of forms, relations, and transformations of subjectivity in interaction. Material and living bodies cannot, in her system, apprehend themselves "in themselves" or be defined by fixed "essences." However, they can assume an unlimited number of identities and activities according

[20] Anthropomorphization functions within a relation of *perception*; personalization is constructed from a network of relations.

[21] Idealistic philosophers maintain that all relations are "internal," i.e., that they are part of a system/code, and that there are no "external" relations in which relations between objects are theoretically independent of each other. Alfred Gell (1999), "deciphering" what he calls a strathernogramme defines the relation as a connection between two terms.

to their articulation within a given relational configuration. Moreover, the relations between terms, according to Strathern, are relations of exchange. Post-Maussian literature shows us that in exchanging gifts, in giving to exchange partners, donors give something of themselves (their persons) to their partners, in the form of an object. It is perhaps possible to consider that Anne and the sensor took part in such a gift/return-gift relation: the sensor afforded Anne a wealth of technical exploration while Anne gave the sensor a poetic potential. The relation (1) that Anne maintained with this motion-sensor technology was dependant on the relation (2) that this technology maintained with potential artistic and poetic expression. From this comes the deduction that the processes that govern the personalization of technologies of motion-sensor concern a relation of relation. And, as Strathern suggests when distributing relations according to a hierarchical order, the relation (1) of Anne to technology was subordinated to her relation (3) to her own "manual poetry."

By considering Strathern's view of relations in regard to our study, we focus on how exchanges implying objects produce person-to-person attachments and connections between persons and themselves. The "living" bodies that we identified in the second part, whether technological or human, exchange competencies through multimodal channels of interaction (Levinson 2006) because they already share certain properties (anatomical, haptic, kinetic); other properties can be activated as their reciprocal "interest" deepens. People are thus nothing but the sum total of their relational components, their connections with other bodies, authorizing them to "be otherwise," influenced by the point of view, sensation, and potentialization of the living and material bodies with which they are interacting. Strathern enables us to illustrate and identify, in the field of technologies, the objectified and externalized relations that give birth to new relations (between people, things and people, people and their immediate environments, and also with oneself). It becomes clear how much people and objects exchange, especially in regard to points of view and points of sensory access to their "respective worlds" — views on bodies and their material malleability, their fluctuating potential of animation or perception and their potentials of personalization: their ways of existing, acting, and connecting, benefiting from all the "hybrid" confused ontologies that make up everyday life.

According to our starting assumption, personalization is a relation of relations between objectification (externalization in a material form) and a relation with an acting entity (self, divinity, deceased, human, Alexis, etc). If objectification is mutually constitutive, personalization is recognition, insofar as one recognizes somebody or something more by sharing a potential than by drawing a physical analogy (the potential of artistic or poetic creativity in our case). This implies that there is no personalization without a concept of situated personhood (here, an artistic creative situation). In our sequence, there was objectification in a form that corresponded to Anne and in which she recognized herself, situated in a sort of material externalization.

Automatic teller machines may afford a valuable comparison to our device. When a user insults (animation and anthropomorphization) a malfunctioning machine, trying to obtain his or her own money, he or she is connected to a person (a banker, an accounts officer) via the objectification of the bankcard. Here, the encounter with the object may be seen as a form of personalization. It is a relation of relations: the user has a relation with the object and a relation with the banker, or in any case with the banker who sold the customer the bankcard or the person who built the machine. This relation of relations becomes obvious in a moment of malfunction; it is experienced as a refusal whose origin is unclear. This indetermination is built on the relation of relations. Anne recognizes herself in her encounter with the objectification of Alexis (in the technological device) and "finds herself" in a situation of construction, elaboration, redefinition of her person, viewed as dynamic and syntactic, "in combination or construction of terms together" (the terms understood here as participants engaged in the interaction). The order of the terms, agreements, dissensions, number, and gender constitute the network of material and living bodies, whose potentialization and potential for animation vary according what they exert on each other — subordination in the case of the ATM, coordination in the case of Anne. In the two cases mentioned above, technologies have a powerful potential contact (no one besides the user touches the bankcard, the capture of a single person's motion), similar internal constitution (the identity chip and the relation to an amount of money, the anatomy of the sensor and dancer), and a potential of interaction (tactile, haptic, kinesthetic, visual, auditory, etc. — cf. Levinson). These three forms of recognition govern the processes of personalization of technologies: these bonds — material and immaterial, technical and identitary, intertwined and indistinguishable during the situation — reconfigure persons, dynamic and syntactic (subordination, coordination) within their network of relations. These devices may bring about the redefinition of positions related to the persons themselves, to others, to things, and to their environment. Shared subjectivities constitutive of personalization is thus based on their technical syntaxes. In this manner, it is possible to trace the paths between persons (human and material), their externalizations (immaterial: sounds, gestures), their objectifications (looped interfaces, technical nodes) and inward returns to oneself (syntactic). It is now manifest how much human and material technicality are syntactically intertwined

5. Conclusions

In the end, our study consisted of attributing a "point of view" to the technological interface embodied by a motion sensor worn by a dancer, in action occurring during an improvisation in which it took part. This means recognizing in it the capability of acting as a subject, in a logical pattern not unlike what governs Amazon perspectivism: "It is not that animals are subjects because they are humans in disguise, but rather that they are human because they are *potential subjects"* (Viveiros de Castro 1998). We took up the risk of this venture because the technological interface showed itself capable of interacting in the tests that the dancer and the engineer set for it, surprisingly "qualifying" as a credible partner in interaction. The originality of our study lies thus first in recognizing a number of human competences in an object, an interface — in particular its capacity to generate ties — and, subsequently, seeing how this interface made it possible to reexamine personalization or more simply to see how personalization is a dual process of objectification and relational loop, by exteriorizing part of oneself that can have a return effect. We had to redistribute competencies between objects and subjects, rendering the two categories indeterminate and propose a new alternative between different "bodies," whether living or material. This alternative, still too discriminating, ceded in the end to a reference to personhood.

Our work thus demonstrates how, from objectification and distributed personhood, (as in the situation of return gift-giving), it was possible to refute the oversimplification consisting of alluding to "augmented personhood" (with all the moral implications that summons up). We suggested, by replacing the duality of subject and object with another, more indeterminate, of living and material bodies, how these bodies acted in their potential of expression, affection, and interaction. The concept of "personhood" made it possible for us to "make more fluid" strict, fixed, essentialist definitions of the entities engaged in the device, considering them all *as persons,* enabling us to enlarge the domain, not limiting them to their mere "bodies" but rather linking them and distributing them within a network of relations. We thus attempted to examine them in regard to everything that bound them together — movement, anatomic analogies, potential for interaction, and ontological indetermination. We can therefore conceive of a person as a technical construction centered on the network of relations ordered by his or her body, respecting a syntax that we have just begun to examine, using the concrete situation of our field of observation. The definition of a person is not established by a pre-conceived network of representations — a culture, a system of signs, a cosmology (that would assign a meaning); rather, it stems from the minute description of respective positions of the entities that *constitute* the fabric of relational configurations, but above all *constituted* as persons by means of the articulation of relations that unite them. This initiative was the result of our desire to relate the question of the meanings of gesture, suggested by the company that commissioned our study, and also of our imperative, identifying the social syntax that motivates all of our efforts, be they methodological, analytical, or theoretical. The importance of movement, action, and emotional processes of multimodal perception renews our view of personhood by allowing it to be perceived in relation to a dynamic process of personalization, with a certain fluidity. This fluidity in practice means that sharing a resource such as movement is, in our study, the smallest elementary unit of potential sociality between humans and non-humans.

The term that Hegel used more than two centuries ago to theorize the process of objectification, *Entäußerung*, guided us. It has often been pejoratively translated as "alienation," quite removed from the positive connotation that Hegel gave to auto-alienation as the essential moment of detachment needed for the historical development of the subject. New technologies, increasingly "intimately coupled" to persons, do not necessarily denote the disappearance of the body chained to them; rather, they may allow for the opposite, a renewed corporality.

References

Bergson, H. (2004), (1896). *Matter and Memory*, Dover Philosophcal Classics. New York: Dover.

Bois, D. (2001). *Le sensible et le mouvement* (Perception and Movement). Saint Quentin en Yvelines: Points D'Appuis.

Boyer, P. (2001). *Et l'homme créa les dieux: Comment expliquer la religion* (And Man Created the Gods: How to Explain Religion). Paris: Robert_Laffont.

Bretm M. and M.-TRAMUS, H (2004). L'interactivité intelligente (Intelligent Interactivity). *Nouvelles de Danse* 52. Bruxelles: Contredanse.

Buisson, C. (2009). « Immersive Theatre #2 – Ethnographie sonore du mouvement dansé: Une écriture performative. Sound Ethnography of

Danced Movement: A Performative Writing", *Ethnographiques.com* 19 (http://www.ethnographiques.org/2009/Buisson).

Casserly L. and G. Lewis. (1997). "Person to… Person?" *in Resonance Magazine* 6(1)

Chateauraynaud, F. (2004). "L'épreuve du tangible. Expériences de l'enquête et surgissements de la preuve." In *La croyance et l'enquête, Raisons pratiques* vol. XV. Paris: EHESS.

Damian, J. (2009). "Des monts et des mondes: Ce que le corps des marcheurs fait à la sociologie." *M@GM@* 7(3). http://www.analisiqualitativa.com/magma/0703/article_10.htm

Deleuze, G., and F. Guattari. (1980). *Mille Plateaux: Capitalisme et schizophrénie 2*. Paris: Minuit.

Descola, P. (2005). *Par-delà nature et culture*. Paris: Gallimard.

Despret, V. (2004). "The body we care for: Figures of anthropo-zoo-genesis." Pp. 111-134 in *Body & Society* 10(2-3). London: Sage.

Despret, V. and S. Galetic. S. (2007). "Faire de James un lecteur anachronique de Von Uexküll: Esquisse d'un perspectivisme radical." In *Vie et expérimentation: Peirce, James, Dewey*, edited by D. Debaise. Paris: Vrin.

Doré, A., (2010). "Le devenir politique des animaux. La circulation des loups dans l'espace public" in *Sociologie et Sociétés* (à paraître)

Gell, A. (1998). *Art and Agency: An Anthropological Theory*. Oxford: Clarendon Press. 1999. *The Art of Anthropology*. London: Cambridge University Press,

Goffman, E. 1991 (1974). *Les cadres de l'expérience*. Paris: Minuit.

Gomart, É. and A. Hennion. (1999). "A Sociology of Attachment: Music Lovers, Drug Addicts." in *Actor Theory Network and After* edited by J. Law and J. Hassard. Oxford: Blackwell: 220–247.

Guthrie, S.E. (1993). *Faces in the Clouds: A New Theory of Religion*. New York: Oxford.

Houseman, M.(2006). "Relationality." Pp. 413-428 in *Theorizing Rituals: Classical Topics, Theoretical Approaches, Analytical Concepts, Annotated Bibliography* edited by J. Kreinath, J. Snoek, and M. Stausberg. Leiden: Brill.

James, W. (1912). *Essays in Radical Empiricism*, New York, London, Bombay, and Calcutta: Longmans (Green & Co.)

Kaplan, F. and P. Y. Oudeyer, P.Y., (2008). "Le corps comme variable expérimentale (The Body as an Experimental Variable)." in *Revue philosophique de la France et de l'étranger* 133(3) : 287 : 298.

Latour, B. (2004). "How to Talk about the Body? The Normative Dimension of Science Studies.", in *Body & Society* 10(2-3). 2005.

Reassembling the Social: An Introduction to Actor-Network Theory. Oxford: Oxford University Press 205: 229.

Leenardt, M. (1947). *Do Kamo, la personne et le mythe dans le monde mélanésien* (Do Kamo: Person and Myth in the Melanesian World). Paris: Gallimard.

Levinson, S. C. (2006). On the Human Interaction Engine. In *Roots of Human Sociality: Culture, cognition and interaction*, edited by N. J. Enfield and S. C. Levinson. Oxford: Berg.

Miller, D. (1987). *Material Culture and Mass Consumption*. Oxford: Blackwell.

Pelsd, S D., K. Hetherington and Vandenberghe. F. (2002). "The Status of the Object: Performances, Mediations, and Techniques." in *Theory, Culture & Society*, 19(5-6) 1–21.

Piette, A. (2008). "L'anthropologie existentiale: présence, coprésence et leurs détails (Existential Anthropology: Presence, Copresence and Their Details)." in *Anthrocom* 4(2) : 131–138.

Ricoeur, P. (1986). "L'imagination dans le discours et dans l'action (Imagination in Speech and Action)" In *Du texte à l'action - Essais d'herméneutique II* (From Text to Action; Essays in Hermeneutics II). Paris: Seuil "Points".

Rubidge, S. (2001) "Action, reaction and interaction", in *Dance Theatre Journal* 17(2).

Severi, C. and M. Houseman, M. (1994). *Naven ou le donné à voir: Essai de l'interprétation rituelle* (Naven or the Gift of Sight: An Attempt at Ritual Interpretation). Paris: CNRS.

Stengers, I. (1997). *Cosmopolitiques - Tome 1 : La guerre des sciences*. Paris: Les Empêcheurs de Penser en Rond.

Strathern, M.,(1988), *The gender of the gift*, University of California Press

Teil, G. and A. Hennion, A.(2004) . "Discovering Quality or Performing Taste? A Sociology of the Amateur." Pp. 19-37 in *Qualities of Food: Alternative Theoretical and Empirical Approaches* edited by M. Harvey, A. McMeekin and A. Warde. Manchester: Manchester University Press,

Viveiros De Castro, E. B. (1998). "Cosmological Deixis and Ameridian Perspectivism." Pp. 469-88 in *Journal of the Royal Anthropological Institute* 4.

Waille, F. *Corps, art et spiritualité chez François Delsarte (1811-1871): Des interactions dynamiques* (The Body, Art and Spirituality in François Delsarte (1811-1871): Dynamic interactions). Thesis in Sciences sociales, Histoire, Histoire religieuse, politique et culturelle.

Defended 17 December 2009.

Zammouri, H. (2008). *Le remplissage matériel et relationnel comme procédé d'animation des objets et ses implications dans les pratiques rituelles* (Material and Relational Filling as a Process of Animating Objects and its Implications in Ritual Practices). Thesis, Master 2 Research en Science Historiques Philologiques et Religieuses, Vème section, Sciences Religieuses. Ecole Pratique des Hautes Etudes. Thesis advisor: Michael Houseman.

Zask, J. (2004). "L'enquête sociale comme inter-objectivation (Social Surveys as Inter-Objectivation)." in *La Croyance et l'enquête: Aux sources du pragmatisme* (Belief and the Survey: At the Sources of Pragmatism), edited by B. Karsenti and L. Quere. Paris: EHESS.

PART TWO:

THE SOCIAL NATURE OF THE BODY: SYNCHRONIZATION, EMBODIMENT AND TECHNOLOGIES

CHAPTER EIGHT

HOMO SAPIENS, THE TECHNOLOGIST: THE RHYTHMIC FUNCTION OF THE TECHNIQUE OF THE BODY

BIANCA MARIA PIRANI

Abstract

Homo sapiens has emerged as a technologist. This is why some social scientists call man a tool-making animal. The evolutionary record affirms this. *Homo habilis* used crude stone implements as far back as 2.5 million years ago, more than two million years before *Homo Sapiens* first appeared on the African savannah. Knowledge of this long overlap of human evolution with tool-making produced significant changes in evolutionary thinking. Earlier views, long held in many disciplines, maintained that evolution first effected biology and only later culture. In the current view, it worked on them together. Tool-making and *Homo sapiens* evolved jointly, tools —their making and use—played a part in the development of the large brain that separates present-day humans from their proto-human predecessors. Man is a species bundled together with technology. The order of tools defines the relationship between the body and things: that is to say, the interfacing mechanisms which enable its orientation in territorial space, its survival and the construction of social life. The deep meaning of technology lies in this pragmatic relationship that allows the body to construct its permanence in the world.

By focusing "bodies in action" as a relational matrix of human sociality, this paper applies the notion of the "technique of the body" classically elaborated by Marcel Mauss to the cognitive activation of the spatio-temporal networks constituting the actor-centered decision making.

Keywords: body techniques, "implicit memories", actor-centered decision making, social networks

1. Homo sapiens, the Technologist

Man is inseparable from his tools. This is why some social scientists call man a tool-making animal. The evolutionary record affirms this. *Homo habilis* used crude stone implements as far back as 2.5 million years ago, more than two million years before *Homo Sapiens* first appeared on the African savannah. Knowledge of this long overlap of human evolution with tool-making produced significant changes in evolutionary thinking. Earlier views, long held in many disciplines, maintained that evolution first effected biology and only later culture. In the current view, it worked on them together. Tool-making and *Homo sapiens* evolved jointly, tools – their making and use—played a part in the development of the large brain that separates present-day humans from their proto-human predecessors.

Man is a species bundled together with technology. The order of tools defines the relationship between the body and things: that is to say, the interfacing mechanisms which enable its orientation in territorial space, its survival and the construction of social life. The deep meaning of technology lies in this pragmatic relationship that allows the body to construct its permanence in the world. Technical relationships are embodied in social relationships and constitute the relational matrix of social intelligence. Werner Sombart (1911: 311-347) considered technology as a great historical modelling force. Technology is a driving force in social change, since it brings about creation and shapes symbolic activities. The development of technical skills, which the body learns by manipulating objects as well as using and perfecting utensils, explains the peculiarity of human cognition compared to animal intelligence as a creative process connecting pre-existing elements in new, and useful, combinations. A "toolbox" fit to invent, after every new interaction, the "able practice" that Marcel Mauss (1936: 394) had already perceived was the "primary tool" of cultural reason. Of all human activities, technique is the only one which does not go back to the starting point. Techniques are learnt and they cumulatively advance with occasional leaps forward. Their aim is to lessen human physical—and lately mental—energy output in under-standing and using the natural world for human purposes.

The development of technical skills, which the body learns by manipulating objects and using utensils, explains the peculiarity of human cognition compared to animal intelligence as a creative process connecting pre-existing elements in new, and useful, combinations. A "toolbox" fit to invent, after every new interaction, the "able practice" that Marcel Mauss (1936: 394) had already perceived was the "primary tool" of cultural reason. The practice of a tool brings about the capacity to repeat a performance with a variation depending upon the changing contexts. This capacity is an issue which implies a sort of correspondence and of stabilization between environmental reactions and conditions. Therefore the technological practice makes the body the *interacting field* connecting the individual with society through the neurophysiological processes generated and regulated by experience. All types of experience and learning modify the neural maps.

Neuroscientists have demonstrated that just three months of activity or passivity can affect the cerebral structure. Indeed, the experiences stimulate the formation of synapsis, activate neural circuits, consolidate them and create new ones.

The human brain is an open and dynamic living system shaped by the changes provoked by the constant variation of the environment, able to receive specific stimulation due to its functional connection to other systems, in particular to other brains. Relational neurobiology (Siegel, 2001) reinterprets individual differences with reference to the interaction between hereditary traits, environmental influences and accidental events, and it shows, with increasing evidence, that the brain functions which are at the foundation of human experiences—memory, affection and emotions—are not functionally separated from cognitive ones. The relational need drives human beings to create connections with the external world in order to keep an adequate neurophysiological balance. The physiological, sensorial and cognitive mechanisms which are put into action in relational contexts mediated by present digital technologies are the same which can be found in the traditional relationships between human beings and technology. According to the Swedish neuroscientist Torkel Klingberg (2008: 10-11) the ability of the contemporary brain is identical to that of the anatomically modern *Homo sapiens*.

The changing elements are represented by the quantity and complexity of the information constituting the "technological environment". Technology is an *activity* rather than a manufact. The definition of technology by D. L. Sills (1968: 58) implies in particular an *activity*.

Technologies are the body of capacities, of knowledge, and of the practice to create, to use and to do useful things.

According to Sills (1968: 58), technology is a modality of creation derived from the *knowing how* (Ryle, 1949), that is to say from the ability to perform specific procedures based on the identification of specific *stimuli*. As a procedural instrument, the basic function of each different technology is to position the body of human beings in their environment, creating those forms of exchange which identify the social space. The adaptive procedures which granted the body its self-preservation and the transmisson of its genetic heritage correspond to the means of intervention on matter. In human beings the nervous system, and in particular the brain, evolved in order to specialize in problem solving, from the ways to prevent drowning to the search for food or for a partner. Each different technological invention is basically an *active manifestation of a new relational product* whose appearance is determined by the unrepeatable nature of its materials, of its events and of the human beings which constitute it. It basically springs from our imperious need to expand, extend, develop the sensorial abilities of our body until this activation enhances and improves it. The main primary justification for technological inventions is represented by this tendency, since the body establishes new relationships with the *habitat* in the unceasing attempt to gain identity in a more complete way.

The *transition to the tool* is functionally justified by the field of relationships which the hand embodies and transforms. The order of tools testifies to the casual evolution of the "repertoire of recalls" which connects the social life of the first *Homo* who inhabited the savannah until the emergence of the symbolic behaviour which constitutes one of the main features of *Homo Sapiens.*

The manufacturing of lithic tools is the most evident expression of the acquisition of a technical intelligence: an aim to attain, the planning, the final performance and then the tool. Although it is true that the use of natural tools such as pebbles, sticks and leaves can be also found in other animal species, from otters to anthropomorphic apes, its systematic use and the resulting technological development are peculiar to the human species.

The *pebble culture* is the oldest existing technology and it is based upon the manufacturing of pebbles. Hominids broke pebbles so as to obtain tools which could dig, cut, cut off and make holes. They soon improved their manufacturing techniques so as to remove splinters on both sides of pebble blocks in order to have sharper edges (bifacial pebble tools). The *pebble culture* presents some primary typologies which summarize its technological evolution.

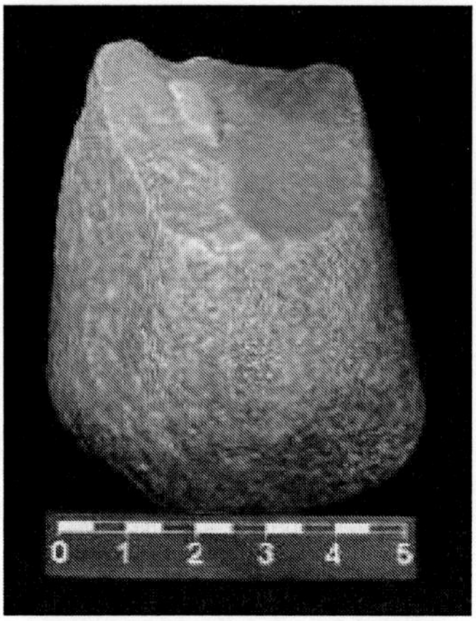

Figure 1. Chopper Tamesrouf, Algerian Sahara

Figure 2. Chopping Tool, Tauz, Algerian Sahara

Chopper [figure 1]:

A very primitive tool consiting of a pebble from which some splinters have been removed with a few carefully aimed blows (on one side only) in order to provide a cutting edge.

Chopping Tool [figure 2]:

Different from the chopper in that both sides were shaped.

Figure 3. Amigdala (Gafsa, Tunisia)

Today we certainly know that our direct ancestors, defined by paleoanthropologists as "anatomically modern" humans, came into being in Africa 200,000 to 165,000 years ago and genetic evidence has demonstrated that all populations living nowadays are connected to those African ancestors. According to the estimate of the team of geneticists headed by Wolfang Enard from the Max Plank Institute of Evolutionary Anthropology in Leipzig, the moment the human form of the gene pool was fixed in the population (through a process of natural selection following a chance mutation) has to be traced back to the last 200,000 years in human history. The "anatomically modern" *Homo sapiens* began with an irradiation to the rest of Africa around 100,000 years ago and was followed by an expansion from Asia probably through a Southern itinerary towards Oceania and a Northern itinerary towards Europe and America. In the course of these itineraries the linguistic abilities of *Homo sapiens* evolved in those referential expressions which are at the basis of symbolic behaviour. In Europe the indicators of a linguistically mediated behaviour suddenly appear forty thousand years ago, when *Homo sapiens* begins to replace the Neanderthal forms and manufactures highly-sophisticated lithic tools and throwing weapons. He establishes even long-distances commercial exchange networks. He begins decorating his own body with painting and carves small female, male, and animal figures. He engraves animal bones to create musical instruments and he paints images of extraordinary beauty on the cave walls of Chauvet, Lascaux and Altamira. This is an unprecedented cultural explosion in hominids' history which takes place during their emigration to Europe on a territorial area going from Asia to Cantabria.

There has not been a commonly shared view on the connection between the African origin of modern anatomy and the origin of modern behaviour and there is no agreement on the meaning of "modern behaviour". Synthetically, there are two opposite theories:

1. Modern behaviour makes its appearance in Africa 50,000 years ago and this event is strictly connected to the behavioural modernity noticed in European Superior Palaeolithic sites;

2. Modern behaviour developed in Africa in an older historical period and it is connected to the anatomic modernity through a gradual process which probably originated 200,000 years ago.

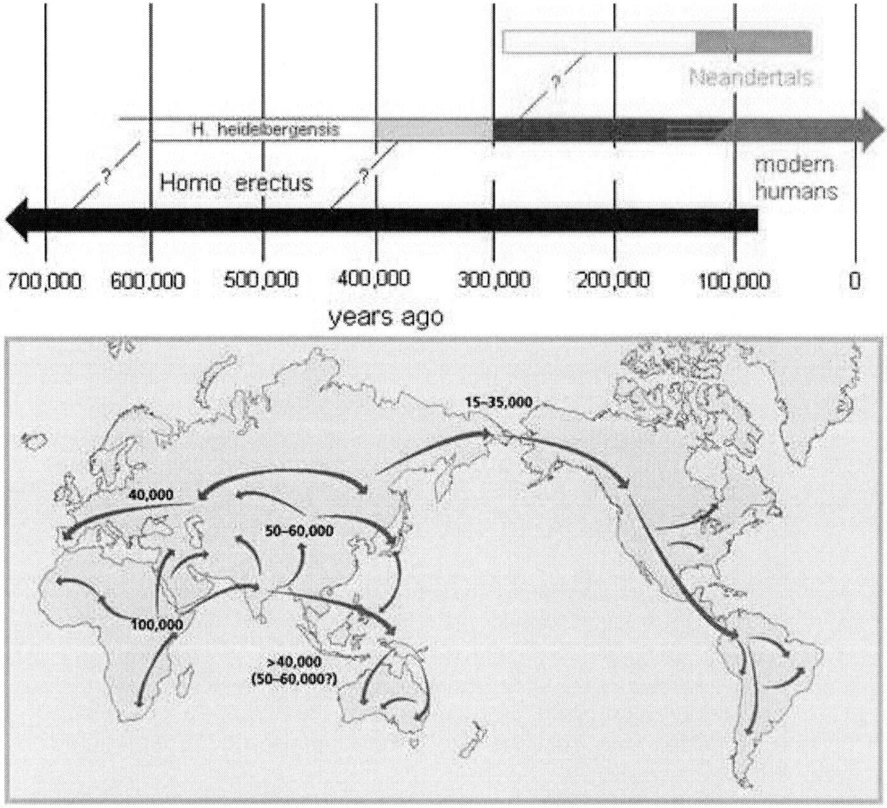

Figure 4. cfr. Feldman & Cavalli-Sforza, Nature Genetics, 2001

Finally, the proof in favour of the "gradualist" thesis discovered on the South African coast 200 miles from Cape Town between 1999 and 2000 at the Blombos cave, the Medium Palaeolithic site which has already revealed to its discoverer—Professor Christopher Hensilwood, from the Centre for Development Studies of the University of Bergen, Norway—the sophisticated bone tools mostly destined to fishing which, on their own, testify to the presence of modern kinds of behaviour. From an undisturbed deposit of ashes and sands—which the luminescent technology subsequently helped in dating to 70,000 b. C.— two fragments of red ochre have emerged, each nearly five inches long, finely chiselled with a grid of intersecting lines, next to the remains of a fireside and to tens of pierced molluscs.

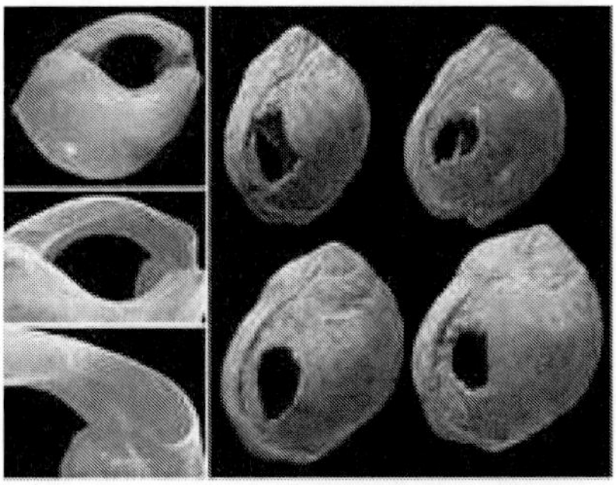

Figure 5. Pierced shells.

Figure 6. Red ochre fragment hand chiselled with a fine grid, Blombos Cave, South Africa

According to Hensilwood, the presence of carved material surfaces denotes cognitive skills and operational capacities ascribing them to modern behaviour. He assumes a mosaic-like "panafrican" developmental model capable of recording ages of innovation and regression in specific contexts. In the whole of Africa there is no possibility of a single trajectory towards behavioural modernity, as well as a single area in which it is possible to find an anatomically modern human behaviour.

Red ochre, a ferrous mineral, has been frequently found in Palaeolithic sites dated less then 100,000 years and was probably used to protect the skin and to tan animal skins. The fragments concerned were first scraped and then ground in order to create flat surfaces, where it was possible to carve the refined grid which leads us to infer the presence of an associative habit. One fact certainly emerges: we are in the presence of a tangible product endowed with an interactive function which has still to be determined, or that it will be impossible to determine because of its strong connection with the spatial and temporal context in which it took place. Carved upon the tangibly present grids in red ochre, this function is a *procedural* one and it is *limited* to its performing context.

The history of Homo sapiens demonstrates that *the body is the empirical area of intersection between the brain and the mind mediated by the technological processes which, in the technical instrumentation and the usage practices emerging from them, make up the spatial and temporal organization of a society and/or a community*. Technological innovation identifies time within the concrete limit. It is within this variable limit, which connects its rhythms *culturally* to the environment, that the body acquires its presence in the context. We define this presence of a specifically *cultural* nature as the place where experience grows.

The "body in action" on the territory is the implicit support for cerebral activity. It defines memory as the fundamental adaptive function through which the human being gradually realizes his potential in the integration between individual local entropies, as the goal of adaptation in space and time, the activity of constant vigilance over sudden changes, the ability to reawaken dormant resources, the voluntary control of action. According to the contents, the memory is transformed and fixed: and the dynamic of these transformations, as Freud guessed, is linked to the emotional contents which connect the sensory organs to the surrounding environment.

Memory is the path of human emotion through the exercise of techniques. Since it is shaped by experience, it constitutes the foundation of human identity. We inhabit the world through our memory, which deposits in our bodies the culture of belonging, the experiences which we have accumulated in the family in which we were born, the emotional expressions which we have acquired and which combine, throughout the eventful course of our lives, to enable each one of us to trace within our biographies the particular features that make us unique and unmistakeable. The implied protagonist of this "inhabiting" is the "body in action" in the environment. Its function can be likened to the unceasing movement of a loom between "inside" and "outside": between the inner surface (in relation to the body as a whole) and the things which constitute the relational boundary between the individual and the environment – "bridges" which link the implicit memory to its contextual connection: *Working Memory*. W*orking Memory* (or operational memory) is the working memory that is useful for resolving contingent problems. Operational memory is produced partly by the prefrontal cortex which integrates perceptions as they occur over a relatively short period of time and combines them with memories of past experiences. It is necessary for many apparently simple aspects of everyday life, such as remembering a telephone number, adding up numbers or driving a car, and it practically organizes our day-to-day activities. In the majority of cases occurring in real life, retrieving information from the memory means having to establish what type of information is useful at that particular point in time, and then having to select it from the vast archive of all the knowledge that is available. This selection requires complex neural calculations which are performed by the frontal lobes (Goldberg, 2001: 92-93).

All the theories that referred to the idea of a more or less mechanical updating of mnestic traces have now been abandoned in favour of more complex conceptions of the specific activities of the brain and the nervous system.

In the post-cognitivist view (Bruner, 1997b; Cohen, Siegel, 1991; Striano, "003), mental processes are considered as transformative and adaptive resources which are defined according to specific contexts which guide their development and design their function on the basis of the fabric of relationships within which knowledge is distributed. A consequence of this is the *multisystemic* conception of memory as a series of different *functional systems* inter-related through zones of convergence. According to this conception, which constitutes the most advanced view among memory theories, by making critical use of the contributions of the neurosciences, from the concepts of plasticity and flexibility to the principles of localization, the notion is suggested of "technique of the body" as a model in which intelligences are brought into play simultaneously as the expression of a *continuous* process of interchange and communication, which is both *transversal* and *synchronous.*

2. The rhythmic function of the "technique of the body"

Sociological interest in the subject of memory focuses on the combinatory platform between the genetic sub-stratum and learning. Far from being considered as an individual process through which the person autonomously preserves essential information, memory is a constitutively relational phenomenon and the formation of mnestic processes is strongly affected by the *distributed* and *situated* nature of knowledge (18). The formative process springs from the synergy of a number of interacting dynamics in which genetic reality and epigenetic development are interlinked through the constant correlation between the biological dimension and relational and socio-cultural matrices. Learning and memory processes touch on one of the distinctive qualities of human behaviour: our ability to acquire new skills through experience and to preserve them in time through the transmission of memory (19).

The ability to codify and transmit memory, in other words acquired experience, has proved a decisive factor in the evolution of human societies, which were able to develop precisely at the time when they began to hand down the testimony of their experience to the new generations. Memory is the fulcrum in the transmission of the human heritage of every society – and of every body, as genetics teaches us.

This chapter will focus on the rhythmic function of the "technique of the body", as developed by Marcel Mauss in 1936 ("Journal de Psychologie", XXXII, 3-4, 15 March-15 April 1936), since it suggests considering the body as an instrument for synchronizing the genetic programme and the practices of action within the topografic combinations brought about by each individual executive phase of the technique. According to our hypothesis, these *unrepeatable* combinations transform implicit memory into *Working Memory* and gradually interweave into the artefacts, practical objects, musical instruments, inhabited spaces, and repertoires of song, music and dance that are handed down over generations, the sources of active strength which make up the *social memory* of a group and/or a society: what Collins (1990: 39) described as "interaction ritual chains". From prehistoric tools in stone, bone or wood to the first ceramic objects, from machines to present-day robots, things, together with our species, have travelled a long way. Changing in accordance with time, place and working processes, arising out of different histories and traditions, they are always, either slowly or suddenly, being invested with new values and given new auras of meaning.

As the Soviet scientist Simonov (2003) pointed out when considering the technique of the actor in Stanislavski , with the technique of the body we affirm the "technique in process", in time. The neuro-scientist compares and translates the elements of the "method of physical action", point by point, into the scientific language of his time. Emotions become electro-chemical exchanges between nerve cells and the "method of physical action" is, in his view, related to psychotherapy. Simonov, in the mid-eighteenth century, was attempting to move "on the dividing line between science and art", as stated in the fourth chapter, and his proposal is all the more significant if one considers how the concept of *adaptability of the emotions* today informs the best analysis of the emotions (Damasio, 2007). Damasio (234) defines "nuclear consciousness" as

> the process which leads to a neural and mental configuration which combines, more or less at the same instant, the configuration concerned with the object, the configuration concerned with the organism and the configuration concerned with the relationship between the two. The emergence of each of these configurations and of their confluence in time

> depends on the contributions of individual cerebral sites acting in close collaboration.

In agreement with Damasio's thesis, we postulate that the technique of the body is aimed at synchronizing, in the executive phase, acquired knowledge with the ability to do; in other words, the combination between the skills which each individual, as a member of the species *Homo sapiens,* carries within himself – including the ability to learn and to remember – and the reciprocal interchanges which characterize agent/context relations. What we place in our individual mnestic systems is a function of our own particular experience; however, our existence and the basic functions of these systems are the product of the genes of our species. The technique of the body regulates this synchrony, which defines in temporal and spatial terms the rapid primary responses of the sensory cortex from which experience originates. In his presentation to the Psychological Society on 17 May 1934, Mauss defined the technique of the body as follows:

> What I call technique is a traditional effective action (and you can see that in this respect it is no different from the magic, religious, symbolic action). It has to be both traditional and effective. There is no technique, and no transmission, if there is no tradition. This is what distinguishes man above all from animals: the transmission of his techniques and, most probably, the fact that it is oral transmission.

Mauss starts from the premise that every act of learning is undoubtedly adaptive, and the only explanation for this, if one wishes to rule out that of a pre-established harmony between organism and environment, which does not seem very reasonable, is that learning presupposes programming based on phylogenetically acquired information. This is demonstrated by the notion of the "technique of the body", already freed from the blissful Cartesian illusion of a fragile body compensated for by the illusion of a "mind for ever". Such an illusion has been fed by the categorial distinction between "environment" and "world" which has been a characteristic of modern philosophical thought and has moreover played a considerable part in most German philosophical culture, including Heidegger himself. The best known and most coherent formulation of this knotty theoretical point is that of Arnold Gehlen (1967; 1983), whose anthropological theory is essential to any reflection on the relationship between technology and human nature. According to Gehlen, man is characterized by a series of morphological and biological peculiarities – premature birth and prolonged childhood, endocrine delay and organic primitivism, tactile hypersensitvity of the body's surface and numerous possible forms of motor coordination and so on – all of which contribute to place him in a form of *exposure to the world* which is without equal in the animal kingdom . Under the weight of this exceptional burden that is his exposure to the world, man is, according to Gelhen, obliged to devise specific ways of reducing the burden, of which technology is destined to be the paradigm. In the decades following its first appearance, this conception of man as the only animal that is "radically open to the world" gave rise to sharp criticism, which was probably justified if being open to the world was unilaterally interpreted as a clear ontological demarcation between man and the non-human animal: an interpretation which appears even more marked in the lectures which Heidegger devoted at that time (1930) to the ontology of life, making a clear distinction between "the world-poor animal", "imprisoned within its environment" and man as "shaper of the world". The idea of "exposure to the world" focuses on the link between technology and human nature and highlights, albeit indirectly, the difference between ancient artisan techniques and modern processes of technologization. In a situation of openness to the world, the wealth of stimuli and the primary uncertainty as to what significant content they may have constitute the indispensable premises for an inventive strategy. In other words, if a biological system is made to construct its equilibrium starting from noise, then noise will be an essential potential resource to be actively sought after and produced .

From the chipped flintstone of the Neolithic Age to nuclear power stations, the technical resources of the mind respond to the primary biological need to dominate the environment, which is an inseparable part of human nature. The arbitrary isolation of modern technology, on the other hand, ultimately produces on the contrary the myth of a possible original relationship with nature which in actual fact has never existed. Nobody is in any doubt that modern technology is different from the technologies of the past in terms of methods and effectiveness, but this progress, impressive though it is, simply confirms a cumulative capability which has always been a normal feature of technological development; and there has probably never been in history a society that did not react with dismay to some technical innovation which was considered abnormal, merely because it broke with consolidated customs. It is not sufficient to observe that man is driven to transform his environment in order to make it more suited to his needs, for the simple reason that that is equally true of a large number of other species, if not of *all* living species. Lewontin (1994: 48) reminds us that practically every living organism is surrounded by a halo of air which is made warmer by the body's metabolism, so that if we take the word "environment" literally to indicate the space surrounding the organism, it is correct to say that every living organism *actively* constructs its own environment. It is conceptually wrong to equate the environment with the "objective" set of data and conditions external to the organism. When, in the late 19th century, Jacob Von Uexküll (1956) intro-duced the concept of environment (*Umwelt*) into biology, his intention was precisely the opposite. Von Uexküll's idea was that if we take a trip into the countryside we are naïve to think that the cricket, the cow, the ants and the farmer simply share the same space. In actual fact, the different set of instincts and the specific sensorimotor apparatus of each of these organisms means that each of them is looking out onto a completely different world, given that one species may be totally blind to the colours and shapes which populate the world-envirnoment of the species living next to it, and that the same blade of grass may have a completely different function for one and for the other. As in the systems theory, in biology too the "environment" does not indicate a specific objective piece of data, but the *correlated specific of a biological system.* In the model devised by Von Uexküll and later adopted by all twentieth-century biology, the core of the organism/environment correlation consists of a series of interactive processes designed to develop the genetic programme that characterizes each individual organism. That is to say that every organism is genetically predisposed to react to a determined range of signals – the colour of a flower, the outline of a predator, and so on. The concept of instinct, subsequently developed mainly by Lorenz (1974), corresponds exactly to this framework. An instinct is in fact a specific motor sequence which is activated in response to a particular signal: it is essential, however, that both the sequence of movements and the quality of the signal be already genetically fixed. The movement does not in fact come about in response to the signal, except insofar as the signal launches an already prearranged programme. Every specific animal skill can be traced back to these "networks" of correlation between genetic programme and contextual action, including the particular constructive aptitudes with which an animal actively intervenes on the outside world.

The technique is the prototype of the human predisposition to transform noise into a *new order.* Each technological invention performs this transformation, by revealing, for example, an efficient weapon in a chip of stone or a powerful remedy against infection in a mould. The craftsman's workshop exemplifies the active search for noise as a resource for survival. Or more precisely, effective mastery of a technique consists of knowing how to exploit every circumstance to one's advantage: like Ulysses, who does not merely escape from the storm but exploits its force to reach his goal more swiftly. It is this *autopoietic ability* that provides the rhythmic function of the technique of the body and its ability to dominate the capital that is time. The body to which Mauss refers concerns the somatic, emotional and behavioural dimension which models implicit memory as a link - observes Leroi-Gourhan (1977, II: 233) - that is

> closely bound to the mastery of a mode of expression that fixes man's true position in a cosmos in which he is placed at the centre and which he does not yet attempt to explore through a process of reasoning in which letters make of thought a penetrating line, long in reach but as thin as a thread.

b

Figure 7. André Leroi- Gourhan, *Le geste et la parole*, 1977, I, fig. 99 b , p. 235

The living body, a difficult and delicate meeting between physiology and activity, is for Mauss a time loom which not only weaves the essential and sufficient base for the group and/or society but also embroiders upon it, through the technique of the body, the rhythmic passing of the days and nights, the seasons, the years and the generations. The technique of the body modulates the body's perceptive and motor activities as an unrepeatable combination between the history of biologically acquired values (the most ancient history in evolutionary terms) and activity aimed at achieving adaptive success over the environmental event. By "techniques of the body", Mauss means "the ways in which man, in each different society, learns to use his body in the environment" (1968: 365). The technique of the body regulates an elementary accounting system which varies according to the regions of the planet, and translates into the appearance of documents serving a memory to which accounting rigour is applied. It is only after the consolidation of urbanized agricultural organisms that social complexity translates into the appearance of documents which testify to men (or to the gods). Mauss identifies in the techniques of the body a means of connecting the biological with the sociological which is strictly related to separation of the sexes, and to age, performance and transmission. The ages of man and his activity in action constitute the criterion of order which, in Mauss's view, forges each different technique of the body. And so he divides the techniques of the body into techniques of birth and obstetrics, techniques of rearing, feeding and weaning children, techniques of adolescence, techniques of adulthood, which he divides between sleeping and waking techniques. Within each one of the various techniques, he links the movements and the stops that characterize it at different times of the day. The interweaving of action with the temporal moment that supports it outlines the "aspects of civilization" which distinguish the social uses of the body. The following is his explanation of the techniques of resting (Mauss, 1965: 401-402):

> Rest may be perfect repose or simply a pause: we can rest lying down, sitting, squatting, etc. Try squatting. You will experience the torture caused, for example, by a Moroccan meal eaten in accordance with all the ritual rules. The way of sitting is fundamental. You can distinguish squatting humanity from sitting humanity and within the latter you can further distinguish between people who use benches and people who use neither benches nor the floor, people who use chairs. The wooden seat supported by crouching figures is widespread – and this is truly remarkable - in all regions of the 15th degree of latitude north and the Equator on both continents. There are peoples who have tables and peoples who do not. The table, the Greek τράπεζα, is far from universal. Throughout the East a carpet or mat is still normally used. All this is quite complicated, because when resting, people eat, converse, etc. Some peoples assume extraordinary positions in order to rest. For example the whole of Nilotic Africa and parts of the region of Chad as far as Tanganyika are inhabited by men who rest in the fields by standing on one leg. Some of them manage to stand on one leg without any support, while others lean on a stick. These resting techniques are true aspects of civilization, common to a large number of individuals, to entire families of peoples.

The practices implemented by the techniques of the body present a temporal order based on the synchronization of biological rhythms with the relational organization of social space. In this "rhythmic scale", the biological rhythms are ordered topographically in the practices which constitute vital programmes for the individual and for the group, such as nutrition and hygiene, work, relational behaviour, the corporeal *habitus*. The technique of the body creates, in short, the conditions of knowledge that serve to construct concrete social phenomena. In his systematic theory of practice, Bourdieu (1980) demonstrated how knowledge, rather than being imported from the mind into experiential contexts, is in turn generated within these contexts in the course of reciprocal involvement with others in ordinary day-to-day activities. He proposes practical knowledge which we associate with skill,– a *know how* which we carry in our bodies and which notoriously does not lend itself to codification in terms of systems of rules and representations. Think of the technique of tying shoe-laces, or breaking an egg, or ironing. Activities such as these are developed not through formal instructions but through the repeated and frequently tacit performance of tasks which entail specific postures and gestures: what Bourdieu calls the *hexis* of the body. The theory of practice suggests that the way in which we perceive depends on the way in which we move. The expert walker adapts his pace to a monitoring process, which may be visual and auditory as well as tactile, of conditions of the terrain which are always variable. In his essay *Techniques of the body*, Mauss sketched a preliminary classification of the extraordinarily varied postures and gestures adopted by peoples around the world in their routine activities, both when resting (sleeping, crouching, sitting or standing) and when moving (walking, running, jumping, climbing, swimming). Every technique of the body, Mauss concluded, is a technique acquired" (1936: 390; 402-403). There is no "natural way" of doing it. How can we reconcile the biology of human nature which underlies each individual technique of the body with "the acquired ability" which shapes its cultural expressions? According to Mauss, the link between human nature and culture can be established by means of a third term: *rhythm*. With regard to rhythm, Mauss (1965: 315) focuses on its function of synchronization between experiential content and the practice of action, not simply on its social nature, which had in any case already been well documented in studies by Wundt (1874; 1920) , Bücher, (1896) and Ribot (1888; 1896). He writes as follows:

> Let us leave aside for a moment the social nature of rhythm. Is it not obvious, for example, if you study dance, even superficially, but from the sociological viewpoint, that it corresponds in one sense to respiratory,

> cardiac and muscular movements which are identical in all individuals and in which the spectators also frequently participate, while at the same time it presupposes and follows a series of images which is itself what the symbol of the dance evokes simultaneously in both groups? Here too we can see the direct link between the sociological and the physiological, not merely that between the social and the psychological.
>
> And if we consider, in rhythm – and also in song – one of its effects: obsession, the way in which it pursues those who have been caught up in it, do we not obtain precisely the same results? In this case there are numerous facts to be quoted, formidable ritual stereotypes: when, for example, in certain dances, often accompanied by a simple cry ululated *ad infinitum*, or simply by some verses of a very simple song for days and nights, groups of people, frequently large groups, strive together to find activity, fatigue, excitement and ecstasy– cause and effect at the same time and reciprocally. Many instances of this kind are to be found in Australia and in South America.

A particularly significant example of technique of the body is dance. As is demonstrated by the history of the human species, the primary function of dance is to synchronize the rhythms of the body with the chrono-spatial maps which shape the spaces for action of each different rhythm. George Leonard (1981), who studied the rhythms of people, is convinced that nothing can happen between human beings that does not correspond to a rhythmic hierarchy. John Dewey was also interested in rhythms. In his book *Art as Experience* (1959), he demonstrates how rhythms pervade all the arts: painting and sculpture, architecture, music, dance. Rhythm is the basis of synchrony. The word "rhythm" is used here to mean the temporal process referring to the various sequences of human behaviour and, more specifically, to the microtime which structures the confines of interactive synchrony. The contextual composition of these confines explains why dance is an unrepeatable combination between the diachronic sequentiality of the genetic code and the synchrony which makes up the maps of adaptation, thus throwing light on the ways in which individuals are linked to one another or isolated from one another by invisible threads of rhythm.

The link between the body and dance is such that in performing it both man and his environment are actively modifying each other. The ethnomusicologist John Blacking (1973: VI-VII) observes:

> The foundation of many, if not all, the essential processes of music is to be found in the human body and in the social interaction systems of human bodies.

Dance makes up the spatial framework in which actions are placed: the appropriate positions for learning and for playing, places for meeting and places for hiding are all established. It coordinates the activities of the group and keeps it together; it keeps human beings within communicating distance so that the presence of food or enemies can be reported. The crucial nature of dance, in short, is the *rhythmic construction of territoriality*.

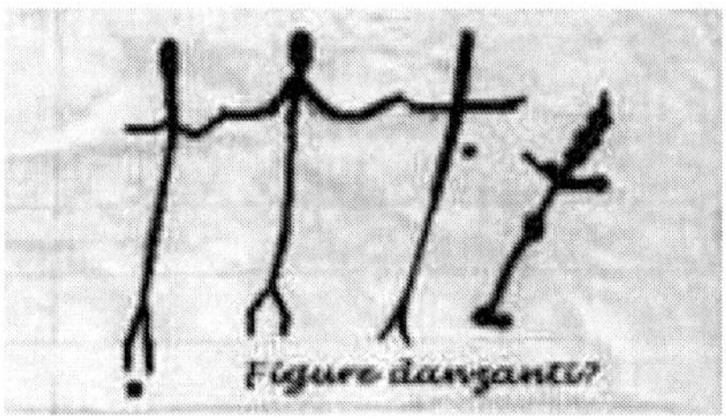

Figure 8. Rock carving dating from the neolithic age (Ciappo delle Conche, Ligurian Appennines. "Ciappo (stone) delle conche" is a prehistoric archeological site consisting of a large stone slab of about 30 m^2 on which 30 or so figures are carved and 7 small basins are hollowed out. The site is in the hills above Finale Ligure, not far from Orco Feglino, just below the top of the southern slope of Monte Cucco (Source: http://it.wikipedia.org/wiki/Incisioni_rupestri#Ciappo_delle_conche)

Dance originated as the *crossroads between heredity and learning* within the framework of biologiclal rhythms. That which has proved itself sufficiently frequently becomes part of the genetic programme and goes to make up a manual of behaviour for all the individuals of the species. The "natural selection" construct refers to forces operating in order to maintain, in the genetic heritage of individuals, the tendency to initiate modes of behaviour which contribute not only to their survival but also to their reproduction and reproductive success. Dance confines and guides the paths of learning within the "musical scores" of action which make up each individual adaptive sequence Let us consider the best known example in the animal kingdom, the dance performed by bees, in order to compare its characteristics with those of human learning. On their return to the hive, after depositing the nectar, the worker-bees perform a characteristic dance on a vertical plane. If the food is located within 50-100 metres of the hive, the worker-bees perform a circular dance in which they rotate on themselves first to the right and then to the left, and so on for about one minute.

If, on the other hand, the food is more than one hundred metres away from the hive the worker-bees perform a belly dance in which they move in a straight line for a short distance, rapidly "wagging" their abdomen before executing a 360° turn which brings them back to move along the same straight line and then turn in the opposite direction. The repetition of these actions creates a sort of figure of eight, which appears to convey information about the direction and distance of the food to the other worker-bees watching the dance (figure 21). Gould (1984) showed that the worker-bees are willing to leave the hive after watching a dance which tells them that the food is on the other side of a lake, but not when the dance indicates that the food is in the middle of the lake. This seems to indicate an astonishing degree of integration between spatial awareness of the territory and communication skills.

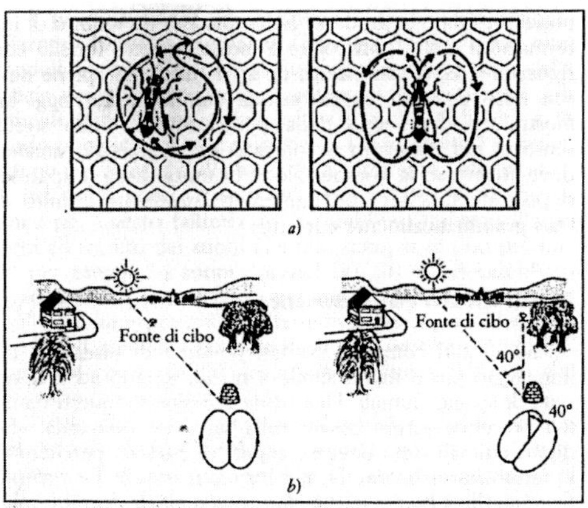

Figure 9. In *a*) the circular dance (left) and the belly dance (right) of the worker-bees; *b)* shows the way in which the belly dance indicates the direction of a source of food. The angle between the direction of the dance and the vertical is the same as that between the source of food and the sun in relation to the hive entrance. (Source: Pearce, 1997)

The physiology of all living beings is characterized by the periodic nature of its functioning in terms of organism, tissues and cells. The systematic study of biorhythms has shown how the traditional conception of the homeostatic balance of organisms is actually in need of critical reappraisal, because it is precisely the periodic variation in the *level of activity* which is the factor that makes it possible to better adapt to changes in environmental conditions. Indeed, the majority of biorhythms appear to be synchronized on natural or astronomic events in order to allow for better utilization of the energetic potential of the organism. The technique of the body regulates the construction of this synchrony, which is in part determined genetically and in part induced, since its purpose is to monitor variations in the level of activity of the task in hand on the basis of variations in the outside environment. The associative effect of the technique of the body is aimed at synchronizing biorhythms and social rhythms. It was demonstrated by the American physiologists Sternman, Mc Ginty and Adinolfi (1971) that certain periodic determinants, such as meal times or sleeping and waking hours, constitute possible social synchronizers. On this topic, one of the most exciting frontiers of thinking on networks is the biological foundation of social networks.

The most significant function of the technique of the body is the production of arousal.This function is designed to improve the performance capabilities of the organism in both behavioural and physiological terms. The state of arousal activates cognitive mechanisms – "be careful!" – which do not require conscious participation. The obvious evolutionary advantage of having basic mechanisms which enable us to rapidly evaluate objects and events in the outside world explains the central role of these processes of our brain. The initial generic response "Be careful" becomes, in a few microseconds: "Act!" (Davidson et al., 1990) through processes of arousal which direct the flows of energy within the system. Recent neurological experiments have shown that the continuous exercise of the state of arousal stimulates neurogenesis.. The first proof of the real growth in size of cerebral structures in relation to environmental factors, considered even on a large scale, was provided only about ten years ago by none other than the London taxi-drivers.

This discovery is remarkable above all for its simplicity and its explicative significance.

It was observed that in London taxi-drivers, whose work requires them to memorize a considerable number of complex routes and localities, the hippocampus is particularly large, larger than in most people. Considering the importance of the hippocampus for the memory, and given that London taxi-drivers have to memorize a particularly large number of routes and localities, it follows logically that their hippocampus has to be subjected to stress, just as a weight-lifter strains his muscles. Moreover, the hippocampus is larger in those who have been exercising this profession for the longest time: the size of the hippocampus is therefore in direct proportion to the number of years on the job. This suggests that there is a direct relationship between the frequency of certain types of mental activitiy and the size of the neural structure involved in that activity. The findings obtained with the taxi-drivers are important in a number of ways. First of all, it was shown that an important neural structure can continue to grow even in adulthood. But even more significant is the fact that the growth of a neural structure is *stimulated by use*. The effects of intense cognitive stimulation appear in this case to have overcome the deteriorating effects of ageing. The findings obtained with the taxi-drivers, spectacular though they are, constituted for some time an exception. And one single research study was not sufficient, precisely because of the exceptional nature of the discovery. A few years later *Nature*, one of the most highly respected scientific journals in the world, published an MRI study on cerebral changes in jugglers. A number of volunteers in excellent health, none of whom had any relevant previous experience, underwent a three-month training period as jugglers. At the end the volunteers were able to keep at least three balls in the air. When their brain MRIs taken before and after the training period were compared, the result was an increase in grey matter in the temporal lobes of both hemispheres and in the parietal lobe of the left hemisphere. This effect gradually faded after suspension of the training, and the advantage acquired in the temporal and parietal lobes was reduced: in particular the third MR examination after suspension of the juggling activity revealed this loss. We can deduce from this that the effects of practical activity on the proliferation of neurons can be observed in specific parts of the brain, even after a relatively short time.

The notion of technique of the body shows, in short, the variables which structure the functions of the body in constructing the memory process and, more specifically, in constructing social memory:

1. *Enrootment in the environment*:

   In the development of thought, doing comes before understanding. No scientific description can equal this enrootment. Especially during the learning process, we acquire experience through the senses and we operate with movement. The enrootment of the body can never be replaced by mere description. It is the ultimate source of our descriptions and the foundation of our way of knowing.

2. *The synchrony between knowing and knowing how to do*:
 Synchrony highlights the body as the space-time vector of the distributed processes which enable human beings, by means of links structured in social networks, to acquire, manipulate, recall and preserve information. Every new *input* – as Hebb (1949) realized – is integrated, thanks to new synaptic links, into an existing network which is reactivated as the information is inserted .

3. *The rhythmic function*:
 Since it is a process *situated* and *distributed* both chronologically and geographically, the technique of the body transforms biological rhythms into social rhythms. As shown by the discovery of the Spanish neurophysiologist Joaquin M. Fuster (1995, 1997, 1999) – the memory systems are believed to be structured in connection networks, whose formation is affected by the elements of experience phylogenetically preserved in the phyletic memory of the primary sensory cortex and transmitted by it to the associative cortex for the development of individual and social memory

Conclusions

For neuro-scientists it is now a known fact that human beings automatically and unconsciously organize their experience in their bodies into spatial units whose confines are established by changes in location, temporality and the interactive elements which determine it. These processes constitute *implicit* (or *procedural)* memory, a form of memory which is not recalled on the conscious level. This memory, conceptualized in French philosophy (Maine de Biran, Ravaisson, Bergson), governs not only simple perceptive and motor skills but also the pirouettes of a Margot Fonteyn or the trumpet technique of a Wynton Marsalis, as well as the leg movements of a cyclist. Descartes himself observed that the lute-player has memory in his hands to play his instrument. Procedural memory constitutes the biological foundation for any performance situation. It is often automatic and is recalled directly through execution, with no conscious effort and no awareness of the fact that we are drawing on memory. Implicit memory constitutes the biological foundation for any performance situation and in its contextual performances it guides us through well consolidated routines, the skills and habits that we have already acquired through experience. If explicit knowledge can be communicated through language with a high degree of awareness, the nature of implicit knowledge is fundamentally concerned with the somatic, emotional and behavioural dimension and refers to the "here and now" of the "present moment" (Stern: 2004).

As already shown in the demonstrations by Goldman-Rakic (1996) and Fuster (1997), the basic function of *working memory* is the *transition from the pre-conscious to the conscious* which gives the procedural (or implicit) processes access to consciousness. These processes (such as body posture, the perception of movement and the dynamics of cooperation between sensory mechanisms and contextual interactive fields) entail mnemonic storage which does not require conscious attention in order to be recalled, usually in the form of habits and perceptive strategies.

These demonstrations, if accepted, radically undermine the traditional Cartesian dogma according to which body and mind would be separate entities. In fact, the brain is a body organ whose characteristics, as aforementioned, are subjected to the history of the evolutionary process. In this respect, it is correct to talk about 'embodied mind'. The noble scientific father of this perspective is, of course, Gerald Edelman. He showed that the brain is embodied and the body is embedded in its environment and this symbiotic intertwining operates in a highly integrated way. Indeed, the deep meaning of this principle is that the brain (and the mind with it) is *'context bound'*. The brain is an embodied *selectional system*[1] that, by recognising patterns, and through creative mapping, produces unique selectional repertoires for every person.

The hypothesis of a 'dynamic nucleus' proposed by Edelman and Tononi (Edelman, Tononi, 2000: 165: 184) explains the conscious experience as an integrated and coherent process, continuous and continuously changing.

> We define a dynamic nucleus as an aggregate of neuronal groups which interact vigorously and have functional confines that distinguish it from the rest of the brain in a temporal scale in the order of fractions of a second. We define it thus to stress, at the same time, its integration and continuously changing composition. A dynamic *nucleus is therefore a process, and not* a thing or a place, and is determined by means of neural interactions, rather than of locating it specifically, of connection patterns or neural activities. Even if it has a spatial extension, a dynamic nucleus, generally speaking, is spatially distributed, besides having a changeable composition.

Edelman considers the body as a *locus of movement*, that is to say, as a *synchronous map* produced at a certain point by the interaction between its own genetically programmed varieties of biological automatism and the

[1] See Edelman, 1993.

action of the external factors which continuously mould its interactive balance. This balance causes the body to take root in the environment through which we consign our biological selves to the network of social interactions. No scientific description is equivalent to the body's taking root. Taking root is the basis of our way of acquiring knowledge.
The hypothesis of a 'dynamic nucleus' shows the capacity limits of consciousness. A state of consciousness can never be subdivided into independent components. In fact we are not able to consciously carry out more than two tasks at a time. The brain faces multiplicity without becoming incoherent. Every conscious event is a process with well-defined confines, and is strictly associated with the coherence of perceptive events. A conscious scene can be created even without external stimulus, and dreams are the most surprising proof of this. Even if the scenes that develop in the dream are quite coherent and sometimes mistaken for real ones (Foulkes, 1996: 609: 24), consciousness is active and functioning only when the state of consciousness is placed in space and time. This dynamic placing is activated by the body understood as a *locus of movement*. As Edelman and Tononi have shown (Edelman, Tononi 2000: 75- 92), conscious experience is associated with the activation of neuronal groups, widely distributed in the cortical-thalamus system. Synchronism between the elementary self-consciousness which regulates somatic processes and the interaction taking place is responsible for the unrepeatable activation that produces the state of active consciousness.

This activation is produced by the movement that the body makes to synchronize its own rhythms with the chemical and electromagnetic flows which, at a certain point, make up the interactive limit. The development of the whole person-organism is carried out in these flows, at the same time body and mind, situated within a continuous relational field. Theflows at issue make up the topographical map-making of bodily experience.

Here we assume that the *topographical construction of the biological body* regulates, by means of the interactive sets produced by used objects and tools, the order of the selective processes within which the evolution of bodily rhythms in the environment is accomplished, thus creating the continuity of a given experience. Proof of this comes from oral cultures. Leroi-Gourhan himself demonstrates that the itineraries of the body construct traditional "domesticity" as the rhythmic result of the progressive balance that man's body manages to negotiate with the environmemtal cycles in the *habitats* in which his social life is contained. In the practices of this body in the various contexts, our nature becomes that of *generating culture*.

Starting from the palaeolithic revolution - but even earlier, with the use of tools – the modern human brain has been irreversibly influenced by interactions between the body and material and symbolic artefacts. Where does the mind end and the rest of the world begin? In their essay *Philosophy of Mind: Classical and Contemporary Readings*, the philosophers Andy Clark and David Chalmers (2002) proposed an "incarnate" vision of the relationships between mind and outside world, which became a subject for debate and is still relevant today. The two philosophers propose an *active externalism*, based on an active role of the environment in guiding cognitive processes. In the cases which they examine, the human organism is linked to an external entity through a bi-directional interaction, giving rise to an associated system which can be seen as a cognitive system in the true sense of the term. All the components of the system play an active role, and together govern behaviour, in the same way that cognition normally acts. If we eliminate the external component, the operational capacity of the system will be reduced, just as if we had eliminated a part of its brain. The position of the two philosophers is borne out today by a growing amount of research in the field of cognitive science. In a number of very different areas, such as the theory of situated cognition (Suchman, 1987), studies of robotics applied to the real world (Beer, 1989), dynamic approaches to child development (Thelen and Smith, 1994), and research on the cognitive properties of collective agents (Hutchins, 1995), cognition is increasingly considered in continuity with the processes of the environment. Therefore to consider cognition as "extended" makes a considerable difference to the methodology of scientific investigation. In practice, explicative methods, which at one time would have been suitable only for analyzing "internal" processes, are now adopted for studying the exterior, and there is good reason to hope that our understanding of cognition may be enriched by this. Some have found this type of externalism impossible to digest. One reason for this could be that many people identify the cognitive with the conscious, and to them it seems implausible that consciousness can extend outside the head. But not all cognitive processes, at least in standard usage, are conscious processes. It is widely agreed that processes of all kinds, beyond the confines of consciousness, play a crucial role in the cognitive process: in recovering memory, in linguistic processes and in acquiring skills, for example. The biological brain in fact evolved and matured insofar as there was the reliable presence of a manipulable external environment. It certainly appears that evolution favoured internal capacities, which are triggered in order to utilize the world environment, so as to rearrange the memory load and even to transform the very nature of computational problems. Our sensory apparatus evolved by relying in various ways on the environment: for example, by exploiting contingent facts relating to the structure of natural scenarios (Ullman and Richards, 1984); or by taking advantage of the computational short cuts produced by bodily movements and locomotion (Blake and Youille, 1992). Once the fundamental role of the environment in conditioning the evolution and development of cognition has been acknowledged, extended cognition is a true cognitive process. Within this process, the body becomes the central point of intersection between memory, technology and context.

The adaptive experience of the individual and of the species develops through the set of *chronobiological maps*, which direct its development and design its functionality on the basis of the web of relationships that form the specificity of a given context. Indeed, the formative process is triggered by the synergy of a number of interacting dynamics, in which genetic reality and epigenetic development are interwoven through the constant correlation between the biological dimension and the relational and socio-cultural matrices. Between time and body there is a very close and strict interdependence, which is rooted in the very lives of cells and molecules. As long as we are and remain a psychosomatic unit, we are possessed by the temporal friction which always accompanies our thoughts, works, days, lives and death. It is the most material experience that restores the history of the body, its density, its temporal resonance. The ultimate originality of this experience is the fact that it is the crossing point between the development of the individual and social experience, individual reference and the collective norm. With regard to the temporal orientation of corporeity, Damasio points out that:

> The whole construction of consciousness, from simple to complex forms, from non-verbal consciousness through images to verbal literary consciousness, depends on the ability to create maps of what happens over a period of time within our organism, around our organism, to our organism, and with our organism – one thing after another, which causes yet another, and so on *ad infinitum*.

The neuro-scientific analysis of the brain and memory is essential for the understanding of the functioning of the brain, of the interrelationship between body and memory, of the nature of memory of the individual. Nevertheless, it has been acknowledged that the brain influences culture but culture also influences the brain, that is, the brain is not immutable but rather adapts to the surrounding socio-cultural environment. This position acknowledges that social factors do play a role in constructing memory as well as its impact on society.

If the body is able to inhabit, and at the same time understand, space-time – and not just be in it– it is because the body itself is this current and conscious space-time. In his fascinating book *Radiant Cool. The strange case of the human mind,* Dan Lloyd (2006: 312), a philosopher at Trinity College in Hartford, Connecticut, has this to say about the relationship between body and time:

> temporality is the skeleton which supports objects and objective reality. Once temporality is understood, the path towards the incarnation of consciouness is wide open.

The "technique of the body" forges the skeleton of human temporality, incorporating the vast accumulation of inventions and creations that make up the cultural aggregate of the species *Homo sapiens.* It is this "incorporation" that constitutes the human nature of technology, from stone-age technology to biotechnology or digital technology.

Every new invention, in order to be accomplished, requires the existence of previous inventions. It is not an unprecedented development: any list of the great inventions of history includes many which are based on the technical limits which generated them: the calendar, the alphabet, the compass, penicillin, double-entry book-keeping, the Constitution of the United States, the domestication of animals, the birth control pill, zero, the germ theory, the laser, electricity, the silicon chip. Each invention constitutes the frame of reference through the constantly expanding network of human technology, from stone-age technology to biotechnology or digital technology.

The technique of the body makes the transition from the implicit body to the executive processes which determine access to consciousness, expanding the fundamental characteristics of life and thus giving rise, within the executive limit, to the transition from procedural memory to the crucial event which we are faced with, be it pleasure or pain, conciliation or conflict. It is what the neuroscientist Eric Kandel (2006: 385) describes as: "*Learning from Memory*".

To combine the physical and biological study of the natural world and its inhabitants with an understanding of the deepest fabric of the mind and of human experience: this is the major task that social scientists, together with neuroscientists, have to face in dealing with the study of the complex bonds that link the body to technology.

The mind is actually the journey that the physiological body undertakes through the relational knots which link the rhythms up with the maps that mould the spatial extension of consciousness. During this journey, the bodily rhythmicity becomes creation, culture and history.

This is the wonder that makes the body the compass of the mind.

References

Ayensu, E.F. and Witfield P. (1982), *The Rhythms of Life*, New York, Crown Publishers.

Barnett, L. (1950) *The Universe and Dr. Einstein*, New York, William Sloane Associates.

Beer, R.(1989), *Intelligence as Adaptive Behavior*, New York, Academic Press.

Bergson, H., (1959) *Oeuvres*, Edition du Centenaire, Paris Presses Universitaires de France.

—. (1972), *Mélanges*, Presses Universitaires de France ;

Bernstein, B. (1965), "A socio–linguistic Approach to Social Learning", in Gould J. (ed.by), *Penguin Survey of the Social Sciences,* Harmondswort, Penguin Books.

Birdwistell, R.L. (1970), *Kinesics and Context: Essays on Body Motion Communication*, Philadelphia, Pennsylvania University Press.

Blake A.,Yuille (ed.by), (1992), *Active Vision*, Cambridge MA, MIT Press

Blacking, J. (1973), *How Musical is Man*?, Seattle, Washington University Press.

—. (ed. by) (1977) *The Anthropology of the Body*, ASA, monograph n.15, London, Academic Press.

Blaking, J. and Kealiinohomoku (eds.) (1979), *The Performing Arts, Music and Dance*, The Hague, Mouton.

Bücher, K. (1896), *Arbeit und Rhythmus*, Leipzig, Hirzel. Boncinelli E. (2002), « Preface », in Ledoux, J., Il *sé sinaptico,* Milan, Raffaele Cortina (orig. Public., *Synaptic Self: How our Brains Become Who We Are*, New York, Viking Penguin).

Clark A. (1989), *Microcognition : Phylosophy, Cognitive Science, and Parallel Distributed Processing*, Cambridge MA, MIT Press

—. (1997), *Being There: Putting Brain, Body and the World Together again*, Cambridge, MA. M.I.T. Press.

Clark, A. and Chalmers D.J. (2002), *Philosophy of Mind*: Classical and Contemporary Readings, Oxford, Oxford University Press.

Cohen, J. (1996), "Does Nature drive Nurture?", *Science*, n.253, 577–578,

Damasio, A. (1993), "The Frontal Lobes", in Heilman E. (ed. By), *Clinical Neuropsychology*, Oxford University Press, New York, Oxford University Press.

—. (1994*), Descartes' Error: Emotion, Reason, and the Human Brain*, New York: Avon Books, Harper Collins.

—. (1999), *The Feeling of What happens: Body and Emotion in the Making of Consciousness,* New York, Harcourt Brace,

Davidson, R.J. Ekman, P.Saron, C. Senulis J., Friesen W. (1990), "Approach–with–Drawal and Cerebral Asymmetry: I. Emotion Expression and Brain Physiology", *Journal of Personality and Social Psychology* , n.58, 330–341.

Davidson, R.J. (1992b), "Prolegomenon to the structure of Emotion: Gleanings from Neuropsychology, *Cognition and Emotion*, n.6, 245–268.

Descola, P.(2005). *Par-delà nature et culture*. Paris: Gallimard.

Dewey, J. (1910/1997). *How we think*. New York: Dover Publication.

—. (1916/1944). *Democracy and education: An introduction to the philosophy of education.* New York: The Free Press.

—. (1922/2002). *Human nature and conduct.* New York: Dover Publication.

—. (1929/1958). Experience and education. New York: Dover Publication.

—. (1938/1998). *Experience and education (The 60th anniversary edition).* West Lafayette, Indiana: Kappa Delta PI.

—. (1897/1997). The significance of the problem of knowledge. In John Dewey, *The influence of Darwin on philosophy and other essays* (pp. 271-304).New York: Prometheus books.

Edelman, G. M. & Tononi, G. (2000). *A universe of consciousness: How matter becomes imagination.* New York: Basic Books.

Edelman, G. M. (1992) *Bright Air, Brilliant Fire: on the Matter of the Mind,* New York: Basic Books.

—. (2004) *Wider than the Sky. The Phenomenal Gift of Consciousness*, Yale: Yale University Press.

Hensilwood, C.S., D'Errico, Vanhaeren M., van Niekerk, K., Jacobs, Z.(2004), "Middle Stone Age shell beads from South Africa", in *Science*, 304: 404.

Gergen, K.J. (1985), 'The Social Constructionist Movement in Modern Psychology', *American Psychologist*, 40: 266-275

Goldman–Rakich P.S. (1995), Cellular Basis of Working Memory , in "Neuron": 477–485.

Goldberg, E (2001). *The Executive brain: Frontal lobes and the civilized mind.* New York: Oxford University Press.

—. (2005). *The wisdom paradox: How your mind can grow stronger as your brain grows older*. New York: Gotham Book.

—. (2009). *The new executive brain. Frontal lobes in a complex world.* New York: Oxford university Press.

Hebb , D.O. (1949), *The Organization of Behaviour*, New York, Wiley.

—. (1954), "The Problem of Consciousness and Introspection", in G. Lindzei (ed.), *Brain Mechanisms and Conscioussness*, Oxford, England: Oxford University Press.

Heidegger, M. (1938), *Die Zeit des Weltbildes* (it. ed. 1968) *L'Epoca dell'Immagine del Mondo,* Firenze: Nuova Italia.

Humprey, N. (1976), "The Social Function of Intellect", in Bateson P. P.G. and Hinde R. A., *Growing points in ethology*, Cambridge: Cambridge University Press, 303-317.

Hutchins, E. (1995), *Cognition in the Wild*, Cambridge MA, Mit Press.

Ingold, T. (1990) An Anthropologist Looks at Biology, in *Man (N.S.)*, Vol. 25, n° 2, 208-229;

—. (1993b), *The Temporality of the Landscape*, 'World Archaeology' 25: 152-174;

—. (1993a) "Tool Use, Sociality and Intelligence", in Gibson K. and Ingold T. (eds.) *Tools, Language and Cognition in Human Evolution*, Cambridge: Cambridge University Press, 429-445;

—. (1993b) Tools and Hunter-Gatherers, in Berthelet A. and Chavaillon J. (eds.) *The Use of Tools by Human and Non-Human Primates*, 281-295, Oxford: Clarendon Press;

—. (1997) Eight themes in the anthropology of technology, in *Social Analysis*, n° 41(1), 106-138;

—. (2000) *The Perception of the Environment: Essays on Livelihood, Dwelling and Skills*, London-NY: Routledge.

—. (2006) "Walking the Plank: Meditations on a Process of Skill", in J. R. Dakers (ed), *Defining Technological Literacy: Towards an Epistemological Framework*, New York : Palgrave Macmillan, pp 65-80.

James, W. (1907/1995): *Pragmatism*. New York: Dover Publications.

—. (1890/1981). *The Principles of Psychology*. Cambridge, Massachusetts: Harvard University Press.

—. (1899/2007). *Talks to Teachers and Students on some of Life's Ideals.* Charleston, SC: BiblioBazaar.

Kagan, J. & Baird, A. (2004). "Brain and Behavioral Development during Childhood, in M. Gazzaniga (Ed.), *The Cognitive Nneuroscience III* Cambridge Massachusetts: MIT Press :93: 103.

Kandel, E. Schwartz. J. & Jessell, T. (2000). *Principles of Neural science.* (4th ed.). New York: McGraw Hill

Kandel, R. E. 2005. *Psychiatry, Psychoanalysis, and the New Biology of Mind*. Washington: American Psychiatric Publishing, Inc.

—. 2006. *In search of memory: The Emergence of a New Science of Mind.* New York: W.W. Norton & company.

Kandel, E. Schwartz. J. & Jessell, T. (2000). *Principles of Neural science.* (4th ed.). New York: McGraw Hill.

Kraut, R. (2002). *Aristotle: Political philosophy.* New York: Oxford University Press.

Klingberg T. (2009), *The Overflowing Brain. Information Overload and the Limits of Working Memory*. New York, Oxford University Press.

Lacan, J. (1999), *Towards a Genetic Theory of the Ego*, in "Some Reflections on the Ego", *The International Journal of Psycho-Analysis* (1953), vol. 34: 12-15.

LeDoux, J. (2002). *Synaptic self: How our brains become who we are.* New York: Viking.

Lieberman P. (1994) "The Origins and Evolution of Language", in Ingold T. (ed) *Companion encyclopedia of anthropology*, London: Routledge: 108-132.

Lynch, G. & Granger, R. (2008). *Big Brain: The Origins and Future of Human Intelligence.* New York: Palgrave.

Lynch, Z. (2009). *The Neurorevolution. How Brain Science is Changing Our World.* New York: St. Martin's Press.

Leonard, G. (19819, *The Silent Pulse*, New York, Bantam Books.

Longo, G.O., (2001), *Homo technologicus*, Meltemi, Roma

Lorentz, K. (1952), *King's Solomon's Ring,* New York : The Thomas Y. Crowell Co.

Maguire, E.A. Gadian, D.G. Johnsrude, I.S., Good C.D. Ashburner, J. Frackowiak, R.S. et al. (2000), "Navigation–related structural change in the hippocampi of taxi drivers", in Proc Natl Acad Sci USA, 97 (8), 4598–4403.

Mauss, M. (1936), « Les techniques du corps », *Journal de Psychologie:* 3–4.

Mauss, M. (1965), Teoria generale della magia, Torino, Einaudi (it.ed.).

Mérleau–Ponty, M. (1945), *Phénoménologie de la Perception*, Paris: Gallimard; English ed. *Phenomenology of Perception* (1962) (trans.by Colin Smith), London: Rout-ledge & Kegan Paul)

Sombart, W., (1919), Der Moderne Kapitalismus historish–systematische Darstellung des gesamteuropäischen Wirtschaftslebens von seinen Anfangen bis zur Gegenwart, 3 voll., München, Duncker & Humblot.

Siegel D.J., (1999), *The Developing Mind*, New York, The Guilford Press.

Simon, H.A. (1996), *The Sciences of the Artificial*, MIT Press, Cambridge (MA)

Simonov, P.V. (1962), *Metod K.S. Stanislavkogo i fisiologija emocij*, Mosca, in Santoro, L.A. (2003) *In corso d'opera*, Lecce, Adriatica editrice salentina.

Suchman, L. (1987), *Plans and Situated Actions,* Cambridge UK, Cambridge University Press.

Thelen, E.& Smith, L. (1994), A Dynamic Systems Approach to the Development of Cognition and Action, Cambridge MA: MIT Press.

Uexküll, J.von (1934), Streifzüge durch Umwelten von Tieren and Menschen, Hamburg Rowolt Taschenbuch Verlag.

Weiss, G. and H. F. Haber. (1999). "Introduction." Pp.143-162 in *Perspective on Embodiment: The Intersections of Nature and Culture* edited by G. Weiss G. and H.F. Haber. H.F. London: Routledge.

CHAPTER NINE

NETWORK SYNCHRONIZATION AND THE SIXTH SENSE: HOW SOCIAL SYSTEMS LEVERAGE INTEGRATION FROM SOCIAL INTERACTION

THOMAS S. SMITH

Abstract

Human behavior is readily entrained by diverse exogenous oscillators—by music, dancing, marching, prayer, rowing, games, circadian shifts, audience participation, and by many other social and biological clocks. These synchronize populations, whether dyads or whole communities, and synchronization, in turn, increases both good feelings and social integration. These outcomes can be understood by studying an innate mechanism standing behind infant-caregiver attachment. We discuss this mechanism as a hyperstructure—a neurosociological arrangement of coupled oscillators that links core hormonal and neurotransmitter systems to attachment and separation behaviors. Distress-dependent, this mechanism recruits socially analgesic resources to sites of human distress, allowing us to understand interaction as an extension of the human immune system and giving rise to a "sixth sense," discussed below.

Falling into step

How fundamental to social life is synchronization? In singing, dancing, worshiping, and making music together, human populations find pleasure, consonance, and rapport. Yet aside from inducing pleasant feelings and rapport, the roots of synchronization seem buried deeper in human nature. Friends walking side-by-side fall into step with one another. Disorganized applause will switch suddenly into a phase of clapping in unison. Two year old children, given drums to play, synchronize their drumming (Kirschner and Tomasello 2009). Even Snowball, the much celebrated sulphur-crested cockatoo, dances to music, bobbing his head to a musical beat (Patel, Iverson, Bregman, and Schulz 2008). All of these examples, whether human or cockatoo, show that behavior is readily entrained by exogenous oscillators of diverse kinds—for humans, by music, by dancing, by marching, by prayer, by rowing, by games, by circadian shifts, by participating in an audience, and by many other social and biological clocks. Steven Mithen (2005) has even speculated that rhythmic vocalization—the "musicking" described by Christopher Small (1998)—probably stood behind the emergence both of speech and of music among Neanderthals. Indeed, observations ranging far back in the evolutionary record suggest that the tendency to synchronize is innate—that evolution has conserved patterns of synchronization in human and nonhuman physiology because they increase adaptive fitness. Some of the most striking evidence of this is very near-to-hand, provided by studies of the interaction of newborns and their mothers. Infant-caregiver synchrony occurs at many levels—through singing and lullaby, through patterns of inter-affectivity mediated by mirror neurons, and, significantly, through oscillatory neural and hormonal controls over attachment behavior and interaction.

This chapter focuses on one set of these hormonal and neural controls. As they appear in infant-caregiver attachment, they can be understood as part of a neurosociological mechanism called a hyperstructure (see Smith 1993, 1997, 1999; Smith and Stevens 1994, 1995, 1999, 2003; Smith and Silon 2005, 2007). In what follows, I want to sketch how this mechanism works, and also to generalize our understanding of the mechanism so as to analyze it as an interpersonal extension of the human immune response. How does this work? The steps in my argument hinge on the following claims:

- Hyperstructures are network-generators: they establish distress-driven links among persons.
- These links serve to modulate distress by stimulating the release of endogenous opioids and other comfort-enhancing neurochemical substances.
- Opioids have a role both in attachment behavior and in the human immune response. This common chemistry is key to understanding attachment as an interpersonal extension of the immune response.
- As distress spreads beyond dyads into larger networks, networks then function as social depots of immune resources. They assemble persons into a sensory grid that recruits socially-analgesic resources to distressed persons or neighborhoods. I will speak of this as a

"sixth sense," following Blalock's (1984, 2005) conceptualization of the immune system as a sensory system.

- For networks to function as sensory systems, they must link multiple oscillators that are synchronized. Synchronization is a network phenomenon.

As we flesh out the steps in this argument, we shall see that synchronization is one of the deep foundations of social integration.

Hyperstructures

The canonical illustration of the hyperstructure mechanism appears in infant-caregiver attachment. When infants cry, the mechanism comes into play. The cry is communicated to the mother, who experiences the infant's distress as if it were her own. She responds with some form of attachment behavior—picking up the baby, holding it, feeding it. Both in herself and in her baby, her behavior has the effect of causing the release of endogenous opioid peptides. Opioids produce soothing, analgesic effects in each of them.

Opioids levels fluctuate in a cyclical pattern. When they fluctuate below a threshold associated with comfort, infants will begin to experience distress, eventually crying. Since opioids have a chemical structure resembling morphine—endogenous opioids were first called endogenous morphine or, in abbreviated form, endorphins—the infant's distress calls are signals akin to withdrawal symptoms—the same symptoms we observe in heroin addicts. That is, they arise as a function of downward fluctuations in opioid levels and in other core hormonal systems associated with comfort. Along with the opioid system, these include activity in the oxytocinergic system and in the HPA-axis. When activity in any one of these systems dips below a comfort-threshold (a level associated with equilibrium levels in each of these systems), newborns will exhibit distress—the sort of distress that begins with mild uneasiness but cumulates into agitated crying. It is such a signal that moves a mother to attach herself to her baby. Depending on the form her attachment takes—whether, for example, merely holding the baby, or holding and feeding it—attachment then stimulates opioid activity, or opioids in combination with increased levels of oxytocin, either of which will have soothing effects. Other forms of anxiety—fussiness, say, or boredom—can indicate diminished levels of arousal, stimulating a mother to separate from her infant, a behavior that has the effect of elevating activity in the HPA-axis. As these two forms of maternal behavior exclude one another—a mother cannot be both attached and separated at the same time—they constitute an oscillatory pattern.[1]

Figure 9-1a and Figure 9-1b are schematics of the opioid cycle and the arousal cycle respectively. For the special case of infant-caregiver interaction, these two cycles can be understood as competing with one another; each cycle depends on one of the two mutually exclusive forms of maternal behavior, attachment or separation. The implications of these cycles can be grasped by reconsidering them in relation to the concept of a depot—a reservoir of neurochemicals (opioids or, for the arousal system, norepinephrine) that can be depleted or can be replenished, depending on the phase of cycle.[2] The symbols *O* and *N* in these graphics refer to opioids and norepinephrine, respectively; *A* and *S* refer to attachment and separation; and the subscripts *i* and *c* to infant and caregiver.

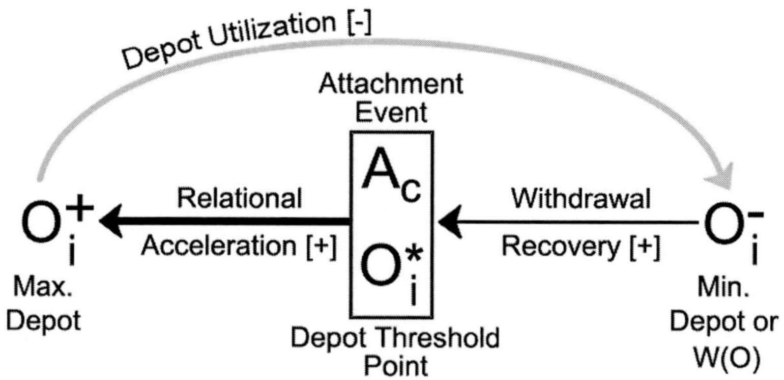

Figure 9-1a. The Opioid Depot Cycle

---

[1] As discussed in various papers: see, for example, Smith, 1994; Smith and Stevens, 1994, 1996a, b, 1999; Smith and Benard, 2002; Smith and Silon 2005.

[2]What is a depot? In abstract terms, a depot is simply a reservoir of resources. In the context of studying complex systems, however, the concept of a depot is important and useful. With depots, or reservoirs of resources, systems have the ability to accomplish work, such as maintaining equilibrium, responding to various demands, or damping and distributing distress. Logically, depots also require associated depot structures that generate and store depot resources for later use and can detect immune cells have a finite life, they must be constantly renewed, or the result would be eventual immune depot decay.

The opioid depot cycle, in Figure 1a, shows that opioid depletion, O_i^-, gives rise to withdrawal pressure in the infant, $W(O)_i$. This indicates a need for depot recovery, and stimulates maternal attachment, A_c, commencing at a depot threshold point we have called O_i^*. When this threshold is passed, maternal attachment accelerates depot recovery up to the depot maximum, O_i^+, where subsequent use of the depot again depletes it.

A comparable logic appears in the arousal depot cycle, shown in Fig 9-1b. Like an empty reading on a fuel gauge, N_i^-, the depot minimum, is associated with the gradual onset of pressures, $W(N)_i$, for depot recovery. When the N_i^* depot threshold is passed, maternal separation S_c begins to stimulate depot recovery up to N_i^+, where use of the depot again gradually depletes it.

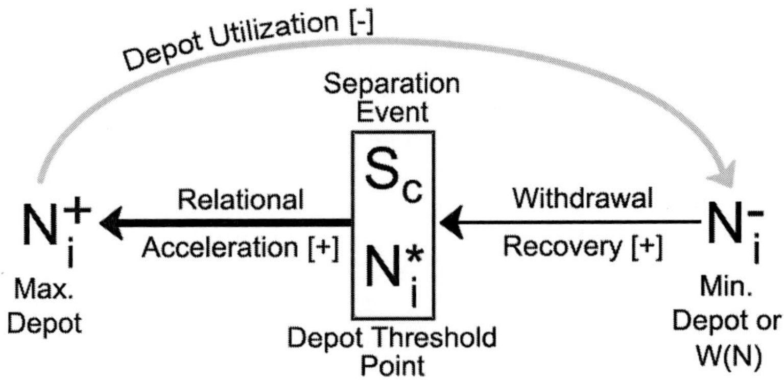

Figure 9-1b. The Arousal Depot cycle

Hyperstructures couple these two neurosocial oscillators, and each then works in oscillatory anti-phase to the other. That is, when the depot associated with either system is depleted, the depot associated with the other becomes active; when the second depot in turn is depleted, the first again becomes active. The coupling functions to synchronize the infant-caregiver dyad, optimizing comfort. The complete coupled hyperstructure is represented in Figure 9-2.

Figure 9-2 is labeled so as to draw attention to various forms of feedback and coupling at work in this mechanism. For example, we speak here of entropy pressures behind the metabolism of resources in each depot, indicating that the mechanism is driven by basic thermodynamic forces that appear both interpersonally and physiologically. This is the significance of the entropy sphere at the center of the mechanism—an acknowledgment of the pressures behind metabolism and replenishment of the depots. For example, "addiction pressure" appears in relation to the replenishment of both depots, and refers to the pressures behind both attachment and separation (or their various substitutes). In the case of attachment, this is an obvious extrapolation from the morphine-like structure of endogenous opioids. There are equivalent addictive liabilities associated with dependence on chemicals of the HPA axis or their surrogates.

The hyperstructure mechanism eventually comes partly under the control of circadian signals as well—transitions between day and night. This circadian control depends on the appearance, several weeks after parturition, of the infant's capacity to process signals of light and dark. But the basic mechanism remains primarily dependent on the comfort that only interpersonal synchronization can create. Comfort, in this sense, is not a property of persons alone but of social systems, whether dyads or larger systems. In the dynamical model that describes the infant and caregiver, accordingly, comfort is an attractor, a social condition that draws persons into itself.[3]

Caregiver Substitutes

At first, of course, babies aren't very good at taking care of themselves. But even in the first months of life, they can be seen to develop weak cognitive and emotional strengths that allow them to decrease dependence for their own comfort on the presence of their mothers. These strengths appear initially in behaviors that babies use to self-soothe or self-stimulate themselves—thumb-sucking and rocking, for example—but they come ultimately under the control of thinking. As thinking acquires increasing coherence over the first two years of life, memory and fantasy begin to become "places"—in effect, new attractor states—a child can turn toward to comfort itself. At first, these cognitive structures supplement in only small and insignificant ways the major direct effect of a mother's attachment. But with increases in the child's capacity to think, the tight coupling of mother and child via the hyperstructure mechanism is loosened.

[3] More precisely, the attractor state is the outcome of combined effects from both the opioid system and the arousal system. When either system shuts down, there is a decay period when its effects continue to be lagged into the cycle of the other system—creating optimal combinations of arousal and soothing.

Eventually children make use of all manner of caregiver-substitutes, and they begin to regulate their own internal states with increasing efficiency.

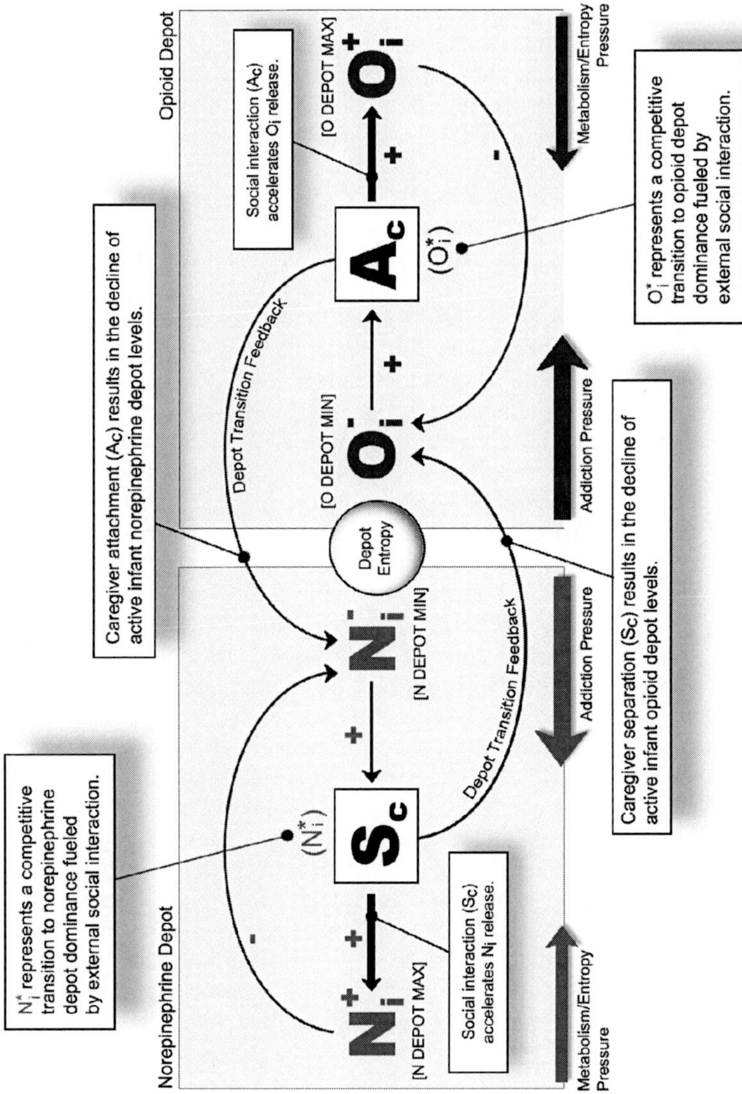

Figure 9-2. The Full Coupled Hyperstructure

Network as nonequilibrium systems

Distress regularly spills beyond the boundaries of dyads. When the demands of infant care overwhelm parents, for example, nearby family members, older children, and even neighbors can be recruited to help manage an infant's needs. When this happens, pre-existing links in a social network are being activated. In effect, the family's network is a depot of potential attachment resources, brought to life by signals of distress, pain, suffering, or anxiety.[4] The propagation of such distress into this depot serves to energize and renew the network, refreshing the links. Without the spreading of distress, the network would eventually collapse, its links unused. Out of synch, lose the link, as with the Hebbian rules describing neural networks.

Networks generated by coupled oscillators, such as those hyperstructures produce, are examples of nonequilibrium systems. In the physical world, the most studied example of such a system is water that is slowly heated. At a certain threshold temperature, the water will suddenly exhibit self-organization. A toroidal convention current will emerge (see Prigogine 1978, 1980; Prigogine and Stengers 1984). This self-organized pattern of movement arises because of the circulation of water molecules—first upward and away from the heat source, and then, as the molecules cool, downward and back toward it. Beneath the temperature threshold where the convection current first appears, the molecules display only Brownian motion; just above it, we find self-organization; heated still more, the water's temperature will eventually pass another threshold and the water will boil. Self-organization is displaced by turbulence and chaos.

Social networks are also nonequilibrium systems. When we study their dynamics, we find similar phenomena of self-organization. The fuel for these dynamics is anxiety and distress—the social equivalents of the heat that produces self-organized features in water. The equivalence is straightforward: No anxiety, no attachment or interaction. By fueling the hyperstructure mechanism, anxiety produces the patterns seen in infant-caregiver attachment, and these are interaction's basic self-organized features—together and apart, attach and separate. Experimental evidence in support of this claim can be found in the literature of social psychology from fifty years ago. In Schacter's (1959) famous experiments with norepinephrine, for example, it was shown that, even in a group of

[4] We refer to this sharing of the burdens of infant care "pooling." Pooling of emotional resources under conditions of stress enables networks to carry higher stress-burdens, and makes them less vulnerable to breakdowns. For computational evidence that supports this argument see (Smith and Stevens 2002).

strangers, increased arousal led to increased interaction. Just as in the case of anxious infants seeking parental attachment, Schacter's subjects looked to one another as their anxiety increased. Dyads formed, and then larger systems of interaction. Through the links in this emergent network spread the anxiety that the experiment had induced. In general, this spreading of anxiety interpersonally is at the core of all interaction's self-organized patterns. These regularly appear so long as anxiety levels surpass the stress-buffering capacity of individuals considered alone. New networks will be generated, pre-existing networks reactivated.[5]

As anxiety-spreading connects persons, it can have dramatic structural effects on networks. Increasing numbers of persons can come into touch, a step or two, sometimes six steps, away from one another, until all are linked into the same world. This is the network structure of the social world—as it turns out, a structure marked by small worlds, or many such small worlds, at least as measured by the average path distance between any two persons. Along the bridges and links connecting these worlds pass information and signals of all kinds.

The social world's network structure thus has a few core architectural features (Smith and Stevens 1999). Most importantly, it is marked by modular organization. Each of its small worlds is a network cluster—persons linked to one another more closely than they are linked to outsiders. When there are bridges between these clusters, these worlds potentially come within reach of each other. Since inter-cluster bridges are typically weak ties, as Granovetter (1973) famously argued in the 1970s, weak ties can be seen to strengthen social integration, chiefly by increasing the connectivity of a society and enabling the diffusion of information.

Yet there is also another face to weak links. Network scientists working in other fields have shown that if we continue to introduce stress into a network, it will eventually collapse. And, perhaps not surprisingly, a network's weakest links are where the collapse will begin. In social systems, there is long-standing evidence that shows this happening. When we raise the stress passing through a community, for example, persons tend to retreat into local clusters—withdrawing from the larger community and retreating into smaller networks of family and kin (Fellin and Litwak 1963).

---

[5] This energy-dependence of networks is a common network property throughout the natural world. Network theorists call this "housekeeping energy"—the level of energy required to prevent a network from collapsing. Fluctuation in energy below the housekeeping threshold will cause networks to collapse. See, for example, Peter Csermely's discussion (2006).

What is important about this? The tendency of weak links to collapse under increasing stress localizes or confines the dissipation of stress, an outcome that adds to network stability. If weak links never collapsed under heavy stress-burdens, networks would be subject to cascading events like panics, even topological phases transitions—the quakes, slides, fractures, and population collapses commonly found in other complex systems. There are many illustrations of this kind of effect in social life. As downturns in the economy gain speed, for example, Stark and Vedres (2002) have shown that business consortiums undergo topological phase transitions. Their networks, once fully connected, suddenly reorganize into star-shaped structures, with one leading firm at the center. The star pattern has reduced the number of links in the network, with each business then connected only to one central firm,. This same star-shape emerges in other situations where stress increases: Where resources grow short, for example, a star network takes shape in communities that were formerly more densely connected. In the military and other "emergency systems" such as police agencies and fire departments—systems in which stress is chronic—a star network is institutionalized in the hierarchy of rank. One α-leader is at the center of each team of responders—each platoon, each engine company, each station house, each barracks. Increases in stress, thus, are associated with the collapse of weak links and a general reduction of the number of links in a network, with hierarchy entering into the picture. But is this the full story? When it comes to social networks, the story has a familiar twist.

Synchronization as a network emergent

Key to this story are regular temporal shifts in network structure. Some of these are calendrical, with weekends and holidays serving as times for relaxation, for de-stressing. Modest increases in the stress-burden of a network, confined to the work week, are followed by days when stress is relieved, as on weekends. Family, religion, and community life—not work—are at the center of these retreats. As in the case of an infant's distress calls, stress creates pressures to attach, to strengthen ties—or, as we have come to understand this in view of the hyperstructure, to service the opioid depot. Once distress is modulated, then the arousal depot will again require servicing, and pressure subsequently supports separation. Oscillation between attachment and separation is the result, equating to the synchronization between infant and caregiver. Yet in other terms altogether we have known something like this for a century or more.

At least since Durkheim's work on religion (1995 [1912]), sociologists have understood the importance to social integration of regular oscillations between communal and individual life. Times of intensified social or communal activity, in Durkheim's famous analysis, reinforced social integration, chiefly through ritual practices that synchronized a community—dance, chant, worship, religious theater, reconstructions of communal memory, and reenactments of communal myth. Again, synchronization is what is at issue here. If we synchronize even weakly connected persons, we strengthen their ties. Synchrony of the sort Durkheim addressed in religious or communal settings, therefore, is far more ubiquitous in social life than he might have imagined. Any performance ensemble—chamber orchestra, dance troupe, choir, theater company, marching band—yields examples. So does the unification of minds engaged in thinking together, the synchrony of a crew in a regatta, the reconstruction of autobiographical memory at a family event, the emotional unification of an audience at a football game—all yield synchrony that can strengthen interpersonal links and deepen personal and social integration. Yet when these moments of synchronization end, persons drift apart and their merger into their shared world slowly attenuates, like a memory that fades. They separate, and the strength of their links weakens.

The mechanics of synchronization always require the coupling of multiple oscillators. Networks, therefore, are where synchronization sets up. No network, no synchronization. Significantly, in settings where synchronization is orchestrated by a community, exogenous oscillators such as dance or music piggyback onto the attachment mechanism. The result is that some of what a mother and an infant feel when synchronized is also felt where others are synchronized—comfort, security, attachment, membership. Examples such as drumming, chant, or communal prayer highlight how this can happen. The beats of a drum or the recital of a chant are clocks, and they slave persons to themselves. When persons are yoked to these exogenous clocks, the strength of their ties to one another increases. Notably, they also find themselves merged into a synchronized system of interaction, whether it is an ensemble making music together, a group of actors performing a play, or whole community celebrating its history.

This pattern of merger-into-a-whole exhibits features of dissociation, not too far removed from what George Herbert Mead (1934) discussed as "taking the role of the other." The person, usually adrift in secular pursuits, is displaced, a figure who is drawn into a story or a collectivity, captivated by the drama and enlivened by the strong feelings—the effervescence, as Durkheim called it—that are driven by synchrony with others. Identity fades, supplanted by membership in a wider world that has history, narratives, roles, and statuses that are usually acknowledged only at a distance, familiar and yet lost to sight when quotidian involvements turn the eye away from the "center" of society (to borrow Edward Shils's [1975] expression).

Perhaps because there is a tendency for weak links to collapse under extreme conditions, communities regularly orchestrate these occasions for stress to be dissipated and for links to be strengthened by synchronization. The oscillation is between a phase of social dispersion—Durkheim's profane time, marked by weaker links to the collective world—and a phase of intensified social and communal activity—Durkheim's sacred time, marked by increasing tie strength.[6] Durkheim saw that this phase of intense synchronization drew persons into contact with the sacred symbols of their social world, and hence augmented moral and cultural integration. But there is still more that synchronization can teach us about moral life, though discovering it requires us to make a conceptual leap—to understand that attachment behavior is actually an extension of the human immune response. How can that be?

Synchronization, the immune response, and the sixth sense

J. E. Blalock (1984, 2005), one of the leading investigators of human immune responses, has described the immune system itself as a "sixth sense," meaning that it can plausibly be likened to organs or structures of the body that perceive threatening stimuli, detecting, in this case, invaders such as bacteria or viruses. Having registered one of these invaders, the immune system deploys its own cellular agents—complement proteins, cytokines, macrophages, neutrophils, natural killer cells, in combination with other chemical and biological players—as part of a concerted biological response to engulf and rid the body of them. Such pathogens are stressors, signaled at the site of their invasion by a growing inflammatory response that is orchestrated by the immune system—increased blood

[6] These correspond to two different phases of moral culture. On the one hand, the secular phase is consistent with the view of culture as a "toolkit" from which, once persons act or make decisions, they can retrofit their decision with explanations or justifications (Swidler 1985). This is a phase of diluted social integration and fragmented morality. By contrast, the phase of intensified synchrony and strengthening ties corresponds to those occasions when exogenous oscillators like dance or worship synchronize whole communities and leverage social and moral integration from interaction. An outcome of this phase is that persons use moral reasoning to guide their activities and reach decisions.

flow, recruitment of macrophages from the blood in nearby tissue, the release of cytokines and other chemical messengers of the immune system, and other effects as well. An important result of inflammation is the release and subsequent recruitment of endogenous opioids—chiefly, β-endorphins—whose effect is to reduce pain at the site of the invasion. Once the body (and the person) has been alerted to the assault, opioids then soothe and reduce the inflammation.

Distress as Inflammation. Attachment can also be seen as an immune response. The argument in support of this hypothesis is straightforward, and indeed generalizes Blalock's concept of a sixth sense (Smith and Silon 2005, 2009). A summary will serve our purposes here. The same way the sting of a bee or an abrasion to the elbow will recruit analgesic substances to inflamed tissue, distress in social life also recruits analgesic resources to sites of social "inflammation" (Bildack 2000). A baby's distress call is the interpersonal equivalent of an inflammation—a signal that first recruits maternal attachment, and that in turn stimulates the release of endogenous opioids. Attachment is thus analgesic—or, more properly put, it stimulates the release of analgesic substances. Alleviation of distress, in this way, is equivalent to the reduction of inflammation. The implication is then obvious: Attachment functions as a social extension of an innate immune response. Whether the cries of a baby, the sting of bee, or the suffering of a friend, attachment makes things better.

When we translate this argument into a network perspective, we begin to uncover some remarkable phenomena. Synchronization extrapolates the effects of attachment across networks, as when an anthem, taken up by an audience, induces a feeling of membership, affiliation, or citizenship—each an extract of attachment itself. If a crew in a regatta rows with perfectly timed strokes, it members have trained to the point where they can feel any deviation from synchrony. When monks sing together in chapel, any monk whose song stands out from the rest has failed his duties by establishing himself apart from the assembly. He is the deviant figure in an otherwise anonymous collectivity. The same is true of musical ensembles—their members are trained to make music together, not as individual performers. If an individual performer surfaces, synchrony is lost. Notably, this is exactly the opposite of how persons behave when they are away from these assemblies, ensembles, or teams. In those other settings, they strive for distinctiveness, achievement, and visibility. But in a synchronized setting, individualism and personal distinctiveness is disturbing. It is the equivalent of attachment loss or abandonment, the worst fear of a child.

Strong integrative pressures therefore exit in synchronized assemblies. Each member is linked to others by synchrony itself, a condition that recapitulates the perfect unity of infant and caregiver. There is family feeling in such assemblies. On the larger boats that compete in regattas, for example, there is always a coxswain who establishes a stroke, and the stroke brings every member into synchrony. The coxswain is the exogenous oscillator, the crew her slave. In these boats and in smaller boats without coxswains—boats where there are only two, four, or six rowers—there is also another form of communication that is not auditory or verbal at all. It occurs through detecting deviations in the balance of the boat. These indicate loss of perfect synchrony, a condition in which the boat loses power and speed. Such deviations are sensed as changes in the boat's balance—feelings to which a crew's training has sensitized them. In maintaining synch in larger teams or performance ensembles of all kinds, exogenous clocks become especially important. As the size of symphony orchestras increased in the second half of the 18th century, for example, orchestras stayed in sync first by relying on cues from concert masters (usually first chair violinists) and then, when orchestra size increased beyond the point where all players could be in visual and auditory contact, finally on the ultimate α-leader, the conductor (see Raynor 1978a, 1978b; Hart 1973). Orchestras thus shifted away from being more densely connected networks to acquiring the familiar star-shaped structure we have discussed above, where stress levels cause weak links to collapse and a central core figure appears.

Coxswains and conductors are metronomic. They are clocks, akin to those we associate with the oscillations in every community between togetherness-and-apartness, attachment and separation. But there is another way to understand how they function, and this arises from the analgesic benefits to collective life network synchronization enables. Networks of persons synchronized to one another extend the comforting immune function of attachment. With synchronization coincides a network-based enhancement to the modulation of distress, akin to the "sixth sense" that Blalock has described. A monk out of sync with is brothers is a disturbance to the whole community. The member of a crew off stroke stresses other persons in the boat. As with these stressors and irritants, other disturbances to communal integration—invasion, illness, deaths, attacks, assaults, accidents, crimes, threats of all kinds—will also spill distress into the links of synched networks from the point at which they are detected. When this happens, the network is activated. An inflammatory signal like a distress call transforms it into an attachment depot. And as in the body, inflammation then recruits analgesic resources from nearby links, sometimes from the whole depot. This is a network phenomenon, and can only occur where networks are periodically strengthened by synchronization.

Lost dogs, missing children, strangers, and thieves

Such shifts and oscillations are important because they reinvigorate local network organization and functioning. In the examples we have reviewed here, synchronization is obvious. But recall, now, the underlying oscillator. The integration of social life does not require ongoing and continuous synchronization. Instead, periodic synch is enough to maintain a network-based social immune response. The rest of the time, networks are latent, waiting to be activated. Their immune response does not even require that all persons in a community be part of a network. But those in it must have their links renewed from time to time.

Socially integrated neighborhoods show their network properties all the time. The best illustrations of this come from studies of crime. Jane Jacobs (1995 [1962]) reported that there were some neighborhoods in Milwaukee where the mere appearance of a stranger on the streets was enough to trigger phone calls to the police. By contrast is the indifference to crime that appears in fragmented neighborhoods, where persons have been observed watching from their windows as crimes take place below them. In general, social control and law enforcement in any society are network outcomes. They can never rest on the shoulders of policemen alone. Civic vigilance and social networks are always involved—reports to the police, witnesses to crimes, concerns for safety and order, and, from time to time, communal action itself. In the frontier settings of the American West, law enforcement was sometimes missing altogether. But taking its place were the posses that would form when news of a crime passed through a community. In European countries, the enforcement of law remained a rotating civic responsibility until the appearance of the first urban police systems at the end of the 19th century, or it similarly depended on the *posse comitatus* (Banton 1965; Buisson 1958).

So networks strengthened by synchronization are responsive to stressors of many kinds. Lost dogs and missing children are enough to mobilize whole neighborhoods into search parties. Strangers are always sources of apprehension. Telephone calls pass from house to house if someone unknown appears on a neighbor's property. If illness afflicts a neighbor, he will awake in the morning to find soup on his doorstep; a death will bring everyone in the network to a wake or a funeral; a house fire will inspire donations of clothes and food. Commonplace though these examples may seem, each illustrates a network response to distress—the recruitment of analgesic resources to sites of social inflammation.

Even in the largest cities, therefore, there are always neighborhoods. People identify themselves as residents of a place with a name—Hyde Park or Greenwich Village or the Latin Quarter or the Trastevere. Such knowledge restores them to home and to the links for which kinship and family are the prototypes. In some of these settings, networks are so synchronized that they function as a kind of social sensory system, akin to the workings of the immune system. They are the sixth sense on which social life depends.

Conclusions

Synchronization in social networks underlies social integration. I have argued in this chapter that the prototype for this is infant-caregiver attachment. In the functioning of healthy families, innate forces always synchronize newborns and their mothers. From their synchrony emerge the remarkable analgesic effects of endogenous opioids, morphine-like neuropeptides whose release is stimulated by attachment behavior. If we extrapolate the functioning of the innate mechanism behind attachment to social networks, we find similar outcomes. Synchronized networks are those Durkheim first described in his great studies of religion, but such networks are ubiquitous in social life.

Synchrony appears wherever networks are activated from time to time by stressors. Powerful illustrations of this are found in ensembles, monasteries, teams, theaters, conversation, and the reconstruction of memories. A few of these illustrations have been developed in detail above. But what they point toward are network phenomena that are far more pervasive in social life. We cannot really speak of neighborhood or community without recognizing how social life leverages social integration from attachment and interaction. No *Gemeinschaft*, no village, no tribe, no congregation, no society can really be fathomed without first understanding how its core properties derive from coupled oscillators—from synchronization that builds and strengthens social networks.

References

Banton, Michael (1965), *The Policeman in the Community*. New York: Basic Books.

Bildack, Jean (2000), "Detection and Function of Opioid Receptors on Cells from the Immune System," *Clinical and Diagnostic Laboratory Immunology*, 7 : 719-723.

Blalock, J. E. (1984), "The Immune System as a Sensory Organ." *Journal of Immunology*. 132 : 1067-70. "The Immune System as a Sixth Sense." *Journal of Internal Medicine*. 257 (2005): 126-138.

Buisson, H. (1958),. *La Police, Son Histoire*. Paris: Nouvelles Éditiones Latine .

Cacioppo, J. and Gary Berentson G. (2002),. "Social Neuroscience." Pp. 3-10 in *Foundations in Social Neuroscience* edited by John Cacioppo, Gary Berentson, et al. Cambridge: MIT Press.

Cacioppo, J., Gary Berntson G.; Ralph Adolphs R., Carter S.,, Richard Davidson R.,, McClintock M.,, McEwen B., Meany M, Daniel Schachter D., Ester Sternberg E., Steve Suomi S., and Shelly Taylor S., (eds) (2002),. *Foundations in Social Neuroscience*. Cambridge: MIT Press.

Csermely, P..(2006), *Weak Links: Stabilizers of Complex Systems from Proteins to Social Networks*. Berlin: Springer-Verlag.

Csikszentmihalyi, M. (1990), *Flow: The Psychology of Optimal Experience*. New York: Harper and Row 1990.

Durkheim, E. (1995), [1912], *The Elementary Forms of Religious Life*. Translated by Karen Fields. New York: The Free Press.

Edwards, D. and . Middleton. D. (1988), "Conversational Remembering and Family Relationships: How Children Learn to Remember." *Journal of Social and Personal Relationships*. 5 : 404-416.

Eigen, M. and P. Schuster P. (1979), "The Hypercycle: A Principle of Natural Self- Organization." Part C. "The Realistic Hypercycle." *Naturwissenschaften.* 65 : 341-369.

Fellin, Ph. and Litwak E., (1963), "Neighborhood cohesion under conditions of mobility," *American Sociological Review*. 28 (1963): 364-376.

Field, T. (1985), "Attachment as Psychobiological Attunement: Being on the Same Wavelength," Pp. 415-454 in *The Psychobiology of Attachment Separation* edited by Martin Reite and Tiffany Field. New York: Academic Press.

Granovetter, M. (1973),"The Strength of Weak Ties," *American Journal of Sociology*, 78 (1973): 1360-1380.

Halbwachs, M. (1980), é1950], *Collective Memory*. Translated by Mary Douglas. New York: Harper and Row..

Hart, Ph. (1973),*Orpheus in the New World: The Symphony Orchestra as an American Cultural Institution.* New York: W. W. Norton 1973.

Heinrich, M. H., Inga Neumann N. I., and Ehlert U. (2002), "Lactation and Stress: Protective Effects of Breast-feeding in Humans," *Stress*. 5): 195-203.

Jacobs, J. (1992), *The Death and Life of Great American Cities*. New York: Vintage.

Kirschner, S., and Tomasello, M.(2009), "Joint Drumming: Social Context Facilitates Synchronization in Preschool Children," *Journal of Experimental Child Psychology*. 102: 299-314.

Mithren, S. (2005), *The Singing Neanderthals: The Origin of Music, Language, Mind and Body*. London: Weidenfeld and Nicholson.

Newman, M, Barabási A.L., and Duncan J. Watts, (eds.) (2006), *The Structure and Dynamics of Networks.* Princeton: Princeton University Press.

Patel, A. D., Iverson J., *et al.* (2008),"Studying Synchronization to a Musical Beat." *Annals of the NewYork Academy of Sciences*.

Pikovsky, A., Rosenblum M., and Jurgen Kurths J. (2001), *Synchronization: A Universal Concept in Nonlinear Science*. New York: Cambridge University Press 2001.

Prigogine, Ilya (1978),. "Time, Structure, Fluctuation." *Science*. 201 (1978): 777-785.

Prigogine, I. and Stengers I(1984), *Order out of Chaos*. New York: Bantam Books 1984.

Raynor, H. *A Social History of Music* and *Music in Society*. 2 vols. New York: Taplinger, 1978. *The Orchestra: A History*. New York: Charles Scribner's Sons, 1978.

Schachter, S. (1951), "Deviation, Rejection, and Communication." *Journal of Abnormal and Social Psychology*. 46 (1951): 190-207. *The Psychology of Affiliation*. Stanford: Stanford University Press, 1979.

Schutz, A. (1978), "Making Music Together." Pp. 159-173 in *Studies in Social Theory*. The Hague: Martinus Nijoff, [1951] 1971.

Shils, E. (1978), *Center and Periphery: Essays in Macrosociology*. Chicago; University of Chicago Press.

Small, C. (1998), *Musicking*. Hanover and London: Wesleyan University Press.

Smith, T. S. (1991), *Strong Interaction*. Chicago: University of Chicago Press, "Catastrophies in Interaction," *Social Psychology Quarterly*. 57 (1994): 274-282. "Nonlinear Dynamics and the Micro-Macro Bridge."

—. (1997), in Raymond E,, Sara Horsfall S., and Mary Lee M. (eds). *Chaos, Complexity, and Sociology* Thousand Oaks, CA: Sage Publications, 1997: 52–63.

Smith. T. S. and Franks, D. (1999),. "Emergence, Reduction, and Levels of Analysis." 3-18 in Franks D.and Smith T. (eds.), *Mind, Brain, and Society* (Volume 5 of Social Perspectives on the Emotions). Stamford, CT: JAI Press.

Smith, T. S. and Silon D., "Neurosociological Foundations of Social Networks: Distress-Dependent Synaptic Architecture in Social Systems." Paper discussed at Plenary Session, International Sociological Association, Interim Meetings, University of Rome, July 12-14, 2005.

Smith, T. S. and Gregory T. Stevens. "Emergence, Self-Organization, and Social Interaction: Arousal-Dependent Structure in Social Systems," *Sociological Theory* 14 (1996): 131-153. [Also: Santa Fe Institute Publication Series, Paper No. 94-08-046.]. "The Architecture of Small Networks: Strong Interaction and Dynamic Organization in Small Social Systems." *American Sociological Review*. 64 (1999): 403-420.

Smith, T. S., Stevens, G. T., and Caldwell, S.. "The Familiar and the Strange: Hopfield Network Models for Prototype Entrained Attachment-Mediated Neurophysiology." in Frank D., Smith T.S. (eds),*Mind, Brain, and Society* (Volume 5 of Social Perspectives on the Emotions) Stamford, CT: JAI Press: 213-245.

Stark, D. (1996), "Recombinant Property in East European Capitalism," *American Journal of Sociology*. 101: 993-1027.

Stark, D., and Vedres B. (2002), "Pathways of Property Transformation." Santa Fe Institute Working Paper No. 20112081.

Strogatz, S. (1994), *Nonlinear Dynamics and Chaos*. Cambridge MA: Westview Press.

—. (2003), *Synch: The Emerging Science of Spontaneous Order*. New York: Hyperion Books.

Swidler, A. (1986),. "Culture in Action," *American Sociological Review*. 51 : 273-286.

CHAPTER TEN

BODY AND TECHNOLOGY: MAN, ARTIFICIALITY AND SOCIAL PRACTICES*

MATTEO NEGRO AND GUIDO NICOLOSI

Abstract

Cartesianism dealt a decisive blow to the concept of human nature, fostering disenchantment and progressively inducing a levelling of the concept of human nature on a mechanistic model. What is natural in man gradually coincides with natural *tout court*, which emerges as a form in itself, like matter.But the body is not merely physical: it includes a topology of one's own body, drawn from the experience of the subject itself. Our body is qualitatively marked as the memory of actions, as the form of space and time, and therefore as the identity of the self, which is grounded upon the original sensori-motor dimension. The second part of the essay presents an analysis of the anthropological relationship existing between body and technique. Different paradigms are critically compared in order to demonstrate that evolution is a *bottom up* process anchored to a circular interaction between body and environment mediated by behavioural choices, phenotypes and organism development. Here organisms are considered as active agents orientated to modify and *co-construct* the world where they live. The author, within this framework, interprets human technique as a *body skill,* an anthropological feature leading beyond the nature/nurture contraposition which usually animates the scientific debate.

1. Man between nature and artificiality

Cartesianism dealt a decisive blow to the concept of human nature, fostering disenchantment and progressively inducing a levelling of the concept of human nature on a mechanistic model. What is natural in man gradually coincides with natural *tout court*, which emerges as a form in itself, like matter. Over time, it becomes increasingly difficult to speak about "human nature", its specificity and uniqueness. Losing the ability to answer the question about the identity of man has led to a pessimism that affects every sphere of activity. The Socratic "know thyself" becomes a huge challenge and, in some respects, an unsustainable one[1]. If we consider that the interpretative diaphragm of human identity that science has established (finding also much support in Cartesianism) fails to account for the inner and spiritual experience, we find the reasons for further pessimism. Certainly, the “machine” model, enhanced by cognitive science, is epistemically powerful and performs the role of minimizing the mystery surrounding human nature well. However, It allows no room for the phenomenological evidence of duality (physicality and innerness[2]), of the dimension of intentionality, of moral evidence (duty, freedom, conscience). Galilean science reached the conclusion that the metaphysical problem of the nature (essence) of things is an evanescent problem from a scientific perspective. As Evandro Agazzi has pointed out, this has not led to a weakening of epistemic scientific conclusions (the scientific image of the world), but has reduced the objectivity of the physical world (including man, as part of it) inside the 'artificiality (experimentation, manipulation, calculation), introducing the possibility of deviation and variation[3]. With

---

* Matteo Negro is the author of paragraphs 1 and 2; Guido Nicolosi is the author of paragraphs 3 and 4. The text was revised for the English language by Stephen Conway.

[1] «L’aspect paradoxal de cette invitation consiste dans le fait que, apparemment, la connaissance de soi est une tâche facile, vu qu’il n’y a pas de diaphragme qui s’interpose entre le sujet qui connaît et l’objet de cette connaissance. Toutefois, cette facilité se transforme en difficulté si l’on considère justement la connaissance comme une opération qui *exige* une altérité, une prise de distance entre le sujet et l’objet. A partir de ce moment, l’ « objectivation» de soi se présente comme une entreprise désespérée et on se sent poussé à la recherche des *moyens* indirects (c’est-à-dire des « sources de connaissance ») pour y parvenir» (Agazzi, 2004b: 49).

[2] «[L]’homme est capable *d’intérioriser* le monde externe tout en ne s’identifiant pas ontologiquement avec celui-ci. Cette « présence interne » du monde constitue l’ensemble des *représentations*, et la faculté de produire des représentations a été indiquée en philosophie par le terme d’*intentionnalité*» (*Ibidem*: 54)

[3] «Scientific observation is *instrumental* observation and, moreover, the experimental set-up is such that, not, so to speak, Nature in itself, but only those aspects of Nature that can be *artificially* isolated by the experimental procedures can really be described and explained. *Contemporary* science has completed this transition: what is studied and discussed in actual scientific practice is not Nature in itself, but previously elaborated *scientific* results and theories» (Agazzi, 1995b: 5).

the advent of modernity, the artificial serves as a mirror which reflects a certain image of man[4].

Western culture has developed the concept of nature since the dawn of Greek thought in a systematic way. Aristotle's philosophy had refined the distinction between natural and artificial. Its conceptual division had persisted for a long time, till the development of modern experimental science. However, a further refinement of the concept of technology and its wider impact on society have exacerbated and also distorted the already ambiguous relationship between the two terms and have forced a semantic revision, whose path is not without pitfalls.

For Aristotle, the relationship between nature and substance is very close and irrevocable. Natural is the substance alone, the artificial belongs to the order of the accidental:

> Of things that exist, some exist by nature, some from other causes. [...] [N]ature is a source or cause of being moved and of being at rest in that to which it belongs primarily, in virtue of itself and not in virtue of a concomitant attribute. [...] None of them [the artificial objects] has in itself the source of its own production. But while in some cases (for instance houses and the other products of manual labour) that principle is in something else external to the thing, in others - those which may cause a change in themselves in virtue of a concomitant attribute - it lies in the things themselves (but not in virtue of what they are). [...] Things 'have a nature' which have a principle of this kind. Each of them is a substance; for it is a subject, and nature always implies a subject in which it inheres (Aristotle, *Physics,* II, 192b).

The definition of nature is immediately applied to humans. And precisely within the context of the debate about human nature, it makes sense to distinguish between substance and accident, as the sphere of human "accidents" is wider than any other: man as a producer, as creator of accidental elements, may be described within a variety of descriptive patterns. As Aristotle states, whatever predicate is used to define the essence of man, it will always refer to an accident. The nature of man lies instead in the "substrate". The essence of man is the substantial ground of his accidents: his specific principle of life and movement[5]. It should not be

---

[4] «What distinguishes the traditional from the modern way of considering the artificial is that in tradition this constituted something intermediate between Nature and man, something which is not natural but is also distinct from man, since it is simply man's *production*. In modernity, the artificial has been more strictly related with man, since it is considered to constitute the "world of man" in a substantial sense» (*Ibidem*: 4).

[5] «For there is nothing to prevent the same thing from being both a man and white

forgotten that the substantial principle animates individuals whose existence is well defined in time and space. The main horizon, the true vanishing point of Aristotelian natural philosophy in examining the status of man, concerns the vital and non-mechanical principle of individuals.

Metaphysical naturalism was also compelling from an axiological point of view and recognized in the metaphysical order of nature the character of perfection. Modern science has emptied the physical world of values and ends, eliminating the risk of the naturalistic fallacy. So, "artificial" changes no longer affect the principle of the order of being, but even fall within the normal course of evolution[6]. These changes ultimately do not bring about the crisis in the framework of teleonomic conception of nature, because natural law is too strong an objective constraint to be equated with any cultural and historical product. Nature, however, lends

---

and countless other things: but still, if one asks whether it is or is not true to say that this is a man, our opponent must give an answer which means one thing, and not add that 'it is also white and large'. For, besides other reasons, it is impossible to enumerate its accidental attributes, which are infinite in number; let him, then, enumerate either all or none. Similarly, therefore, even if the same thing is a thousand times a man and a not-man, he must not, in answering the question whether this is a man, add that it is also at the same time a not-man, unless he is bound to add also all the other accidents, all that the subject is or is not; and if he does this, he is not observing the rules of argument. And in general those who say this do away with substance and essence. For they must say that all attributes are accidents, and that there is no such thing as 'being essentially a man' or 'an animal'. For if there is to be any such thing as 'being essentially a man' this will not be 'being a not-man' or 'not being a man' (yet these are negations of it); for there was one thing which it meant, and this was the substance of something. And denoting the substance of a thing means that the essence of the thing is nothing else. But if its being essentially a man is to be the same as either being essentially a not-man or essentially not being a man, then its essence will be something else. Therefore our opponents must say that there cannot be such a definition of anything, but that all attributes are accidental; for this is the distinction between substance and accident-'white' is accidental to man, because though he is white, whiteness is not his essence. But if all statements are accidental, there will be nothing primary about which they are made, if the accidental always implies predication about a subject. The predication, then, must go on ad infinitum». (Aristotle, *Metaphysics*, IV, 4, 1007a).

[6] «With the constitution of modern science, as we have seen, ontological necessity is removed from Nature, but a weaker condition is introduced in its place, that of natural *determinism*: natural laws are perhaps contingent (in the sense that they might have been different from what they actually are), however since they exist, they strictly command natural phenomena and it is impossible to escape them» (Agazzi, 1995b: 8)

itself to an interaction with rational and free subjects, who can, if they wish, foster its development in harmony with the pursuit of other rational ends, or try to impede it by introducing types of theoretical and practical incompatibilities.

Accordingly, the crucial issue from a philosophical anthropological viewpoint consists in the agreement between the respect for human beings as such and the enhancement of the artificial, as documentation of the human operation that continually creates something to adapt the world to himself. The "principle of natural compatibility" (PNC), highlighted by Agazzi, summarizes on one hand the recognition of the teleological foundation of nature and on the other the use of contingent means, fit for the pursuit of natural ends. In summary: the natural is expressed as "realized" and not inherently fictitious compatibility between the intrinsically natural and the contingently natural, i.e. the artificial[7]. Obviously, though, as mentioned, the image of the scientific world does not provide the concept of teleology, human experience in its complexity would be opaque outside any consideration concerning the ends (ends of action, etc...). Aims and purposes which are not objectively determinable by the method of experimental science, but only through a careful philosophical analysis.

The artificial world is the natural environment of human beings and, so, the respect for nature does not imply the rejection of the artificial, the technological, but the right consideration of the relationship between the region of the artificial in its various forms, and the other ontological regions[8]. Accordingly, it is not possible to superimpose the one region to

---

[7] «[E]ven if we admit that the artificial is legitimate provided it respects the natural ends, it is by no means logically founded that we must also keep faithful to the means contingently employed by Nature for attaining such ends» (*Ibidem*: 11).

[8] «[L]e monde de l'*artificiel* (c'est-à-dire, le résultat de la manipulation humaine de la Nature) est actuellement l'environnement *naturel* de l'être humain (c'est-à-dire l'environnement qui est proportionné à sa nature spécifique). C'est pourquoi l'exigence morale de respecter la nature [...] entraîne le respect du droit humain à utiliser et à modifier la Nature et considère l'artificiel comme une partie de la Nature qui provient de la nature particulière de l'être humain. Seule une perspective cosmocentrique [...] semble contenir les éléments nécessaires pour éviter de glisser de l'appréciation positive (légitime) de l'artificiel à l'affirmation que *chaque* forme de l'artificiel est moralement légitime. [...] la justification de la légitimité de l'artificiel, qui est une défense particulière de la *liberté d'action* humaine, doit être inscrite à l'intérieur d'un contexte moral de *respect* aussi pour les régions ontologiques qui sont différentes de celle de l'être humain, de son monde et de ses besoins. En particulier la réalisation de l'artificiel doit respecter non seulement un étalage d'obligations *morales* concernant les humains [...], mais aussi la Nature» (Agazzi, 2004a: 84-5).

another and operate some kind of reductionism. What fundamentally marks the ontological differences between the human and the artificial, and somehow seals their compatibility and continuity, is the sphere of intentionality[9]. Intentionality is the true principle of individuation of human action, which is different from any reproducible activity not by virtue of its effectiveness, of the mere result, but of the rational attitude that inspires and marks it in a unique and definitive way. Action is therefore an operation which is perfectly appropriate to the rational attitude of the subject, who is irrevocably connected to it (*operari sequitur esse*). By contrast, the evaluations of the operations of automata or machines are in general based on the results, as outcomes of emulating processes[10]. The intention, the "ought" dimension of each typically human act is an ideal-factual criterion (open to intersubjectivity)[11].

---

[9] «[P]ossiamo dire che l'anello mancante, per usare una famosa locuzione cara ai teorici dell'evoluzione biologica, fra l'intelligenza artificiale e quella naturale (che si mostra già a livello animale, prima che a livello umano), pare indubbiamente essere l'intenzionalità. Il riconoscimento di questo fatto non si limita a confutare ogni riduzionismo fisicalistico, ma mostra l'inadeguatezza anche di quelle forme di emergentismo che sono capaci di vedere i livelli più elevati della realtà soltanto come complessificazioni di quelle inferiori, lasciandosi sfuggire la vera natura del nuovo e del diverso» (Agazzi, 1991: 13).

[10] «[T]he technical attitude focuses its interest on what the different beings (machines, animals, men, etc.) are able to *do*, while the philosophical attitude aims at establishing what these beings *are in themselves*. Now, the arbitrary extrapolation of the technical attitude consists in the pretention of claiming that two beings are *essentially* of the same nature if they are both able to *do* a certain number of identical or very similar things (underestimating the question about *how* these two beings do the same things, and also that of the existence of *other things* which one of them may be able to do while the other is not). On the other hand the extrapolation of the philosophical attitude consists in pretending that these two beings do actually perform *different* things, when they seem to perform the same things, because of a difference in nature which is considered to exist between them (and this is clearly false from a strictly technical point of view)» (Agazzi, 1981b: 196-7).

[11] «La presenza di una dimensione di dover-essere è infatti accettabile a *tutti* i livelli dell'agire *tipicamente* umano, ossia di quell'agire in cui *consta* che l'uomo si comporta diversamente da un animale [...] l'esistenza presso l'uomo di una dimensione del dover-essere è qualcosa *che si constata*, ha cioè una radice fattuale, non comporta un illecito "passaggio dall'essere al dover-essere", bensì rispecchierà il fatto che *qualunque* azione che sia autenticamente umana (indipendentemente dal "livello" in cui si colloca) è sempre guidata da una "intenzione" che ne assegna lo scopo e le offre un criterio o un ideale di "perfezione". Ma, appunto, questo criterio non è più fattuale, bensì ideale, esso non si trova nell'esperienza stessa [...]

The man-machine analogy tends to ignore (although it favours some interesting local applications) that there are behaviours which are not only aimed at one or more goals, but which are also oriented by values, through intentioned ideal entities which become models of our activity[12]. Moreover, the effects on the moral and even legal sphere are not long in coming. The machine that is programmed for a particular purpose is "valid" to the extent that its program is running; when it stops running it, it loses all its value. However, the machine falls fully within the framework of the behaviour of those subjects who produce and program it. Such behaviours are accompanied by normative requirements and reasons, that is by logical and rational judgments on relations subsisting between states of affairs; in other words, human action requires an understanding of relations between states of affairs, the ability to judge, which is also a normative capacity, because judgment is the norm of individual and social action.

From the philosophical analysis, human ontological superiority is ensured by the indisputable fact that man as a person has a non-material or spiritual dimension. But this is a property of the human being in its entirety, according to the totality of his experience. And the inviolability of the person is the inviolability of the whole person and not of a single part, disconnected from the whole. If someone loses an arm in an accident, the entirety of his being a person is not weakened. The amputated arm does not share such a property, since it is hopelessly separated from the person to whom all parts belong. On the other hand, if after a while a prosthesis is implanted (like any other limb implanted in similar cases) to replace the arm and restore partial functionality, this does not infringe upon the principle of inviolability, but it becomes part of the personal unity. So, man pursues the ultimate (also naturally founded) goals, that the rational nature first identifies and highlights, and his physical parts with their particular functions interact with those goals.

In addition, compared to other living beings, man has a sense of "ought to". He does not respond to reality only through the recording of data, but also by putting in place a sort of moral intentionality, abstracting the image of the accomplishment of an action, a desire, a thought, and recognizing and establishing values, motivations, meanings. This is an instance which goes beyond the scope of scientific experience. The

---

tuttavia esso è tale che, dimenticandocene, non riusciremmo più a qualificare come "umane" le "azioni dell'uomo"» (Agazzi, 1987: 333-4).

[12] See Agazzi, 1992: 128-9

deontological dimension has its own ideal cogency, which is asymmetric to the naturalistic domain[13].

Nature is indeed a multifactorial reality and cannot therefore be studied through a single method of investigation. The study of nature is compositive and meets two requirements. The first requirement involves the application of a compatibilist criterion. Each particular method must tend to play up its own nontrivial internal dynamics (metaphysical and regulative) of reconstruction of the entire experience and not of reduction of the experience itself. The second requirement concerns the philosophical perspective of the whole, which does not derive from the sum of specific investigations, but rather from the unavoidable gap that exists between any theory and nature as a general problem[14].

What does it mean then to regard man as a natural being? Certainly it means taking account of factual knowledge deriving from the natural sciences; it also means including observation points upon the intrinsic normativity of human experience, to an objective eidetic-constitutive phenomenology. The prescriptive, the values, the "ought to" as a structural dimension of agency, intentionality, consciousness, the inner experience; they all are also factors of experience, to which the empirical sciences should devote their hermeneutic efforts, but according to a more compatibilist mode than a reductionist one. It is one thing to investigate the intentionality of consciousness by applying a computational or biological matrix chasing the perspective of compatibility, it is another to investigate the intentionality of consciousness through a unique computational

---

[13] «When acting according to this ideal picture, we act morally, in the sense that our actions (including our creation of the artificial, the manipulation of Nature, the use of our body, the intervening on it and on that of other persons) are guided by our conscious intention not to passively adhere to an alleged given Nature, but to promote what is *good* for ourselves and others» (Agazzi, 1995b: 17).

[14] «[L]'ineludibilità del "punto di vista dell'intero", non soltanto nell'interpretazione dei risultati teorici delle scienze, ma anche nella costruzione delle stesse teorie scientifiche, dal momento che quella che di solito viene indicata come esigenza di rispettare in esse la *coerenza logica*, non ha quasi mai i caratteri banali di una richiesta di evitare le contraddizioni formali, bensì quella di essere compatibile con certi criteri molto generali di *intelligibilità* [...] il punto di vista dell'intero è proprio quello che caratterizza la prospettiva filosofica in senso lato, e quella metafisica in particolare: filosofia e metafisica, infatti non si caratterizzano tanto per il fatto di occuparsi di certe cose o di certi tempi, ma per il *modo* di occuparsene, ossia per la prospettiva di globalità con cui investono *qualunque* oggetto di cui si occupano (anche se non intendiamo dire che lo stile della filosofia sia soltanto questo). Diremo allora, in sostanza, che le scienze contengono parecchia filosofia e metafisica *implicite*» (Agazzi, 1995a: 5).

or biological platform[15]. Therefore, it makes no sense to categorically separate "body" and "mind," says Agazzi, if to these categorizations we? cannot contribute different and compatible ontologies. The human dimension of embodiment, then, documents in a special way the incontrovertible fact of being oriented towards the realization of “non-physical” tasks[16].

---

[15] «Questo insieme di conoscenze [*factual knowledge*] contribuisce potentemente a farci delineare un'*immagine dell'uomo*, a dirci in una certa misura “che cosa effettivamente siamo”. Ma la nostra immagine dell'uomo contiene altresì, e in misura non meno essenziale, la delineazione di “ciò che vorremmo essere”, ossia l'insieme dei *fini* o degli ideali che, secondo noi, un uomo o un'umanità veramente realizzati *dovrebbero attingere*. Possiamo quindi tornare a parlare di *natura umana*, purché ci rendiamo conto che questa natura, molto di più che qualcosa di dato, è un *compito da realizzare*, è un insieme di condizioni di fatto e di fini [...] In tal modo risulta anche possibile superare la dicotomia tra descrittivo e prescrittivo, dicotomia che si esprime spesso dicendo che, se anche dovessi trovarmi in presenza di una chiara tavola dei valori, non ne scaturirebbe per me il *dovere* razionale di attuarli. Ciò è vero se i valori vengono ipostatizzati, ossia concepiti come entità metafisiche disincarnate, ma se i valori vengono considerati come facenti parte dell'immagine dell'uomo (ossia della natura umana quale effettivamente e razionalmente ho accettato di rappresentarmela), allora essi assumono la natura di *caratteri costitutivi* e diventa pertanto *razionale* che l'uomo si comporti in conformità ai suoi caratteri costitutivi. Come si vede, il tradizionale concetto metafisico della morale che si ancora nella natura umana è di fatto un'immagine storicamente in evoluzione, nella quale l'umanità iscrive anche i fini e i valori che ritiene meritevole, o addirittura doveroso, promuovere» (Agazzi, 1992: 293).

[16] «Something of which we are becoming more and more aware nowadays is that we must recover a unitary view of man, as man *is* a complex and *many-sided* reality, which cannot be understood by *separating*, e.g., body and spirit (or body and mind, if we understand “mind” in a sufficiently rich sense). If we keep this need of unity in view, we can accept to a large extent the implicit intention of dualism, in as far as it highlights that the complex reality of man includes aspects which may be grouped together under two main headings, for which the names of *mind* (or soul, or spirit) and *body* may be used, owing to the historical weight of their meanings; but it would be dangerous to “substantialize” these headings. As a matter of fact, what is becoming increasingly clear is that the human body iself is a peculiar body indeed, which is, as it were, “oriented” toward the realization of performances which we still call “spiritual” [...] to the delineation of the unitary and complex image of man no source of information and insight may remain alien» (Agazzi, 1981a: 19-20).

2. Embodiment, intentionality and action

In human beings, the body, which obeys very precise working and empirically verifiable mechanisms, is subject to the execution of non-corporeal operations. In the natural universe, it is certainly a unique phenomenon: the human machine pulsates and moves on the basis of something that does not belong to its functioning but that is structurally and inseparably connected to it, so as to constitute its potential energy, as Simondon wrote. Eating and living are not the only goals for man; they are indeed related to the affirmation of non-material goals (projects, ideas, beliefs, etc..)[17]. Not only that, but also the experience of objective interiority is the focus of intersubjective life, languages, social and cultural traditions and plays a role that is more often even denied to much more prosaic existential events. It is an experience of the person, within the person, but it is not reducible to an epiphenomenon of a pure subjectivity, to the manifestation of the monadic autonomy of individuals.

The "normal" man is therefore the one that carries everything to a unity of sense, through its intentional sphere. As Canguilhem wrote, the ordinary man is the normative man, always able to institute new rules, even organic. What does "normative man" mean? The term denotes those who are able to internalize in an intentional way the objects of their experience, through rules of use, cognitive procedures, issuing from perception and action. This internalization is of particular value in adapting to situations and contexts. However, everyone tends to create or restore a space in which to find "his/her" normality. This is the core problem. Life is a polarity, said Canguilhem. This indicates that life tends to polarize around itself all the necessary conditions to ensure its own subsistence and development. Compared to this polarization then it makes sense to speak of biological or physiological normality.

---

[17] «Cette sphère de l'intériorité est loin de se réduire purement et simplement à une sorte de *point* élémentaire insondable, obscur et inexprimable: elle constitue un véritable *monde intérieur* qu'on peut analyser, approfondir, explorer, même en utilisant des *méthodes* appropriées que l'humanité s'est efforcée de préciser et de *pratiquer* dans les différentes cultures, y compris la culture occidentale, avant que l'impérialisme de la technoscience ne fasse tomber en désuétude une pareille préoccupation [...] Nous devons reconquérir l'idée d'une *intériorité objective*, car celle-ci est une réalité dont tous les hommes ont l'*expérience* (même s'il ne s'agit pas de l'expérience sensible), qu'ils peuvent comprendre, dont ils peuvent parler réciproquement, qu'ils peuvent sonder et dont ils peuvent enseigner l'analyse à d'autres personnes» (Agazzi, 2004b: 57-8)

The common concept of normality seems to refer to the ability of adapting to any situation, but this is, obviously, an unlikely potentiality, which has no basis for its verifiability. In spite of that, the myth of normality as universal adaptation resists and influences markedly the negative judgment on exceptional or pathological conditions. Instead, from another point of view, all human conditions are unique and largely inevitable. It is not sufficient to adopt statistical criteria in an attempt to determine the normal states, or simply to narrow the horizon of the possibilities of adaptation to the conditions which are contemplated by laboratory researches. Canguilhem already saw the limits of an approach to the problem of normality and pathology of human behaviour that does not take into account the habits and customs of each individual. With regard to the

> relationship of adjustments to environments, we must not forget that the laboratory itself constitutes a new environment in which life certainly establishes norms whose extrapolation does not work without risk when removed from the conditions to which these norms relate. For the animal or for man, the laboratory environment is one possible environment among others (Canguilhem, 1978: 85).

The category of the normal institutes the pathological as its subspecies. The biological abnormality is antinomian to the biological normal but not to the normal of life.

If life is thus the polarity that aggregates any manifestation, even the artificiality, if human life implies the perception of an organized whole which is the condition of agency, then the brain has no magical power. The living self instead weaves a net of intentionality, in the words of Merleau-Ponty[18], and in doing so it constitutes/institutes the time as its time. Qualitative changes of intensity emerge in our life which have nothing to do with the mathematical quantity, with the continuum. The real, normal and normative self is free from the blackmail of quantity.

As already noted, concerning the theme of embodiment, Cartesianism introduced a misunderstanding in the scientific and philosophical discourses. As remarked by Marc Richir, «..."psychological" means "intangible" or "incorporeal" only in a mindset where the body has been surreptitiously reduced to the materiality or the physical (spatial) embodiment of an instrument, of a friend who is more or less indulgent or malicious»[19]. This amounts to saying that the Cartesian body-matter is

---

[18] «Time is not a line, but a network of intentionalities» (Merleau-Ponty, 2002: 484)

[19] Richir, 1993: 13, our trans.

merely a passive instrument of crossed causal forces, while the psychological sphere maintains a sort of unadulterated autonomy. The work of not reducing consciousness to matter, undertaken by most cognitivists, in fact did not render, contrary to appearances, a great service to mentalism, because the mental properties and physical properties, although qualitatively and phenomenologically separate, remain both "natural" properties (in a scientific way). But the provocation of Richir goes further and urges us to rethink, in other words, the "body", as no longer synonymous with "material", nor, therefore, opposed to "intangible", but as a "synolon" of form and matter, of *morphé* and *hyle*. The body is not merely physical: it includes a topology of one's own body, drawn from the experience of the subject itself, altered and amended by the perceptual field[20]. It is therefore purely phenomenological evidence, which is equally confirmed within a certain type of scientific research, neurophysiological research in particular. The historic discovery of mirror neurons significantly validates the improbability of the divergence of the body schema from the perceptual processes and the intentional sphere. This is because our body is qualitatively marked as the memory of actions, as the form of space and time, and therefore as the true identity of the self, which is grounded upon the original sensori-motor dimension[21]. The world is not the terminus of a physical and computational relation, but it is a great case for action. And the identity of the self forms from the history of the endeavoured or accomplished actions. Michel Henry writes:

---

[20] Salvatore Natoli correctly observes: «Se dunque l'esperienza del corpo la facciamo nell'agire, i confini del corpo non possono che essere altrimenti tracciati se non nell'ambito dell'azione. Bateson ci fornisce in proposito delle precise indicazioni. Se voglio, ad esempio, tagliare un albero, in primo ho certamente l'intenzione di farlo, Ma chi sono *io*? Sono certo il mio corpo, ma sono anche la sequenza indicata. Il proprio corpo non può essere mai percepito al di fuori della rete di relazioni del mondo. Quale è, dunque, il confine del corpo? ... il nostro sentirci corpo è definito nella e dall'azione e per questo la mente, tornando a Spinoza, non è mai separabile dal suo schema corporeo» (Natoli, 2004: 63).

[21] A very good remark by Fontanille : «Le *Moi* de référence est la *chair* animée par le *sensori-motricité* ; le *Soi* en devenir est le corps propre dont les limites et l'individualité sont progressivement définies par l'accumulation et la mémoire des réactions de saturation et de rémanence. En somme, et conformément au principe d'interdéfinition des deux instances, le *Soi* prend forme dans les traces imprimées par les tensions et pressions dans la chair du *Moi* ; en bref, le *Soi* (corps propre) serait, entre autres, la mémoire du *Moi* (la chair), et la somme des empreintes qu'elle conserve» (Fontanille, 2004 : 27)

> the existence of my body is practice, i.e. a general possibility of the ego, it is its ontological relevance, it is the identity of the body, and also, as Maine de Biran said, it is "the very durability of our personal uniqueness". The body is not immediate knowledge, it is the permanent knowledge that is my existence itself, it is memory. (Henry, 2001: 134-5; our trans.)

Habit-based memory is the very form of the body. *Habitus*, i.e. unity of the possibilities which are expressed by the corporeal self in the endeavored and performed actions[22]. Unity is therefore not given by an external time, by the time of a neutral and independent observation, by the time of the laboratory. Instead, action is incarnate (embodied) since it does not match the simple execution of a performative rule. The self is the rule of action:

> By using the term action we mean to emphasize once again that sensory and motor processes, perception and action, are fundamentally inseparable in lived cognition. Indeed, the two are not merely contingently linked in individuals: they have also evolved together (Varela-Thomson-Rosch, 1991: 173).

The enactive approach here involves two concepts, «that perception consists in perceptually guided action and that cognitive structures emerge from the recurrent sensorimotor patterns that enable action to be perceptually guided»[23].

What connection is there, then, between intention and action? If we conceive intentionality as an intrinsic property of mind, a state of consciousness that causes a bodily movement, we can assume that the link

---

[22] «On ne peut donc songer à faire reposer l'unité de notre être originaire, l'unité du corps ou celle de l'ego, sur la mémoire. Ce n'est pas par le souvenir, c'est-à-dire par la médiation du temps, que se rassemble dans une unité l'ensemble des actes et des mouvements qu'accomplissent ou exécutent les pouvoirs fondamentaux de mon être propre. Cette unité de mon être à travers le temps que constitue la mémoire requiert un fondement. Ce fondement, c'est l'habitude, ou, si l'on préfère, l'être même de mon corps qui rend précisément possible l'acte de mémoration et qui n'est rien d'autre que l'être de la connaissance ontologique entendu comme l'être réel de la pure possibilité. Expliquer par la mémoire l'unité de l'ego ou du corps, c'est commettre un paralogisme, c'est expliquer l'unité originaire par une faculté qui trouve au contraire son fondement dans une telle unité. En d'autres termes, l'unité de notre être n'est pas constituée, elle n'est pas arrachée au temps par le pouvoir de nos protentions et de nos rétentions, elle est immanente, au contraire à l'être de notre connaissance ontologique, elles se confond avec lui, elle est précisément ce qui fait de cet être l'être d'une connaissance ontologique» (Henry, 2001: 138-9).

[23] Varela-Thomson-Rosch,1991: 200.

is heavily subject to the variability of counterfactual conditions. We also assume that the body is a mere tool, however useless in the case of an ascertained cognitive deficiency, or equivalent to any artificial object. Instead, if intentionality is the form itself of action, namely its being a movement towards an accomplishment, the relationship is manifested in an unequivocal way: the body is itself intentionality in action. Even the newborn shows to have a sensori-motor proprioceptive sense, which is not a representation of her own body, but, as Shaun Gallagher noted, the original and pre-reflective sense of herself as "embodied". Varela remarked suggestively with Thompson and Rosch:

> The child begins with only her sensorimotor system [...] the newborn infant is neither an objectivist nor an idealist; she has only her own activity, and even the simplest act of recognition of an object can be understood only in terms of her own activity (Varela-Thomson-Rosch,1991: 176).

Intentionality as a way of being and not merely as a property of mental representations: the discovery and understanding of such evidence could also provide a new insight even into the ethical issue of the dignity of the human body.

3. Hominization: a 'matter' of (or) mind?

It would be a serious mistake to consider human phylogeny as a process leading to the enfranchisement of thought from the material environment. It is true, as many anthropologists claim, that we are essentially 'thinking apes', but, at the same time, in a very fundamental epitome we equally should acknowledge that our 'thinking brain' doesn't place us *above* things. We are not privileged animals, and our destiny is, like all other non-human animals, living our life embedded within matter.

3.1. Different paradigms of hominization

At the beginning of the XX century, physical anthropology was dominated by a 'brain-centred' view, assuming that the hominization process started 'from the head'. In fact, anthropologists shared the idea that the brain evolution was the first step in explaining the whole of Man's evolution. Interpreting literally the Cartesian paradigm and the rationalist belief of the illuminist, their expectations were focused on the evolutionary process of intelligence. For this reason, Warnier reminds us correctly (Warnier, 2005: 53), that diggers looking for fossil records at the time were expecting a 'missing link', idealistically represented as a human head on an ape's body. The empirical evidence collected over time showed exactly the contrary: a (non-linear) converging tendency towards a modern human body with the ape's head. That is, out of metaphor, all this striking evidence showed that as a fundamental evolutionary step the pre-human body changed its configuration, in particular, assuming a new bipedal physical condition. The brain, instead, at the very beginning of this new process was the same as their ape ancestors. The brain was to change later, 'rapidly' growing in size, absolute and relative, and in its functional morphology. What is commonly accepted is that the new bipedal posture caused many important anatomical transformations: facial architecture, teeth, cranium and cranio-vertebral alignments, pelvis modification, curvature of thigh bone (femur), structure of the knee joints, foot and toes, etc. (Tobias, 1965). At the same time, it seems quite evident that the bipedal revolution freed the proto-human hands provoking their consequent re-functionalization. These scientific acquisitions led to a new wave of studies within which many distinguished anthropologists symmetrically reversed the brain-centred paradigm. Gradually a new paradigm was created focusing on locomotion: the bipedal paradigm (Warnier, *ibidem*). Probably, we may consider André Leroi-Gourhan and Sherwood Washburn the most famous exponents of two different approaches within this latter. Both consider the erect posture as the most important feature which can explain the evolutionary process leading to *Homo sapiens*. But where Leroi-Gourhan emphasized the subsequent development of a body (hand, face, brain, etc.) equipped for technique (Leroi-Gourhan, 1983), in which technique is considered almost deterministically, as a sort of logical and inevitable trend, Washburn tried to explain this development (bipedal posture) within the traditional Darwinian perspective (natural selection, genetic inheritance, mutations, fitness, randomness, etc.). Today, claims Warnier, we should consider this perspective old-fashioned, and he suggests adopting a different approach which he calls the 'cultural-bipedal paradigm' (*ibidem*: 65). According to this latter, while accepting the fundamental importance of biological dimensions like bipedal posture, we need to render the frame more complex by introducing culture, communication, tradition, learning and social organization, probably already active within higher primates.

For example, the bipedal posture itself could not be the Darwinian effect of a genetic mutation[24], but the outcome of a behavioural innovation achieving a 'selective prize' provided by the re-functionalization of hands

[24] That is, a genetic random mutation positively selected by environmental pressure.

which can now be used for objects and tools manipulation aimed at social and practical goals. Thus, within this frame, Warnier states we find a behavioural innovation concerning 'body techniques' and 'instrumental techniques' (Mauss, 1966). Here, behavioural choices are the 'interface amongst society, culture, and environment'. Moreover, free hands can be used now not only for technical gestures, but even for the whole *armamentarium* of gestures aimed at constructing and strengthening social relations (*grooming*, artefacts uses, body manipulation, etc.). In Warnier's words:

> Ludic and instrumental manipulation of body and matter develops thumb opposition, the prehensile hand, manual ability and development of the cortical areas controlling the hands which, compared with brain weight, are three times more developed in man than apes. Finally, as already noted, the 'couple' formed by hand and artefact replaces teeth for grasping, carrying, grinding, ripping, boring, menacing, immobilizing and fighting, in an overall system together a reduction in facial mass and its musculature, cortex development and enhancement of manual ability and bipedalism. The re-functionalization of the hands is systemic with the teeth, the brain, the artefact and walking within a social context. (Warnier, *ibidem*: 67, *our translation*).

It is now evident how hominization and the hand-brain binomial, within this perspective, is not an issue concerning only tools or technique. We have to avoid falling into the trap of reducing Man to a sort of rationalist *Homo faber* (like Marx or Leroi-Gourhan himself). Free hands represented a wonderful opportunity even for the 'construction of the Self and the socialized body'[25]. Warnier uses a very effective expression: 'the artefact is the material pin which holds together the assembly bipedalism / hand / motor conduct / sociality / culture / language / partition / brain / hunting' (*ibidem*: 80, *our translation*). We will try to explain the meaning of this expression in the following paragraphs. Before, however, we wish to discuss the meaning of the central concept we used here: the behavioural choice.

---

[25] Let's consider how important hands are for seducing or performing sex or caressing children, etc. We could propose similar considerations about language. The erect posture made anatomically possible the functionalisation of the basic respiratory components involved in speech: lungs, larynx and the supralaryngeal vocal tract (Lieberman, 1994), but the articulated word is the consequence of a 'contingent choice'. This choice, again, is not only a technical one. Words are, at the same time, technical instruments and gestures orientated towards construction of the Self and social relations.

3.2 Behavioural choices, exaptation and evolution

Robert Foley, a distinguished anthropologist, in a very intriguing book entitled *'Humans before humanity'* (1999), argued that when we talk about natural selection, if we want to understand the real process in action we have to focus our attention on the active role played by individuals rather than on anonymous populations. That is, individuals with different and specific behavioural patterns. This principle is valid for all organisms, but it is particularly so for complex organisms like apes or humans. This perspective is very important because it acknowledges an active role of the individual organism that puts it at the very centre of the evolutionary process. Organisms are not passive entities only waiting for random mutations, and evolution is not the passive response provided by the species to environmental challenges. Organisms have different survival strategies[26] which become specific behaviours. These behaviours can be innovative and radically transform[27] the ecological niche within which the organisms live. Human evolution is, following this idea, somehow related to the behavioural strategy adopted by some ancient hominids and that changed some basic biological traits: erect posture, brain size, etc. Foley describes a feedback system where behavioural changes come first, causing biological changes and dictating the selective conditions causing, again, new biological changes.

Patrick Bateson (1988), explicitly quoted by Foley, goes further on this fundamental aspect. He shows that the traditional Darwinian image of selection[28] leads to represent evolution with the very fascinating metaphor of the 'external hand', downplaying the role of organisms within the evolutionary process. Bateson, on the contrary, considers this metaphor an 'encumbrance'. He describes four interacting processes:

a) Organisms make choices which could influence the course of evolution[29];

---

[26] The term strategy can be misleading. It doesn't mean a rational, intentional and conscious thinking. It refers to an opportunistic search for evolutionary gain.
[27] If they are successful in the long run: that is, if by adopting a specific behaviour individuals will achieve a higher reproductive success than those who don't.
[28] Here, it is Nature that elects organisms fitting best within their environment.
[29] For example, within a wide range of environmental conditions they could inhabit, they can choose which one to actually inhabit (Waddington, 1959). Moreover, they can move to another environment, if that of their ancestors is no longer adaptive for molecular changes previously intervened (Dover, 1986).

b) Organisms alter actively their physical and social environment condition[30];
c) Adaptability[31];
d) Alteration in gene expression[32].

For all these reasons, Bateson talks about *'behaviourally induced evolutionary change* (*ibidem*: 198), and Foley himself used this highly illuminating sentence: 'behaviour, particularly the social one, which usually goes beyond the scope of paleo-biology, is essential to understand evolutionary events' (Foley, 1999: 233, *our translation*).

Moreover, these 'processes' could also be explained with the help of another but related and important concept of post-Darwinian theory: *exaptation*. This concept has a long history dating back to 1872, when Charles Darwin answered the zoologist George Mivart's objections to the presumed inability of natural selection to account for the incipient stages of particularly complex biological structures. Darwin used the neologism *'pre-adaptation'* and introduced the possibility of *redundancy* in nature within the relationship between organs and functions. This redundancy, for Darwin, exists in order to allow that a trait developed for a certain reason could be co-opted or converted towards a function completely independent from the previous one. This functional co-option was renamed by Gould and Vrba (1982) 'exaptation', to indicate how organisms often readapt already available structures in opportunistic ways for new functions. This concept challenges the adaptationist view and confutes all the interpretations (also neo-creationists) which assume the existence of an 'intelligent design' guiding Nature. Conceptually, as shown by Ingold (1997), this concept highlighted an interesting analogy between organic and technical change. It shows, in fact, that in Nature new structures do not appear from nowhere. They are not designed for a specific aim *'de novo'*. Ingold says: 'natural selection can only work on the stock of materials that are already

---

[30] Plants shed leaves thereby changing soil characteristics for their descendants; beavers dam rivers; social competition can cause unexpected properties, etc.

[31] See the so-called Baldwin effect, where the cost of a phenotypic adaptation transmitted by learning can be reduced by random genotypic change (thus reducing learning costs).

[32] See 'genetic assimilation' (Waddington, 1953). According to Waddington, laboratory empirical research showed that some somatic non-inherited changes provoked on phenotypes by environmental stimuli can be genetically assimilated and transmitted to the following generations, even when the stimulus which generated them is no longer active. What is fundamental here is the fact that genetic mutations are not random, but environment-oriented or pheno-centered.

available' (Ingold, 1997: 119). In this perspective, changes within an ecological niche can lead the organism to co-opt one structure, evolved for a specific aim, to do a different job. The history of artefacts is very similar. Also the technical sphere involves exaptation whereby old instruments are co-opted for new uses every time.

From this point of view, we could consider exaptation as a post-Darwinian theory which is able to go beyond the sclerotic version of Darwinism interpreted by Modern Synthesis. Evolution is the interaction between external (environmental changes, ecological niche, etc.) and internal (genes, cellular structures, etc.) forces; here Nature is able to recycle available organic materials. An/The Organism is considered a sort of *bricoleur* which faces change through the best use of what it already has, and ontogenesis plays a crucial role. Exaptation shows that evolution is also the consequence of an opportunistic assembly which uses redundancy as the main source of creativity. Adaptation here is a contingent side effect more than a direct optimization and flexibility is the organism's capacity to react creatively to environmental change. In this sense, flexibility is the main tool at the service of the organism to guarantee its survival and this happens thanks to imperfection, multiplicity and redundancy (Pievani, 2004).

This principle of *bricolage* is, of course, valid also at the micro-level, as claimed by Jacques Monod. He clearly stated (1977) that the living world is characterized by diversity and unity at the same time. The genetic multiplicity which evolution produces is not based on absolute novelty every time, but the same biological elements are combined in new ways. It seems that, from a genomic point of view, life is based on the alteration and recombination of existing molecular structures. In this direction, the work of Susumo Ohno on *'gene duplication'* (1970) proved fundamental. Ohno argued that gene duplication is the most important evolutionary force since the emergence of a common ancestor. The duplication of a gene means that an additional copy, free from selective pressure, is created. This new copy can mutate without deleterious consequences to the organism. This freedom from consequences allows the mutation of novel genes that could potentially increase the fitness of the organism or codify for a new function. This process[33] produces a huge amount of temporary genomic redundancy (neutral) but which represents and important potential stock for future evolutive needs[34].

---

[33] See also the so-called *'selfish gene'* or the *'junk DNA'*.

[34] This argument helps us accept a wider degree of phenotype *plasticity* (Pigliucci, 2001; West-Eberhard, 2003).

However, it is quite evident that complex forms of behavioural choices are, then, related to complex neural processing. So it is quite predictable that complex animals, like mammals, with higher relative brain size have higher rates of behaviourally induced evolutionary changes. In this argumentation, Bateson implicitly suggests we look at evolution increasingly as a *bottom up* process, where the main issue is the particularly circular relationship between organism and environment, mediated by behavioural choices, phenotypes and organism development[35]. However, says Bateson, if we compare the total amount of DNA of mice with that of bacteria we see that mice have 6000 times as much DNA as bacteria. Apparently, this explains the different complexity between them. But humans have no more DNA than mice. So, if we compare mice and humans and consider differences within the nervous system and behavioural complexity, we can plausibly infer that 'the gradual emergence of behavioural complexity was not achieved by accumulating genes' (Bateson, 1988: 201). Probably, relative brain size is a more reliable indicator of complexity.

Going back to human evolution, it was a behavioural choice that induced hominoids to adopt erect posture, a new body technique, to better adapt to their environment. It took around two or three millions of years and many psychological and anatomical adaptations (see above) to achieve a good verticality and, consequently, free the hands for making and transporting tools, but also for taking care of sons and developing a strong sociality. Probably, for Foley, the starting point was ecological: bipedal gait guarantees energetic savings within given time-space restraints (Foley, 1999). What is really relevant, here, is that at the end of this process, the bipedal gait made an inexorable growth of the brain size possible (absolute and, more importantly, relative size). Humans have the highest relative Encephalization Quotient (EQ) within the animal kingdom. That makes humans more intelligent and intelligence is a fundamental trait for ecologically 'sophisticated' animals and, the key trait for animals whose most important characteristic is their *sociality*.

4. Towards a socio-anthropology of technique

Hegemonic scientific thought, probably still anchored to the traditional dualistic Cartesian philosophical paradigm, persists in representing evolution as a *top down* process. Many critical authors define this representation gene-centred (Ho & Fox, 1988; Oyama, 1998; Jablonka &

---

[35] Again, in this 'game', two distinct but related 'non-random' processes play a central role: *exaptation* and *emergence* (Tattersall, 2009).

Lamb, 2007; etc.). Genocentrism considers the genome as a 'barrel' containing discrete information units, linearly linked to a unique determination chain from genotype to phenotype. In this vision, there is a close relationship between natural selection and genetic inheritance. Natural selection cannot operate on acquired traits because they are not transmitted to offspring. Differential proliferation of new genetic combination is the evolution essence. Here, evolution produces genetic programmes (a sort of *design* or informational software) controlling ontogenetic development. Between ontogeny and phylogeny there is an unbridgeable divide.

Instead, we have shown that the hominization process has been a *bottom up* process linked to the specific feedbacked relation between a behavioural choice, which took the form of a particular body technique (erect posture) not previously genetically programmed, and a specific environment. This phenotypic body technique, which radically transformed the environmental niche of these hominoids and consequently the kind of selective pressure operating on them, over time has been genetically assimilated, following a 'behaviourally induced evolutionary change' pattern (Bateson, *ibidem*). That is, a kind of dynamics clearly discussed and postulated within epigenetic approaches (Nicolosi & Ruivenkamp, *in press*).

We focused particularly on the fact that this evolutionary process had, as its main consequence, brain size growth and the brain's new functional neuro-anatomy. This is an important issue we would like to discuss in greater depth. Indeed, in our view, this point may lead us to discover a fundamental interconnection amongst three basic dimension of our life: body, technique and sociality.

4.1 Brain development: ecology, sociality or socio-technical skills?

Today, the idea that the mind depends on the size and morphology of the brain, from a quantitative and qualitative point of view, is quite acceptable. This assumption, if accepted, radically undermines the traditional Cartesian dogma according to which body and mind would be separate entities. In fact, the brain is a body organ whose characteristics, as aforementioned, are subjected to the history of the evolutionary process. In this respect, it is correct to talk about 'embodied mind'. The noble scientific father of this perspective is, of course, Gerald Edelman. He showed that the brain is embodied and the body is embedded in its environment and this symbiotic intertwining operates in a highly integrated way. Of course, this relationship is not considered deterministic. Indeed, the deep meaning of this principle is that the brain (and the mind with it) is *'context bound'*. It is not an abstract machine, like a computer or clock, elaborating programmed information; and it does not produce preordained outcomes. The brain is an embodied *selectional system*[36] that, by recognising patterns, and through creative mapping, produces unique selectional repertoires for every person.

Now, simplifying for the sake of argument, anthropological analysis of brain evolutionary development can be polarized into two opposing factions. One faction claims that pressures favouring intellectual advances are ecological, the other claims they are social. The first faction, which enjoys a remarkable circulation, is prevalently traceable to the work of Leroi-Gourhan and Washburn. In this perspective, feeding strategies linked to ecological (territory and food) constraints required increasingly complex adaptive responses based on tool-making and tool-using and able to stimulate selection of higher intellective faculties. As a consequence of this view, many scholars try to find evidence showing the link between technique and specific cognitive and linguistic skills. For instance, some authors argued that tool construction requires some motor skills based on a linear, sequential and causal connection, a kind of 'logic of movement' and suggested that specific capacity of generating this kind of movement sequences was progressively developed by motor and pre-motor cortex strongly influencing even the Broca's area. In other words, in this perspective, speaking that is to 'articulate a sequence of syllables', is very similar to splitting? a piece of wood (tools production) as well as throwing objects (e.g. a spear) (Castelli and Peretto, 2006). Further, language and object manipulation are both hierarchical and based on the combination of different hierarchical subunits. Some scholars emphasize that Broca's area, particularly large in modern human brains, is involved for both hand motility and language and provides a common circuitry for hierarchical organization of tool-using and linguistic skills (Lieberman, 1994). This relationship seems to be influential even in ontogenetic processes (Langer, 1993)[37], and considered confirmed by the relationship between *aphasia* and *apraxia* (Kempler, 1993). Indeed, relationships between technique and specific cognitive and linguistic skills do exist, but we are far from finding wide acceptance of this assumption. For example, Thomas Wynn, an

---

[36] See Edelman, 1993.

[37] According to Langer, object manipulation helps in human maturation of cognitive, linguistic and social skills. He argues that object manipulation, in fact, develops concepts successively integrated into classificatory, mathematical, linguistic and social constructs.

authority in this field, clearly denies this relationship. He explicitly affirms that if tools played a role in human intelligence then evolutionary development was 'a fleeting one':

> I am not arguing that tools were not important in human evolution. They clearly played a role in changing subsistence patterns, geographic expansion, and so on. However, the question here is intelligence, not ecological success, and we cannot assume that the two are tied directly to one another through the agency of tools. (Wynn, 1988: 283)

In conclusion, we are probably far from being able to say that we are in the presence of a linear and solid phylogenetic relationship between tool-using and intelligence. This relationship seems rather complex and more interdisciplinary research is needed (Gibson, 1991).

The other faction claims that this selection was stimulated by sociality. This claim is based on the fact that 'brains are exceedingly expensive both to evolve and to maintain and […] it is difficult to justify the claim that primates, and especially humans, need larger brains than other species merely to do the same ecological job' (Dunbar, 1998: 178). For this reason, during the late 1980s, they elaborated an alternative hypothesis to explain primates' large brains as the evolutionary answer to the 'computational demands' of complex social systems within which they live. This assumption is even related to the specific evolutionary history of the brain (Dunbar, 2009). It is said, for primates, that the social group dimension is directly correlated, within each species, to the size of the neo-cortex, which is phylogenetically the latest 'layer' of the brain cortex. The evolutionary expansion of the brain took place in the anterior part, in particular the frontal lobe, where intelligence within humans and anthropomorphic apes is prevalently 'located' [38]. What is really interesting for us is that, compared with apes, the frontal lobe in humans is bigger proportionally to the average number of social relations they are able to manage. Moreover, it seems that this development of the frontal lobe has been more relevant than that of the other parts of neo-cortex.

Developing arguments put forward by primatologists such as Michael R. A. Chance and Allan P. Mead (1953), Alison Jolly (1966) and Nicolas Humphrey (1976), this suggestion became celebrated with the title of 'Social brain hypothesis', better known to a larger audience as 'The Machiavellian intelligence hypothesis' for being prevalently focused on tactics and coalition formation processes and abilities. This focus is

---

[38] Of course, this is a simplification, because the brain is an integrated organ, but it is nowadays ascertained in neurology that the frontal lobe has a very important role in social abilities (Damasio, 1995).

probably the main limit of this argument, because it ends up by reducing the idea of sociality to a very poor niche.

Tim Ingold (1993a), who claims that the contraposition between ecological and social interpretations of brain and intelligence development is a serious impediment to the full understanding of both phenomena, helps us overcome it. He argues that this contraposition is based on a dichotomy between technique and sociality which mirrors the fundamental dualistic western contraposition between nature and society. Further, he says masterfully:

> […] this same tradition of thought, whose penchant for constructing dichotomies is one of its main defining characteristics, has given us a distinction between intellect (as a property of mind) and behaviour (as bodily execution), along with the idea that all purposive or intentional action is preceded by an intellectual act of cognition, involving the construction of representations, the consideration of alternatives, and the formulation of plans. I dispute this, and with it the very idea of intelligence that informs (often implicitly) both 'technical' and 'social' accounts of its origins and evolution. (Ingold, *ibidem*: 431)

The very relevant issue Ingold discusses is the concept of intelligence itself. He strongly refuses the idea according to which, intelligence would be a sort of internal cognitive *device* which, elaborating information, is able to guide our actions. At the same time, it would be folly to think of locomotion as produced by an internal 'engine'. Thus, no intelligence, just like no locomotion, exists separate from the whole organism. Cognition and locomotion are qualities of the whole animal, and they are strongly intertwined. But Ingold, in this essay, emphasizes another important related issue. Intelligence among primates and, above all, among humans is often represented by *planning* abilities, that is, the power of thinking about our past experience to imagine future alternative scenarios and construct plans. But, Ingold says, people plan actions or scenarios intelligently just as the same people make artefacts intelligently. The significance of this is that they are both intelligent activities, where constructed plans are not 'necessary preludes but rather its products' (Ingold, *ibidem*: 431). Plans, for Ingold, are 'guidelines' which are not able to produce actions or behaviours deterministically; and intelligence is, here, considered as 'immanent in the action itself'. Tool-using, as well as speaking, for instance, obviously require a particular neural organization, but this latter, in harmony with Edelman's teaching, is wholly 'immersed within a total context of bodily activity in an environment' (Ingold, *ibidem*: 432). Tool-using or speaking, thus, are not the pre-programmed output of an internal device located within our brain.

For all these reasons, Ingold distinguishes technique from technology and considers technique a *skilled practise.* In fact, the form of a constructed physical object doesn't issue directly from an acquired or self-authored formal representation. This is because even when the agent has the plan (the design) in mind, this latter doesn't contain all we need to enucleate the form. Construction (the substance) is not the mechanical transcription of the design (the form). In his view, the object *emerges* from a 'complex pattern of skilled movements'. More precisely, Ingold says:

> The forms of artefacts are not inscribed by the rational intellect upon the concrete surface of nature, but are rather generated in the course of the gradual unfolding of that field of forces and relations set up through the active and sensuous engagement of the practitioner and the material with which he or she works (Ingold, 1997: 112).

For this reason, thinking about the 'making' of Man should lead us to think about *processes* and not *structures* (or programmes); where there is no room for representations functioning without any reference to practical application in specific contexts. In this perspective, in tool-making processes, there is less of a gap between higher animals and men than usually believed. Both don't execute programmes mechanically (either cultural or genetic). Moreover, in both cases, we witness skills that are 'neither innate nor acquired, but developmentally incorporated into the *modus operandi* of the body through practise and experience in an environment' (Ingold, *ibidem*). These ideas become very clear when we are able to distinguish conceptually, with Ingold, *tools* from *artefacts.* Tool is an object extending the agent's capacity to operate within a specific environment. Artefact is an object with a specific shape orientated by a pre-existing mental form. There are tools which are not artefacts and vice versa. As Marcel Mauss showed well, says Ingold, believing that every technical action needs a tool (an instrument) shows a cultural prejudice according to which the technical quality of action lies in the instrumental object rather than the human agency. This prejudice is the consequence of the introjection of the mechanistic perspective at the root of the modern concept of technology. The 'conflation of the technical with the mechanical', which marked the birth of modern technology, transformed the figure of workman into that of an operator. From that point on, technique has been considered as residing within the tool, outside the agent, whereas technique is embedded within his experience in a specific (ecological and social) environment. For these reasons, Ingold considers the split between the technical and the social unacceptable.

4.2 Technique, technology and the 'law of the irreducibility of skills'

As aforementioned, Ingold explicitly refers his analysis to the 'theory of practise', that is, to a *corpus* of philosophical and socio-anthropological knowledge whose best known authors are Wittgenstein, Heidegger, Merlau-Ponty, Pierre Bourdieu, etc. According to this tradition of studies, human beings are (in philosophical terms) *being-in-the-world,* a phenomenological figure opposed to that of 'self-contained individuals' facing an external world, current in methodological individualism as well as cognitivism. Following this approach, human beings, inhabiting the world, make it a 'meaningful environment'. This is essential, because it refuses the idea according to which the world is meaningful because we are able to assimilate it to a formal design. For this reason, referring to technique, Ingold points out that precisely because we are embodied presences we cannot build, operate technically as it were, without *dwelling*. We live *within* the world not on its surface. Thus, acting technically implies a personal involvement within the world. Hence, a tool is not merely an object. We talk about an object when we refer to its intrinsic characteristics, but we talk about a tool when we refer to the user characteristics and purposes (Ingold, 1993b). An object can become a tool when it is linked to a technique, that is, a technique of a skilled subject. In this fundamental sense, a tool is a body extension which is, at the same time, a projection of a personal power and character. That is because with tools the agent brings into play an intention issuing from his personhood. But this strong interconnection makes tools, techniques and skills *social entities*. They mediate personal relations and so they are *embedded* in social relations.

> Indeed the distinctive features of human technique are essentially the same as those of human sociality: its reflexivity, its creativity, its monitoring according to ideal standards, and its temporal orientation. All have their basis in the properties of consciousness, self-awareness, and intersubjectivity. (Ingold, *ibidem*: 284)

This means that technique is inseparable from the particular and specific concrete experience of real organisms who apply it within a specific bio-social environment, when they give life and shape to particular objects. That is why *'isolated objects do not tell us anything'* (Sigaut, 1993: 383). The meaning of a tool depends on the purpose of those who created it or use it. If this purpose changes, even the meaning of the tool changes. At the same time, the tool can be modified to better serve the old or new purpose, etc. For this reason, Ingold stresses the idea that an object becomes a tool only when it is tied to a technique, whose specific *'locus'* is the subject. Thus, in the end, we have a tool when an object is embedded within a technical system based on the strong relationship between skilled agents and their ecological and socio-cultural environment (Ingold, 1993b).

Separating the domains of technology and society as we do in the Western world is a serious misunderstanding. An 'error of perspective' caused by the gradual process of *disembedding* from the social which has affected technique at least from the Neolithic on, even if this process has exponentially multiplied its effects in modern society, after the advent of the industrial revolution (Giddens, 1991). But the crucial point, here, is that the tool-making process, originally embedded in an experienced human agent able to perform a skilled hand-eye co-ordination, has gradually been replaced by machines. With machines, technique becomes technology, where we witness a shift of relevance from the skilled practise to a *corpus* of rational, mechanical and objective principles encapsulated within the technological object. This shift is a fundamental anthropological stage implying a transition from the subjective-knowledge based skill, whose main feature was (social) practise and (personal) intuition, to mechanical functioning:

> Machines are indifferent to the personhood of their operators, as is technology to the personhood of its bearers. Whereas tools and techniques place the human subject at the centre of productive practice, both technology and the machine affirm the independence of production from human agency. Thus persons have been withdrawn from the centre to the periphery of the productive process, so that the technical forces activated in the latter have become *external* to relations between persons [...]. This is tantamount to the *disembedding* of technical from social relations [...]. (Ingold, 1993b: 286)

But to understand well Ingold's view about what modern sociology defines a *disembedding* process, we have to add that he considers the concept of skill as strongly intertwined with that of *task*. Task is a socially embedded activity, which can be performed only by persons, not by machines. Tasks require a technical skill, they are objective-oriented (no reference to formal rules or procedures), they are index of social identities and are not isolated, but in a never ending relationship with the performances of the other members of the community, that is, are part of a wider *taskscape* (Ingold, 2000).

Now, following the classical and influential Marxian interpretation of the rise of industrialism, we are led to believe that modern society has definitively swept tasks away from our horizon. Because in pre-modern societies there is no meaningful distinction between work and social life. Tasks were 'technically skilled activities' of persons who have specific social identities. In this frame, time was a flow intrinsic to work, and consequently to tasks and social life. In these societies, time was *social time*, because it was embedded within local community rhythms in close relation with the cyclic rhythms of the environment (night and day, seasons, etc.). But, it is often argued, with the rise of modern industrial societies, this qualitative and irreducible task-oriented time was dramatically replaced by a new concept of the abstract *sidereal* time of the clock, used to measure an equally abstract and quantitative concept of labour (labour-power). Together, modern time and modern labour, became the main units to measure the value of goods by currency ('time is money'). In this social condition, we witness a radical splitting of every dimension of life. The necessity/freedom dichotomy seems to pervade every domain of human experience: time (clock/free), tasks (work/leisure), tekhnē (technology/art), prestation (market/gift), etc. (Ingold, 2000: 329). In this (very pessimistic) frame, there is no room for task-orientation and technically skilled activity, except for the so-called 'home activity' (social reproduction domain as a residual of the pre-industrial era) as opposed to the 'workplace activity' (production domain). In this perspective, workers are alienated and 'strangers to the world whose forms and meanings are created through this activity' (Ingold, *ibidem*: 331).

Nevertheless, Ingold, in recent years, has proposed a more complex and interesting perspective. In his view, it is not entirely true that tasks and consequently skilled practises disappear from the horizon of the modern worker. His words are illuminating:

> [...] in their concrete presence, machines are substantial components of the immediate environment, and engaging with them is an inevitable part of the business of everyday coping in the world. Thus rather than simply operating a technology, the activity of industrial workers consists in *coping with machines*. And viewed in this light, such activity not only belongs to them, but also calls for a good measure of skill, of a kind that can only be acquired through experience on the job. Moreover it is trough the development of skills of coping that workers are able to resist the impositions of a regime of command and control [...]. It is true that the machinery that workers are required to operate may – on account of its noise, heat, vibration of whatever – strain the human body to its limits of tolerance. However, despite, Marx's claim to the contrary, the worker does not cease to dwell in the workplace. He is 'at home' there. But home is often a profoundly uncomfortable place to be. (Ingold, 2000: 332)

In other words, Ingold emphasizes that, in the end, it is the person who is engaged with the machine and the environment surrounds them both. Thus, even in the 'workplace' we can find task-orientation and skilled practises. Explicitly referring to Sigaut's teaching, Ingold argues that in Western society, technology has pursued a project of appropriating the craftsman's skills, transforming his social practises into a set of mechanical rules and procedures aimed at marginalizing the role of human experience. Today, in the technological world, it seems that the practitioner agency is *determined* by the operations encapsulated within the design and machine. With the effect of replacing 'the rhythmic pulse of dextrous activity, governed by the coupling of perception and action, with the metronomic oscillations of a mechanically determining system' (Ingold, 2006: 80). What Ingold suggests is that this is a 'deceptive appearance', because task orientation is 'indestructible' and consequently socially situated skills and prestations are still the fundamental existential condition of our life as human beings and the indispensable pre-condition of sociality and social identity with which modernity has to 'come to terms'.

For this reason, in modern technological society we have a dialectical tension, that is, a dynamic oscillation between the two poles of the 'necessity/freedom' dichotomy illustrated above. These two poles co-habit within all the spheres of the modern technological world, they are not sharply split into two different areas (working time and leisure time), and so we have to achieve a temporal compromise between the sidereal time of the clock and the intuitive time of our body. This negotiation, namely the coordination of our activities and movements with the clock-time, is an acquired task-orientated skill in itself. The most important to be able to live in a technologically advanced world. For this reason, Ingold suggests that to announce the end of the 'social production of technical skills' due to the rise of modern technology goes against the evidence, that is,

> the entire history of technics might be interpreted as a constantly renewed attempt to build skills into machines by means of algorithms, an attempt constantly foiled because other skills always tend to develop around the new machines (Sigaut 1994: 446).

Sigaut calls this principle *'the law of the irreducibility of skills'*. It is based on the fact that skill is inherent to the creative ability of practitioners in de-assembling the technological artefact to then re-embed its parts or procedures into their locally situated social experiences.

References

Agazzi E. (1981a) Mind and body: a philosophical delineation of the problem, in *Epistemologia*, IV (n.s.), 3-20.

—. (1981b) Intentionality and artificial intelligence, in *Epistemologia*, IV (n.s.), 195-228.

—. (1987) I fondamenti filosofici dei diritti umani, in *Cenobio*, 4, 325-338.

—. (1991) Operazionalità e intenzionalità: l'anello mancante dell'intelligenza artificiale", in in S. Biolo (ed.), *Intelligenza naturale e intelligenza artificiale*, Marietti, Genova, 1-23.

—. (1992) *Il bene, il male e la scienza. Le dimensioni etiche dell'impresa scientifico-tecnologica*, Milano: Rusconi.

—. (1994) The Human Being as a Person, in *Forum Trends in Experimental and Clinical Medicine*, 4, 221-232.

—. (1995a) *Filosofia della natura. Scienza e cosmologia*, Casale Monferrato: Piemme.

—. (1995b) Nature and the Natural: Some Philosophical Reflections, in *Studies in Science and Theology*, 3, 3-19.

—. (2004a) L'être humain et la nature dans le discours bioéthique, in P.Kemp (ed.), *Le discours bioéthique*, Paris: Cerf, 71-85.

—. (2004b) Sciences de la vie, image de l'homme et progrès, in *Biologie moderne et visions de l'humanité*, Bruxelles: De Boeck, 47-59.

Aristotle, *Metaphysics*, transl. by W.D. Ross (http://classics.mit.edu//Aristotle/Metaphysics.html).

—. *Physics*, transl. by R.P. Hardie and R.K. Gaye (http://classics.mit.edu//Aristotle/physics.html).

Bateson P. (1988) The active role of behaviour in evolution, in Ho M and Fox S. (eds.) *Evolutionary processes and metaphors,* New York: John Wiley.

Canguilhem G. (1978) *On the normal and the pathological*, Dordrecht: Reidel Pub. Co, .

Chance M. R. A. and Mead A. P. (1953) Social behaviour and primate evolution, in *Symposia of the society for experimental biology,* VII, 395-439.

Castelli O. and Peretto C. (2006) The phylogenesis of language: the grammar of gestures and the manipulation of words, in *Human Evolution*, 21, n° 1, 45-49.

Damasio A. (1995) *L'errore di Cartesio. Emozione, ragione e cervello umano*, Milano: Adelphi.

Dover G. A. (1986) Molecular drive in multigene families: how biological novelties arise, spread and are assimilated, in *Trends Genet.*, 2, 159-165.

Dunbar R. (1998) The social brain hypothesis, in *Evolutionary Anthropologist*, 6, 178-180.

—. (2009) *La scimmia pensante. Storia dell'evoluzione umana*, Bologna: Il Mulino.

Edelman G. (1993) *Sulla materia della mente*, Milano: Adelphi.

Foley R. (1999) *Gli umani prima dell'umanità*, Roma: Editori Riuniti.

Fontanille J. (2004) *Soma et séma. Figures du corps,* Paris: Maisonneuve & Larose.

Henry M. (2001) *Philosophie et phénoménologie du corps,* Paris: PUF.

Gibson K. (1991) Tools, language and intelligence: evolutionary implications, in *Man (N.S.),* vol. 26, n° 2, 255-264.

Giddens A. (1991) *Modernity and self-identity: self and society in the late modern age,* Cambridge: Polity Press.

Gould S. J. and Vrba E. S. (1982), Exaptation a missing term in the science of form, in *Paleobiology,* 8 (1): 4–15.

Ho M and Fox S. (eds.) (1988) *Evolutionary processes and metaphors,* New York: John Wiley.

Humprey N. (1976) The social function of intellect, in Bateson P. P.G. and Hinde R. A., *Growing points in ethology*, Cambridge: Cambridge University Press, 303-317.

Ingold T. (1990) An anthropologist looks at biology, in *Man (N.S.),* Vol. 25, n° 2, 208-229.

—. (1993a) Tool use, sociality and intelligence, in Gibson K. and Ingold T. (eds.) *Tools, language and cognition in human evolution*, Cambridge: Cambridge University Press, 429-445.

—. (1993b) Tools and hunter-gatherers, in Berthelet A. and Chavaillon J. (eds.) *The use of tools by human and non-human primates*, 281-295, Oxford: Clarendon Press.

—. (1997) Eight themes in the anthropology of technology, in *Social Analysis*, n° 41(1), 106-138.

—. (2000) *The perception of the environment: essays on livelihood, dwelling and skills*, London-NY: Routledge.

—. (2006) Walking the plank: meditations on a process of skill, in J. R. Dakers (ed), *Defining technological literacy: towards an epistemological framework*, New York : Palgrave Macmillan, pp 65-80.

Jablonka E. and Lamb M. (2007) *L'evoluzione in quattro dimensioni*, Milano: UTET.

Jolly A. (1966) Lemur social behaviour and primate intelligence, in *Science*, 153, 501-506.

Kempler D. (1993) Disorders of language and tool use: neurological and cognitive skills, in Gibson K. and Ingold T. (eds.) *Tools, language and cognition in human evolution*, Cambridge: Cambridge University Press, 193-215.

Langer J. (1993) Comparative cognitive development, in Gibson K. and Ingold T. (eds.) *Tools, language and cognition in human evolution*, Cambridge: Cambridge University Press, 300-313.

Leroi-Gourhan A. (1983) *Mécaniques vivant. Le crâne des vertébrés du poisson à l'homme*, Paris : Fayard.

Lieberman P. (1994) The origins and evolution of language, in Ingold T. (ed) *Companion encyclopedia of anthropology*, London: Routledge, 108-132.

Mauss M. (1966) Le tecniche del corpo, in *Teoria generale della magia e altri saggi*, Torino: Einaudi, 385-409.

Merleau-Ponty M. (2002) *Phenomenology of Perception,* London: Routledge.

Monod J. (1977) Le frontiere della biologia, in AA. VV., *La biologia molecolare. Storia e ricerca*, Roma: Newton Compton, 33-47.

Natoli S. (2004) Essere nel mondo, in *The Body and the Sense of Self*, Milano: Fondazione Carlo Erba, pp. 59-64.

Nicolosi G. & Ruivenkamp G., The 'epigenetic turn'. From info-genes to body-organisms, *in press*.

Ohno S. (1970) *Evolution by gene duplication*, Berlin: Springer-Verlag.

Oyama S. (1998) *The evolution's eye. A systems view of the biology-culture divide*, Durham: Duke University Press.

Pievani T. (2004) Exaptation: la biologia dell'imprevedibile. Un approccio costruttivista all'idea di "adattamento", in Barbetta P., Capararo M. and Pievani T., *Sotto il velo della normalità*, Roma: Meltemi, 13-142.

Pigliucci M. (2001) *Phenotypic plasticity*, Baltimore: JHU Press.

Richir M. (1993) *Le corps. Essai sur l'intériorité*, Paris: Hatier.

Sigaut F. (1993) How can we analyse and describe technical actions?, in Berthelet A. and Chavaillon J. (eds.) *The use of tools by human and non-human primates*, 381-400, Oxford: Clarendon Press.

Sigaut F. (1994) Technology, in Ingold T. (ed.) *Companion encyclopedia of anthropology: humanity, culture and social life*, London: Routledge, 420-459.

Tattersall I. (2009) *Il mondo prima della storia*, Milano: Raffaello Cortina Editore.

Tobias P. V. (1965) Australopithecus, Homo Habilis, Tool-Using, And Tool-Making, in *The South African Archaeological Bullettin,* Vol. 20, No 80, 167-192.

Varela F. J., Thompson E. and Rosch E. (1991), *The Embodied Mind. Cognitive Science and Human Experience*, Cambridge, Ma.: The MIT Press

Waddington C. H. (1953) Genetic assimilation of an acquired character, in *Evolution*, 7, 118-126.

—. (1959) Evolutionary systems – animal and human, in *Nature*, 183, 1634-8.

Warnier J.P. (2005) *La cultura materiale*, Roma: Meltemi.

West-Eberhard M. J. (2003) *Developmental plasticity*, New York: Oxford University Press.

Wynn T. (1988) Tools and the evolution of human intelligence, in Byrne R. and Whiten A. (eds.) *Machiavellian intelligence. Social expertise and the evolution of intellect in monkeys, apes and humans*, Oxford: Oxford University Press, 271-284.

Thanks to Stephen Conway for revising the English language of this text.

CHAPTER ELEVEN

FLESH ON MEMORY: AN EMBODIMENT OF DROUGHTS AMONG THE TURKANA, THE PASTORALIST IN NORTHWESTERN KENYA

SHINSUKE SAKUMICHI

Abstract

Background: The Turkana people are pastoralists of a semi-arid area in the northwestern Kenya. Since the 1980s this area has been hit by successive droughts, which consequently cause social changes. While at the same time, an illness, called *Ewosin-a-ngacin* (EAN) with abdominal problems as its main symptoms, has emerged. Massage has been used as the remedy.

Purpose: The purpose is to show the resourcefulness of the body that can reconstitute the social order, transform cultural meanings and mediate between the self and the world in the case of EAN. We focus on the massage séances in which the body of EAN with an innate logic is materialized from the massive body.

Field & Method: The 14 massagers and the 17 patients were interviewed and videotaped in the 4 towns of Turkana district from 2002 to 2005.

Findings: The interviews and videos show: the massagers produce the body of EAN by their fingers on the patients' body, like an artisan moulding a clay figure. It is not only explained in the new combination of the anatomical knowledge of livestock and the folk medicine of human illnesses, but also interpreted in the context of droughts and the consequent hardships; loss of livestock, change of staple food from milk to maize etc.

Conclusion: EAN is a new type of illness based on the common body produced between a massager and a patient, who commonly experienced the droughts. The emergence of EAN means the effort for the Turkana people to adapt themselves to the droughts and social change by the elaboration of the common body, the inter-corporeality: by the embodiment of droughts.

The Turkana people are pastoralists of semi-arid area in the northwestern Kenya. Since the 1980s this area has been hit by the successive droughts, which consequently cause the social changes. While at the same time, an illness, called *Ewosin-a-ngacin* (EAN) with the abdominal problems as main symptoms, has emerged. The main symptom is constipation. A massage has been used as the remedy.

The purpose of this chapter is to illustrate that EAN embodies the social changes caused by droughts. We would like to stress the importance of three features. First, the Turkana people achieved the embodiment by locating droughts and its hardship in the pastral-physical understanding. Secondly, as its result, they could make a new pathology based on syndrome concept. Thirdly, the body of EAN became the common body for the Turkana people to remember and transmit droughts experience: flesh in memory.

We intend to show the resourcefulness (Sugawara,2005) of body that can reconstitute the social order, transform the cultural meanings and mediate between the self and the world in the case of EAN. The focus is put on the massage scenes in which the body of EAN with an innate logic is materialized from the massive body.

1. Purpose

The Turkana people are pastoralists of a semi-arid area in northwestern Kenya. Since the 1980s successive droughts in the area have brought about social change among the population. During this same period, an illness called *Ewosin-a-ngacin* (EAN) has emerged. The illness is characterized by constipation and is treated with massage.

The purpose of this chapter is to illustrate that EAN embodies the social changes brought about by the droughts. Three important features of this embodiment process are highlighted. First, the Turkana people achieved the embodiment by locating droughts and the associated hardships in the pastoral-physical understanding. Second, as a result, they could create a new pathology from the standpoint of the syndrome concept. Lastly, the EAN body became the common body for the Turkana people to remember and transmit their drought experience: flesh on memory.

I intend to show here, in the case of EAN, the resourcefulness (Sugawara, 2005) of the body that can reconstitute social order, transform cultural meanings, and mediate between the self and the world. Focus is placed on the massage scenes in which an innate logic about the EAN body has materialized from the physical body.

2. Theoretical Background

In the realm of medical anthropology, illness-coping refers to the process by which illnesses become embedded into preexisting meaning structures, such as an indigenous cosmology mediated by diviners (Kleinman, 1980) or modern discourses circulating in the mass media, from the standpoint of socio-cultural construction. As typical examples, Nichter (1992) analyzed the epidemiological process of Kyasanur forest ***disease*** in India, and Farmer (1992) followed historically the socio-political and cultural construction of AIDS in Haiti. These research studies focused on the cultural representations and discourses, but in doing so failed to grasp the physical body in which patients experience symptoms and pain (Csordas, 1994a). As a result, the physical body was taken as a passive recipient of a social process. Moreover, such studies were still rooted in the Cartesian separation of mind and body.

What is important about embodiment is to understand the body not merely as subject to external agency, but as an agent in its own world construction: creating “a way of living or inhabiting the world through one’s acculturated body” (Weiss & Harber, 1999). Researchers have highlighted the unconscious interaction with other human bodies from which cultural meanings form: in healing rituals, divination, trance, and culture-bound illness (Csordas, 1990, 1994b; Roseman, 2000, 2002; Jenkins & Valiente, 1994).

Notwithstanding, studies on embodiment have paid little attention to the relation of physical structure and bodily process. The approaches adopted had two related problems: an ignorance of folk knowledge of physical structure and of embodiment by conscious exploration. With illness, we usually start with primary care, not praying to ancestors, but touching and identifying the affected parts, dialoguing with the body to decide treatment.

Kawai (1998, 2008) described the development of anatomical and organ-physiological knowledge in the ethnomedicine of the pastoral Chamus of northern Kenya. They derived such knowledge from their daily experiences of dissecting livestock and examining illness. They not only explain illness in reference to folk anatomy and physiology, but also experience illness as it happens in the same folk anatomical and physiological bodily structure. I will refer to the embodied ethnomedicine specific to pastoral societies as pastoral-physical understanding in this chapter. Cast in such terms, the Chamus embody the pastoral-physical knowledge as the lived experience. Similarly to the Chamus, the Turkana have developed ethnomedicine based on their experiences of livestock keeping.

I intend to show EAN as an ethnographic case of embodiment based on a pastoral-physical understanding, focusing on massage scenes in which the person massaging interacts with the patient's body and elaborates the bodily structure of EAN. In this chapter, we are concerned with the pastoral-physical basis of bodily experience. Our interest lies not merely with the culturally inscribed body but also with the body as the seat of agency and intentionality. The findings of this case study are expected to contribute to the literature on the cultural variations of the embodiment from the perspective of the bodily structure of a pastoralist society, as well as to provide an empirical basis for theorizing the possibility of the acculturation effected by the bodily change: the transformation of intercorporeality.

3.Changing Field

Demography

The Turkana people are pastoralists living in a semi-arid area of northwestern Kenya.

Most of the 450,000 population live in the Turkana District that covers an area of 68,000 square kilometers. Their land is dry, receiving only 200 to 400 mm annual rainfall on average <Fig. 1>, and most of the people have lived a pastoral way of life, keeping cattle, camels, goats, sheep, and donkeys (e.g., Little and Leslie 1999, MaCabe 2007)<Fig.2>.

The Turkana District is one of the remotest areas in Kenya, and is continuously ignored by national development policies. Our research team has been conducting anthropological research since 1978. We are based in the town of Kakuma, located 100 km northwest of Lodwar, the largest town in northwestern Kenya <Figure.3>. I joined the team in 1993.

Fig. 1 Typical landscape in research field; dusty dry plain, shrubs and hills.

Fig. 2 Herdboy started for one-day pasture. Herd was less than twenty small livestock.

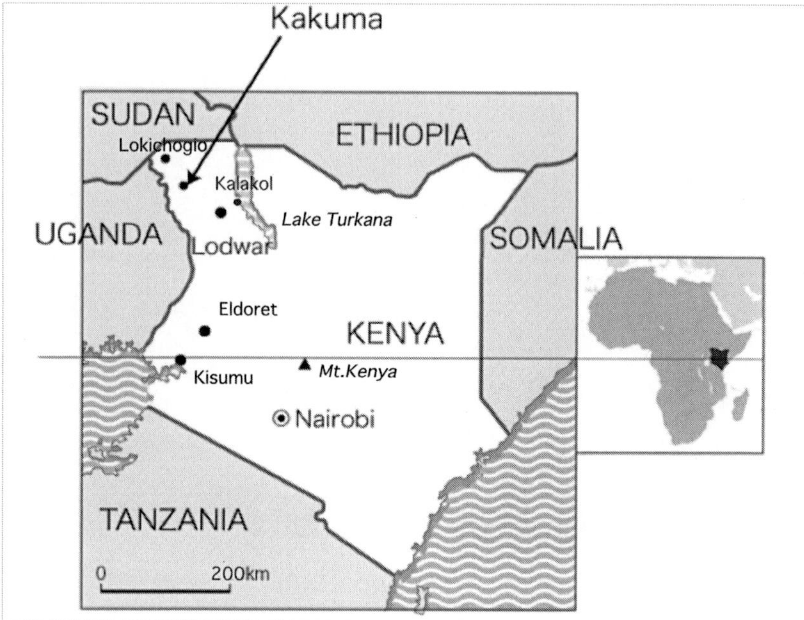

Fig. 3 Map of research field, Kakuma.

The Aftermath of the Droughts

The District suffered its first major drought in 1981. These major droughts are referred to as *Lopiar*, meaning "sweeping away all livestock". The social impacts of the *Lopiar* can be summarized by the following three points. The first is the tendency to migrate to and settle in townships because of famine, food aid, security, and medication. The second is the tendency to change staple food from milk products to maize grain (flour). This change has meant that the Turkana now depend on the foods available in the markets and on aid, which supposedly has deteriorated their health. They themselves recognize that increasing numbers of people are sick after *Lopiar*. The third is a tendency to leave behind reciprocal social relationships which acted as a safety net and support network when they suffered deprivation of their livestock (Gualiver, 1955; Spencer, 1997; Broch-Due, 1990, 1999). These tendencies have accelerated since the Kakuma Refugee Camp was established in 1992. At that time the Turkana people in the Kakuma area were confronted with an unprecedented situation: a huge influx of refugees from different ethnic groups, from Sudan, Somalia, Ethiopia, etc. In response, the UNHCR set up the refugee camp in order to accommodate the refugees. The camp population had grown to more than 50,000 by 2009. This swelled the numbers of people in the camp and in Kakuma to more than 100,000: a massive increase over the estimated population of 35,000 in Kakuma in 2005, and the mere 9,000 in 1999. The refugee camp created a huge market, which continues to attract the surrounding Turkana people (Ohta, 2005).

"The swept away"

Eight of thirty-one informant families gave up the pastoral life, settling near the township to ensure a supply of food and earn a wage. I shall present here the main life events of the Atukan family members as an example of the many people displaced by drought.

The life history of the Atukan family

1) Locodo in 1970: Locodo, later to become the wife of Atukan, lived a pastoral life with her ex-husband and five children around Lopuri, 20 kilometers north of Kakuma. Her husband died in 1970 from cholera. She refused a marriage to her brother-in-law and moved to her blood brother's land in Kalakol, facing Lake Turkana (*see* Fig. 3).

2) Atukan in 1975: Atukan, later to become the husband of Locodo, was a member of Locodo's ex-husband's family. They were personally acquainted. The family had remained in Lopuri after Locodo left, but moved to the outskirts of Kakuma in 1975 following the loss of all their livestock to disease. They planned to live by selling firewood and charcoal in Kakuma. Atukan remembered that an airplane dropped some bags of maize and the struggle that ensued among the people on the ground for them. The year 1975 is locally called "the year they could get a large quantity of oil because so many livestock died".

3) Atukan and Locodo in 1981: The famine forced the Atukan family to head for Kalakol where aid from "white men" was available. There, Atukan and Locodo met again and started a new life together; Atukan engaged in fishing, using a boat and net provided by the aid group, and Locodo made handicrafts such as hemp mats and bags and wooden sticks. The aid group backed them by buying their products.

4) Migrations in 1984 and 1992: They migrated to the periphery of Kakuma with livestock they had bought from savings made. They planned to earn a living from their livestock and to receive food distribution. They moved at least five times in this area, always in the vicinity of the food

distribution points. In 1992, they moved close to the refugee camp to seek work. Atukan worked as a blacksmith for local neighbours, sold firewood and provided physical labour for the camp.

5) Self-help in 1993: Following a dream that he could get cows in Uganda if he smeared his body with white soil, Atukan put together a traditional raiding party of six men. They found a lightly-guarded herd of cows belonging to the neighbouring ethnic group, the *Jie* of Uganda, and successfully captured over 40 cows. However, the *Jie* recaptured all of the cows. Atukan recognized he had a talent for divination and started as a diviner (casting sandals), although he hardly had any clients come and sometimes gave a divination to only a family member.

This family's history shows the vulnerability of pastoralists and the importance of towns for the rebuilding of lives. The pastoralists in droughts thought that they could survive in towns, with their markets selling livestock, firewood, and handicrafts as well as their food distribution stations. Hence towns were seen as a starting point for future life as well as an evacuation point against famine for the pastoral people who had lost their livestock (Little & Leslie, 1999). The year of the first *Lopiar*, 1981, not only swept away all the livestock, but also swept the pastoral people into the towns. The tendency for them to migrate to the Kakuma area was increased because of the refugee camp's establishment. The life history of Atukan family thus reflects the difficulties the Turkana people have commonly experienced since the onset of the droughts.

4. Pastoral-physical understanding

There are two types of illness recognized by the Turkana; illnesses of god and illnesses of humans. The former refers to illnesses for which the cause and mechanism are unknown (only God, *Akuj* knows), but remedies are available. The latter refers to illnesses for which the causes are known (witchcraft and sorcery), but there are no remedies. Our focus in this chapter is on the illnesses of God for which primary care is given.

"The human body is equal to livestock", the Turkana commonly say. They, in principle, think that illnesses are caused by abnormality on the pastoral-physical body constituting the folk anatomical and physiological knowledge derived from the experiences of livestock keeping. The first step in illness-coping is to localize and materialize an illness as an afflicted part.

Vessels full of blood

One day, I asked a research informant to show me the organs and tissues of a male goat he was dissecting. The men started cutting the throat and held the esophagus out of the way with their fingers in order to cut the white artery behind it; they knew well that the contents of the stomach would regurgitate if they cut the esophagus. The blood dripped out in container and they proceeded to dissect the animal while drinking the blood.

The informant kindly pointed out the organs and tissues. I asked the location of the *cirai* (main artery) and he sought it between spine and organs, picking it up to show me. The other informants explained that there was no blood when the animal was sick.

Towards the end of the dissection the informant asked me if I liked blood. He had noticed that I did not participate in the blood drinking, despite their recommendation. I replied in the negative. In reality, I do not like it, because I usually experience as if blood flow from the nose, after drinking. He tried to persuade me, saying that blood gives me a life. He also indicated his blood vessels in his arms and hands, proudly saying "The vessels did not have blood before drinking. Now they are full of blood. I am energized."

They not only understood and expressed the physical condition by using pastoral-physical terms, but also embodied it as lived experience. Moreover, the episode fully expresses that their knowledge is analogically based on the observation of their illnesses as well as on the bodily structure of livestock. They are used to eating the meat of sick livestock after removing the affected parts. As a result, they could make a correspondence between tangible, visible disorders and hidden internal lesions. This experimentalism enabled them to enrich the folk pathology. *Cirai* is a main artery, which the Turkana view as a significant organ for explaining the link between cause and the affected part, as exemplified in the following case.

Headache caused by blood in the womb

I took the wife of my informant to a healer as she was suffering from chronic malaria and severe headache. The healer looked at her general appearance and touched her abdomen, feeling a "rugged womb" (swelling of the uterus). Her diagnosis was that blood remaining in the woman's womb after childbirth had been carried via blood vessels to the head, causing the headache. She prescribed a herbal remedy to collect together and excrete the illness (i.e. a laxative regimen). Herbs in Turkana are commonly used as emetics and laxatives. The effects of decoctions of herbs are explained pastoral-physically: the herbs collect the illness into the large intestine and induce excretion from the mouth and anus.

Swollen spleen and liver

The diagnosis and treatment of *etid-loriwo* (a swollen spleen and liver) is another good illustration of the Turkana's understanding of illness based on the pastoral-physical body. In many instances the spleen and liver swell together as the blood overflows from both organs; the cause is unknown, but the Turkana know from dissections that both organs contain blood. The symptoms of this illness are languor and fever in the afternoon, and are treated by phlebotomy (blood-letting). The treatment I observed involved the practitioner locating the spleen and liver by touching the body, resetting the misaligned positions of the organs, and making several cuts in the skin over the swollen organs with a razor. He pointed out the blood loss, saying "The blood streams like a river. You can see the black sticky mud that's left after the clear water flowed away. This is the illness."

The basic agenda for the treatment was to locate the illness in the pastoral-physical body and remove the affected parts in the same manner as the affected parts are removed prior to eating diseased livestock. As the Turkana typically eat the meat of livestock that have died from disease after removing the affected parts, their dissections out of the meat in the field serve to refine the pastoral-physical knowledge. In other words, they are likely to identify illness itself in the affected body parts: thus the practitioner confirmed the existence of illness in the affected blood in the present example.

Pluralism of illness

The Turkana people take it for granted that the one illness masks other illnesses lurking in the body. They believe that certain illnesses occasionally came to the fore, and that if the treatment fails (for which no-one is to blame), the other masked illnesses will manifest themselves.

We will take one example: In the 1990s a young adult male returned from a mountainous area because the refugee camp had opened. Within a one–month period, he suffered from eye pain, headache, leg pain, and ulcers on the hands and feet. He claimed that the illness was made up of *ngakonen* (eye disease), *lokou* (severe headache expressed as "the brain is scrambled"), *amoding* (boils), s*ir* (sickness of the blood) and *lobai* (bone and joint sickness). The man demanded a sheep for the healing ritual. Later he was diagnosed at the hospital with a parasite syndrome.

Each of the illnesses has specific symptoms to be diagnosed and the interrelationship between the illnesses is not taken into account. This view of the distinctive nature of multiple illnesses contrasts with the syndrome concept in modern medicine.

5. Symptoms and Perceived Etiology

Spread of EAN

In 1997-1998 we conducted research with eighty-eight persons (thirty-one families, aged in their 20s-70s), asking them to remember the history of their illnesses. The results shown in the table reveal that EAN appeared in the 1980s and the number of cases has increased since the 1990s.

In 2004 we investigated the history of illness in 10 families composed of 149 persons (56 males, 82 females, 11 of unknown gender; 39 of whom were aged under ten years) to determine the number of EAN patients. The table shows that only two of the ten families did not report members having had EAN, and EAN accounted for 34 per cent of the 100 patients who reported any illness.

Table 1 Prevalence of illnesses eighty-eight informants experienced during 1970-1998. Triangle marks show the illnesses of which the number increased twice more than previous period (1980-1989). Two illnesses but EAN are psychological problems. Social change might worsen mental sates of inhabitants.

Illness	Symptoms/Year	1970-79	1980-1989	1990-1998	Sum
autro etau	loss of heart	6	20	□44	70
Kuwam	Epatiti	6	24	23	53
lomezekin	swelling of body	13	21	15	49
Lokou	severe headache	7	20	15	42
Sir	pain in blood vessels	8	15	18	41
etid-loriwo	swelling of spleen and liver	6	16	7	29
eramatau	wandering heart	2	3	□18	23
Sikina	fat of nipple	5	8	10	23
Agule	paralysis of legs (polio)	3	8	11	22
EAN	Constipation	0	3	□**17**	20
lomagali	pain in ribs	5	10	5	20
Lobai	pain in bones	1	7	7	15
Puuru	Measles	8	4	2	14
ngijyokio	mental disorders	1	0	3	4
Total episodes		71	159	195	405
Informants		63	83	88	
Average of episode		1.1	1.9	2.2	

Table 2 Number of EAN patients among ten families in 2004.

Family ID	Members	Patients	EAN patients
1	45	36	4
2	11	5	0
3	15	13	2
4	6	4	1
5	7	6	0
6	10	5	1
7	11	9	9
8	8	6	1
9	11	11	11
10	5	5	5
Total	129	100	34

Symptoms of EAN

The symptoms of EAN include constipation, diarrhea, difficulty defecating, bloody stool, difficulty swallowing, nausea, heartburn, difficulty breathing, chest pressure, heart and chest pains, cough, difficulty walking, and paralysis of the legs.

We can classify the symptoms into three types. The first type concerns digestive, intestinal symptoms such as constipation, difficulty defecating, diarrhea, and bloody stool. The second type concerns gastric symptoms like abdominal swelling, nausea, heartburn, and difficulty breathing. The third type is pain throughout the body and paralysis of the legs (a*gule*), of which they are very much afraid.

Three EAN patients explained their symptoms, course and cause of illness, and the interview data they provided is informative to delineate the sickness pattern.

Patient 1

Patient 1 worked for an international aid organization. In 1992 he showed malaria-like symptoms and had sticky, bloody stool. The stool became firmer owing to his antiviral medication. He recovered after a healing ritual and drinking soup followed by massage for one week, which helped to excrete the illness. The same symptoms recurred in 1995.

In 2001 he experienced symptoms of joint stiffness, painful ribs, hard things in the stomach, and abdominal swelling, and he started massage treatment. He was identified as having EAN. The following were considered as evidence of EAN: the pain was emanating and rising from the toes; he felt as if the pain were expanded to his legs. He feared that he would suffer from *agule*, paralysis of the legs. Patient 1's theory regarding his illness was that EAN is the father of other serious illnesses as well as *agule*. EAN is held to be contagious by most of the Turkana, particularly among those living in townships. They associate the onset of EAN among the population as the result of famine: food, if insufficient in quantity, becomes stuck in the large intestine.

Patient 2

Patient 2 was a 30-year-old woman. She felt as if her bowels' rising up just to the gorge and as if she gagged on certain foods. Therefore, she suspected EAN. Although she had not suffered much illness in her youth, she had suffered *agule* from an early age. She had two healing rituals done, following the massage.

Patient 3

Patient 3 was a 50-year-old man who experienced rib and abdominal pain in 1998. Someone identified EAN as the cause, where grains of maize had adhered to the wall of the large intestine and hard pieces of feces of maize grain had damaged the wall. The treatment recommended was 1) intake of digestive foods, particularly maize flour with fat and milk, and 2) massage followed by intake of herb soup to induce excretion. The intake of oil is also considered effective for treating intestinal wounds.

Patient 3 did not know the existence of EAN when he lived and worked at the livestock camps in his youth: it was not until he returned home in the 1980s because most of his livestock had been lost in the drought that he heard about it for the first time.

The features of EAN can be summarized as follows: 1) the illness emerged following the onset of the droughts of the 1980s, 2) the famine and intake of maize grain (flour) without fat and milk causes the illness, 3) the physiological mechanism involves injury to the intestinal wall by feces consisting of hard pieces of maize, 4) as the cause of many illnesses it is referred to as the father of illness, and 5) it is an illness of townships (settled areas), not of the mountains.

6. Analysis of Massage

Let us move to the massage of the body of a person with EAN, an event which involves the practitioner, patient and audience. Twenty-one massage sessions performed by 17 practitioners were studied in 2004-2005 at Kakuma and three other towns (Logichogio, Lodwar, and Lokichar). The massage sessions were observed and video-recorded, and interviews with the practitioners, patients, and audience were conducted at the events. The research focus was the correspondence between practice and interpretation.

The practitioner, K, was a 50-year-old woman. She had migrated to Kakuma from the mountains, giving up a pastoral life in 1995. She had been practising massage since 1993. The patient was a 40-year-old man who presented the following symptoms: difficulty defecating, lower abdominal pain, painful ribs and hips, upper limb pain, and difficulty walking. He was warned that there was a risk of leg paralysis (*agule*).

Massage trails and thrust

In order to ascertain the complicated movements of the practitioners' hands during massage, we analyzed the general trails of the hands. Moreover, we differentiated between thrust massage and other massage techniques. In thrust massage, the hands stop or progress slowly to the specific affected spot, then the practitioner makes a thrusting movement and vibrates the affected spot with the fingers (see Fig.4 and 5).

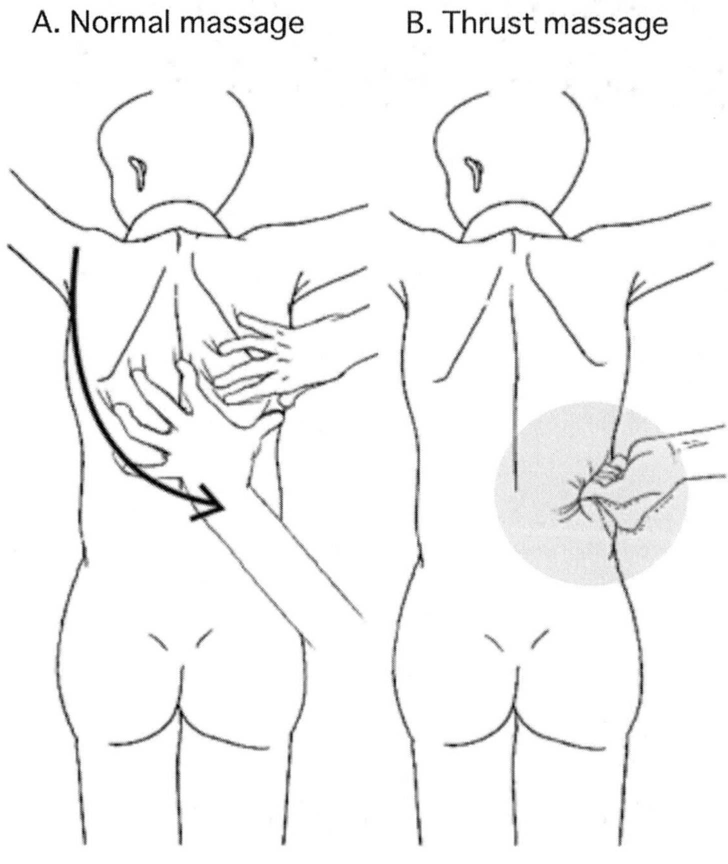

Fig. 4 Two types of massage; thrust massage thrusts and vibrates fingers in affected parts, intending to work on organs and tissues in the depth of body.

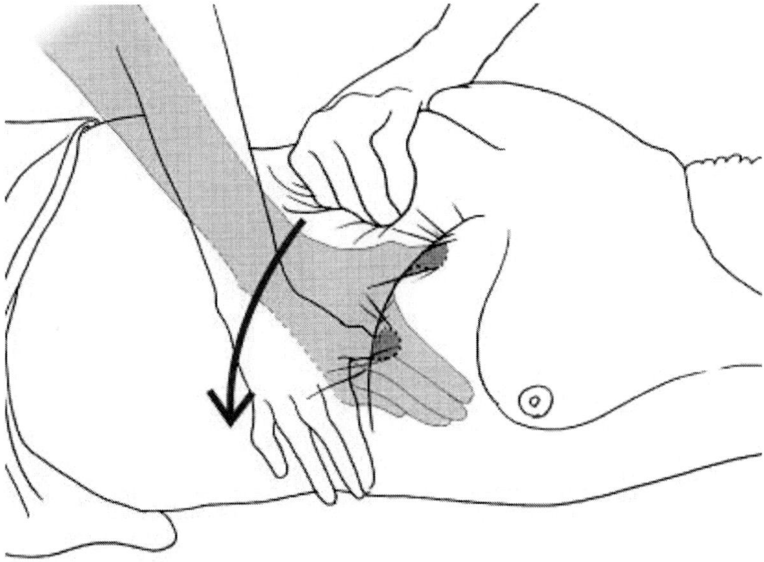

Fig. 5 Thrust massage in abdomen; fingers of right hand of massager, thumb specifically, thrust under left ribs of patient and move slowly to flank, vibrating. Patient cries in pain.

The practitioner uses thrust massage to work on affected parts that are located deeper inside the body.

An example of a massage healing session is given below, in which the practitioner focused on the abdomen and was sensitive to the physical reaction of the patient's body. The massage necessarily routinely starts with the patient lying in a prone position. The start from this position was significantly related with the pastoral-physical understanding of EAN.

Prone Position

The first four trails of massage in the session (*see* Fig. 6) are described and subsequently explained below and shown.

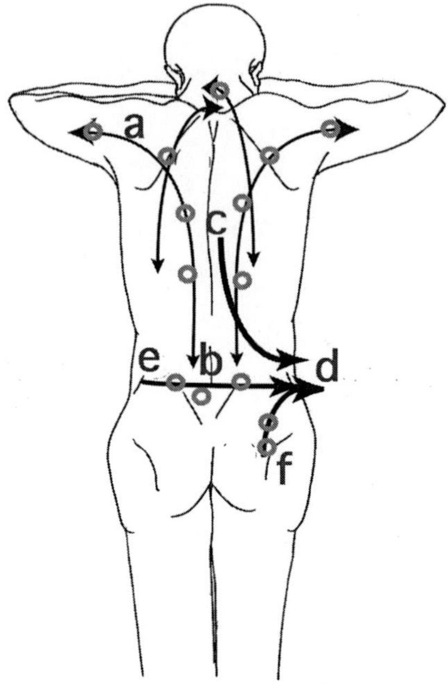

Fig. 6 Trails of massage in prone position. Small circles are "hard round things".

Fig. 7 Massage started from lower back upward to ribs, shoulder blades and neck. The massaged part gleamed wetly with oil.

Trail a-b

The practitioner massaged the lower back upwards along the spine to the neck, extended out to the shoulders and ribs, then finished in the reverse direction (Fig.7).

Trail c-d

She then moved her hands down from back to both sides of the flank, thrusting the right flank (Fig.8). The patient gasped with pain.

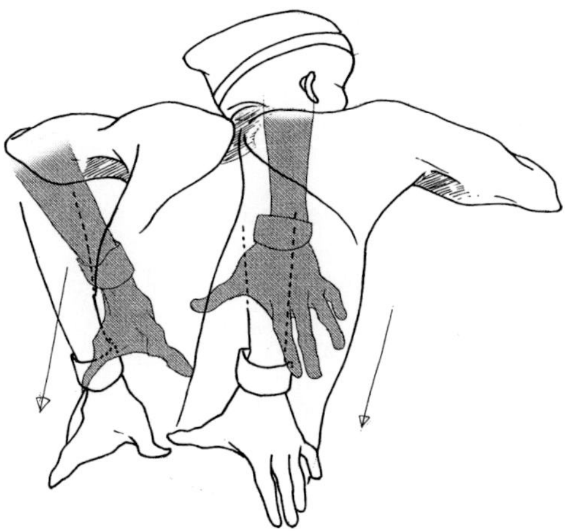

Fig. 8 Practioner gave thrust-massage to flank, massaging along spine down to lower back.

Trail e-d

The practitioner massaged the lower back (loin), thrust-massaging a few areas and ended at the right flank with a thrust. The patient gasped with pain.

Trail f-d

She then massaged from the right buttock to the flank, thrust-massaging several areas. The patient cried, tapping his hand on the floor with pain.

The massage steps shown in Fig.6, f-d were performed to collect hard round things scattered throughout the upper body and return them to the bowels. A woman in the audience pointed out several sites on the back, ribs, spine, and right hip and buttock, saying "There are many hard round things here and there" (circles in Fig.6). The hard things (illnesses or the origin of pain) had originally "flowed" from the large intestine through the blood vessels to the rest of the body, while hard pieces of feces clung to the intestinal wall.

The woman also pointed out the hard round things on the hip and buttock, recommending that I confirm their stiffness. I then traced them towards the flank, tapping with my fingers. I identified them. The patient's body was treated as if it were a shared body. The patient was highly concerned about the risk of *agule*. The woman's gestures showed that the illness spread out from the flanks to the hips and legs; the massage traced the "infection route" in the reverse direction. The hard round things in the bowels as well as in the hip and buttock were haptically moved. She called attention to the sounds of moving bowels (bowel peristalsis) when she massaged from the buttock to the right flank. She took sounds as evidence of the return of the hard round things to the bowels.

Supine position

We will next discuss the trails shown in Figure 9.

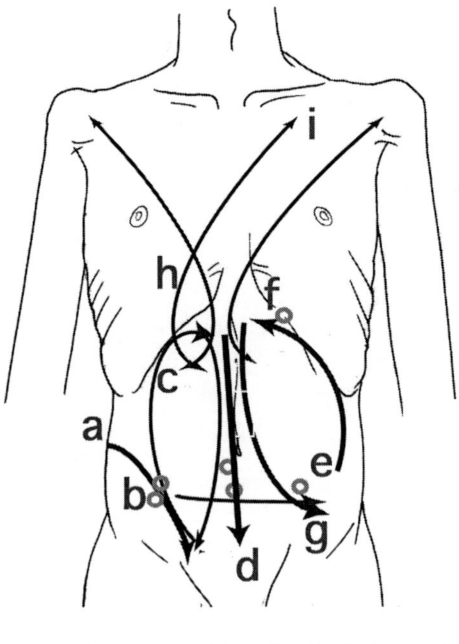

Fig. 9 Trails of massage in prone position. Small circles are "hard round things" and affected parts thrust-massaged.

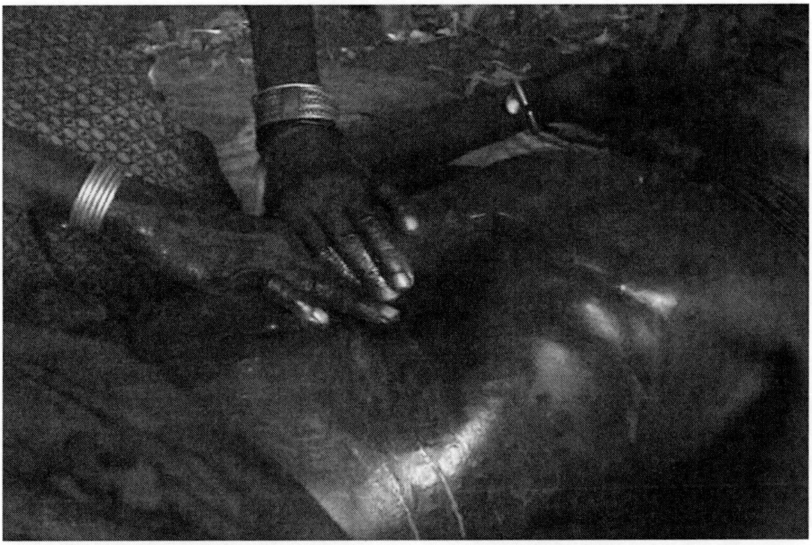

Fig. 10 Massager thrust-massaged from pit of stomach to navel and lower abdomen along right ribs.

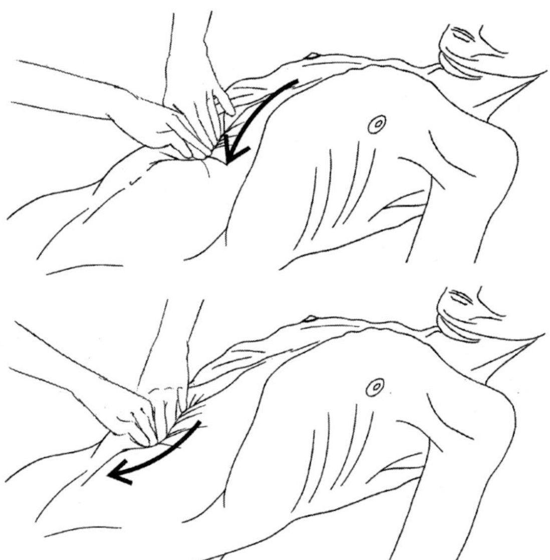

Fig. 11 The thrust massage from pit of stomach to navel in illustration.

Trail a-b

The practitioner massaged and thrust the right flank, then descended to the lower abdomen where fingers touched hard things (circles in Fig.9, b). She thrust and vibrated them.

Trail b-c-d

She then massaged up to the pit of the stomach along the edge of the right ribcage before descending to the lower abdomen in which she felt hard things. She thrust and vibrated the pit of the stomach through the navel to the lower abdomen (circles in Fig.9, d).

Trail e-f-g

The practitioner massaged from the left flank along the left rib cage up to the pit of the stomach, and then descended to the lower abdomen via the navel. Her fingers touched hard things at the edge of the left rib cage (e) and lower abdomen. With her finger she indicated e as *Korian* (hard pieces of feces) and said "Here and there, we can touch inside" while touching the wider area of ribs. Her fingers descended from the pit of thestomach to the lower abdomen, where she thrust-massaged in Fig.9, g, saying "Let them go out".

Trail h-i

She massaged downwards from the collarbone, shoulders, arms, and ribs to the pit of the stomach (in Fig.9,h). She massaged from the left shoulder to the right ribs across the chest, saying "Collect to abdomen" (in Fig.9, i).

The trail of the right flank to the lower abdomen (trail a-b) was to return the collected illness to the abdomen in the large intestine. The trail from the pit of the stomach to the lower abdomen (trails c-d) was intended to drag down the bowels which had risen up and create a space again between the heart and bowels. The stomach had been pushed upwards by the bowels rising, ultimately making both organs press against the heart. The hard things (hard pieces of feces, in Fig.9, f) in the large intestine were identified at the left rib cage and were evidence of swollen bowels. At the same time "something was boiling up" in the stomach when pressed and it descended with the sound of *chyu* (sounds of cooling boiled water) if massaged.

The thrust-massage to displace hard pieces of feces was performed to excrete softened feces and to make twisted bowels straight. The large intestine had swollen to become rugged.

The trail from the upper limbs, neck, shoulders, armpits and ribs to the pit of the stomach (trail h-i) was intended to return the pain scattered via the blood vessels to the abdomen.

The massage in the prone and supine positions described above constituted one sequence. The basic purpose of the massage was to collect and return to the large intestine the illnesses that had been spread throughout the body by the blood vessels and tissues. The flanks were gateways linking the back with the large intestine, which is why the massage always starts from the back and finishes at the abdomen.

7. The body with EAN

Let us now discuss the pathogenic process of EAN as explained from the pastoral-physical standpoint, using the informants' words. The findings from the other massage scenes can be summarized in the following.

EAN as a syndrome

1) EAN emerged as an illness with the change from the staple food of milk products to that of maize that was caused by the severe droughts.

2) The insufficient supply and quality of maize resulted in pieces of feces becoming hard in the large intestine and adhering to the intestinal wall. As a result, the large intestine became swollen and painful. Detachment of the hard pieces of feces injured the large intestine, causing blood to appear in the stool.

3) The large intestine became dry owing to the lack of fat and milk in the diet, causing the stomach to roll up; expressed in terms of "a dried up strip of leather". The stomach was compressed by the large intestine, causing nausea and breathing difficulties, just like "foam rising in a simmering pot." The large intestine and stomach rose up together towards the heart into the pit of the stomach, touching some of the ribs. The compression of the heart caused heart and chest pains and cough. Informants said there was no space between the heart and bowels or that he or she felt the intestine rise up to the gorge. Moreover, the swollen large intestine directly touched the womb and bladder causing *apot* (infertility) and difficulty urinating.

The swollen part of the large intestine transmitted pain in two directions: towards the upper half of the body and towards the lower half.

4) It transmitted pain via the main artery and other blood vessels to the upper half of the body (hip, waist, back, shoulders and shoulder blades,

arms, neck, temples, and head). These affected parts had been categorized as symptoms of other discrete illnesses before the emergence of EAN.

5) It transmitted pain to the waist and legs through blood vessels or by direct contact between tissues. The illness in the affected parts might progress to *agule*, paralysis of legs.

The causes of various symptoms and illnesses were attributed to EAN or specifically to the large intestine affected by EAN. Thus, it was possible to explain all symptoms and illnesses with one cause. It follows from what has been observed that EAN was the first illness in Turkana to be explained by the syndrome concept. It is speculated that the health condition of the Turkana people worsened so much that the pluralism concept could no longer suffice.

In the last analysis, two important accounts are related, drawn from the case of practitioner K. It was useful to facilitate discussions about memory and transmittance of culture.

Flesh in memory

In the interview with practitioner K after the massage session, she complained: "The low quality maize cooked without fat and milk is the cause of EAN … we lost our livestock, so nobody visited or stayed here … because guests would eat nothing but the worst quality maize.... Who visits a person who has lost their livestock?" Both the practitioner and the patient then demanded a sheep for a healing ritual from me.

Practitioner K expressed a sense of dropout from the reciprocal social support network as well as loss of livestock. The practitioner and the patient as "the swept away" shared a common experience of droughts, which elaborated the EAN body. It became clear that they not only massaged the body, but also materialized the lived experience as flesh in memory.

Discovery and contamination

Practitioner K lived with her family in a mountainous area, keeping a small number of livestock. She had experienced receiving massage for *Agule* in her childhood (1960s). Her family decided to migrate to Kakuma in 1992 hoping that her son would find employment in the refugee camp's shops.

She performed her first massage in 1992 for a female patient in Kakuma. As soon as she started the massage, she not only recognized it as the same illness she had suffered in her childhood but also noticed a difference: the hard things in the abdomen gradually disappeared as the massage proceeded. The second opportunity for her to perform massage was for her husband who showed symptoms of abdominal swelling and bloody stool. She performed the same massage treatment as for the female patient she had treated. She again found that the hard pieces of feces disappeared as he recovered. She learned that it was EAN from a neighbour. Her third patient was her son's boss, a Sudanese. She identified his EAN and performed massage, receiving money and maize as payment. She then began to take on patients as a practitioner.

Practitioner K's story is a good example of how practitioners often come to know of EAN. Taking other cases into consideration, we identified three routes to this realization: 1) women found out about EAN by themselves, 2) women who knew other practitioners found out about EAN from them, as in the case of practitioner K, and 3) women who experienced EAN massage as a patient started as a practitioner: patients who acquired the EAN body might be prospective practitioners. In sum, the EAN body is culturally transmitted through contact between bodies as if it were contagious. At the same time, the EAN body transforms the relationship of the self to the world for the patient. In other words, the intercorporeality (Csordas, 2005) might form among bodies, so that the Turkana people recognized themselves as afflicted victims of droughts.

8. Conclusions

Turkana society has accumulated and developed a pastoral-physical understanding of the body to cope with illnesses, in place of cosmology in which supernatural agents work or socio-political discourses in which economic discrepancy is blamed. Their pastoral-physical knowledge is mainly based on their experiences with livestock, as we can find in their saying "the human body is equal to livestock." The Turkana people were committed to materialize illness in the pastoral-physical bodily structure.

EAN is defined as an illness of the large intestine, caused by the change of staple food from milk to maize cooked without fat and milk. The swollen large intestine affects all parts of the body via blood vessels and tissues, because of hard pieces of feces adhering to the intestinal wall. Thus, EAN is referred to as the "father of illness."

We arrive at the following implications. : First, the Turkana not only located and identified the symptoms of newly emerged EAN in the existing pastoral-physical body, but also reorganized the pastoral-physical body to adjust EAN into a new body that has a specific connection between the blood vessels, organs, and tissues. In other words, the

Turkana people adapted to various hardships brought about by successive droughts in terms of the bodily transformation, and this entailed an elaboration of the syndrome concept and its underpinning bodily structure. Second, the pieces of hard feces functioned in their narratives as metonymy for the droughts and hardships they have experienced: indigestible maize, food without fat and milk, loss of livestock, disruption of the reciprocal social support network, and migration into townships. We therefore concluded that the emergence of EAN was the result of the embodiment of droughts, or the Turkana people's efforts to bring some order to the social change.

The significance is that both the understanding of EAN as a syndrome and the flesh-on-memory of droughts were made by "hands and eyes" of massage practitioners. Through their cooperation with the patients, they identified and created it by seeking out the affected parts in organs and tissues, sensing and listening to the reactions of the touched body, based on knowledge of the pastoral-physical body.

The massage practitioners elaborated the EAN body through massage, in the manner of an artisan making sculptures out of mud. They listened to the body's voices in the bowel sounds and touched its gestures in the movements of the bowels. They touched pulses in the blood vessels and kneaded stiffness in the muscles and bowels, tracing the infectious route. They took the patient's crying and gasping as the body's reply. Bodies replied, murmured, and made statements: the communication of the practitioners and the bodies (patients) through massage was vital. EAN has highlighted for us not only the agency and intentionality of the body, but also its resourcefulness in the sense that it is a flexible instrument of developing knowledge, on which they remember droughts, and which identifies the self and the world.

References

Broch–Due, V. (1990). "Livestock Speak Louder than Sweet Words: Changing Property and Gender Relations among the Turkana." *Poverty and People: Changing Rights in Property and Problems of Pastoral*: 147: 163. Baxter P.T.V. (ed.by),*Development* edited Manchester: Department of Anthropology, University of Manchester.

—. 1999. "Remembered Cattle, Forgotten People: The Morality of Exchange & the Exclusion of the Turkana Poor." Pp. 50-88 in *The poor are not us: Poverty & Pastoralism in Eastern Africa* edited by D.M. Anderson, D.M. and V. Broch–Due. Ohio University Press.

Comaroff, J. and Comaroff J. (1992), *Ethnography and the Historical Imagination*. Boulder: Westview Press.

Csordas, T. J. (1990). "Embodiment as a Paradigm for Anthropology: The 1988 Stirling Award Essay." Pp. 5-47, *Ethos* 18: 5–47.

—. (1994a). "Words from the Holy People: a Case Study inCultural Phenomenology." Pp. 269-290 in *Embodiment and Experience:The Existential Ground of Culture and Self* edited by T. J. Csordas. Cambridge:Cambridge University Press: 269: 290.

—. (1994b). "Introduction: The Body As Representation and Being in-the-world." Pp.1-24 in *Embodiment and Experience: The Existential Ground of Culture and Self* edited by T. J. Csordas. Cambridge: Cambridge University Press: 1–24.

—. (1999). "Embodiment and Cultural Phenomenology." in *Perspective on Embodiment: The Intersections of Nature and Culture* edited by GAIL, W. and HONI, F.H. London: Routledge: 143–162.

—. (2005). "Intersubjectivity and Intercorporeality." Pp. 18-30 in *Construction and Distribution of Body Resources: Correlations between Ecological, Symbolic and Medical Systems* edited by Sugawara, K. Tokyo: RILCAA, Tokyo University of Foreign Studies: 18: 30.

Farmer, P. 1992. *AIDS and Accusation: Haiti and the Geography of Blame*. Chicago: University of Chicago Press.

Gulliver, P. H. (1955). *The Family Herds*. London: Routledge & Kegan Paul.

Jenkins, J. H. and Valente, M. (1994). "Bodily Transactions of the Passions:El Calor among Salvadoran Women Refeugees." In Csordas, T.J. (ed.by) *Embodiment and Experience: The Existential Ground of Culture and Self*, Cambridge: Cambridge University Press: 163: 182.

Kawaj , K. (2008). "The Epistemology and Ontology of the Chamus in
Kenya:The Human Body as Nature." *African Study Monographs* 29(3): 119–131.

KawajI, K. (1998). *Pastoralist's View of Body and Therapeutic Practice* (in Japanese). Tokyo: University of Tokyo Press

Kleinman, A. (1980). *Patients and Healers in the Context of Culture: An Exploration of the Borderland between Anthropology, Medicine, and Psychiatry*. Berkeley: University of California Press.

Little, A. M. and W. P. Leslie. (1999). *Turkana Herders of the Dry Savanna: Ecology and Biobehavioral Response of Nomads to an Uncertain Environment.* New York: Oxford University Press.

Maccabe, J.T. (2007). *Cattle Bring Us to Our Enemies: Turkana Ecology, Politics, and Raiding in a Disequilibrium System.* The University of Michigan Press.

Nichiter, M. (1992). “Of ticks, Kings, spirits, and the promise of vaccines.” Pp. 224-256 in *Paths to Asian Medical Knowledge* edited by C. Leslie and A. Young, Berkeley: University of California Press: 224–256.

Ohia, I. (2005). “Multiple Socio-Economic Relationships Improvised between the Turkana and Refugees in Kakuma Area, Northwestern Kenya.” Pp. 315-337 in *Displacement Risks in Africa: Refugees, Resettlers and their Host Population* edited by I. Ohta and Y. D. Gebre. Kyoto: Kyoto University Press and Transpacific Press.

Roseman, M. (2000). *Healing Sounds from the Malaysian Rainforest: Temiar music and medicine*. Berkeley: The University of California Press.

—. (2002) “Making Sense out of Modernity.” Pp. 111-140 in *New Horizons in Medical Anthropology: Essays in honor of Charles Leslie* edited by M. NICHTER and M. LOCK. London: Routledge 11:140.

Spencer, P. (1997). *The Pastoral Continuum: The Marginalization of Tradition in East Africa.* New York: Oxford University Press.

Sugawara, K. (2005). “Introduction: Toward The Anthropological Understanding of Body Resources.” Pp.18-30 in *Construction and Distribution of Body Resources: Correlations between Ecological, Symbolic and Medical Systems* edited by Sugawara, K. Tokyo: Rilcaa, Tokyo University of Foreign Studies.

Weiss, G. and H. F. HABER. (1999). “Introduction.” Pp.143-162 in *Perspective on Embodiment: The Intersections of Nature and Culture* edited by G. Weiss G. and H.F. Haber. H.F. London: Routledge.

CHAPTER TWELVE

AN INVITATION TO EMPOWERMENT-ORIENTED NEUROSOCIOLOGY: A CONTRIBUTION TO INDIVIDUAL/SOCIETY AND THEORY/PRACTICE CHALLENGES

ALIREZA MOULA

Abstract

Wieviorka, Joas and Björnberg the leading organizers of the international congress of sociologists, 2010 wrote: "Determinism is dead in the social sciences… The world changes and this change, to a large extent, depends on human action and imagination. If sociology is to be useful, it has to contribute to an understanding of change – and it has to change itself." The aim of this chapter is to encourage dialogue between different voices of the individual/society and theory/practice disputes by presenting a new perspective, neurosociology. Instead of arguing for the significance of one or the other voice that already exists *within* sociology, the "imaginative voice" of empowerment-oriented neurosociology can contribute to the solution of these longstanding challenges and hopes to lead to dialogue with the "critical voices". The necessary building blocks of such a neurosociology and its application through a person-in-environment formula and an empowerment model will be presented. The chapter proceeds in the following order: (1) what is sociology's and neuroscience's main concerns, (2) what is, or what could be considered as neurosociology, (3) the four necessary cornerstones of an empowerment-oriented neurosociology, and (4) a concluding section. The writer concludes that neuro-sociology can study individuals' interactions with their environments. Empowerment will take place by a kind of social intervention based on the biopsychosocial knowledge of human beings. With the help of the knowledge about our brains, we can cognitively empower individuals toward intelligent interaction with their environments.

Key words: sociology, neuroscience, neurosociology, brain, empowerment, RAHYAB.

Introduction

> A number of philosophers and scientists have been attracted to a picture of science as a dialogue between an imaginative voice and a critical voice… In contrast to Popper, Hull argues that there is no need for individual scientists to take a cautious and skeptical attitude toward their own work; others will do this for them (Godfrey-Smith, 2003, p. 165).

Berger and Kellner (1982) in *Sociology reinterpreted* encourage sociologists to return to the big questions. They ask for "revitalization of sociology" meaning "a revitalization of a certain perspective, of a way of seeing the world." Sociologists should retain some ability for cross-cultural, global comparisons. The "calling of sociologist is by its very nature a cosmopolitan one." At the same time, these two sociologists raise the issue of "makability of the world" and add that the world "could be different from what it is" (pp. 13-17). Wieviorka, Joas and Björnberg (Website of International Congress of Sociology, 2010) write:

> Determinism is dead in the social sciences. Despite a strong interest in social structures, social mechanisms, forms of reproduction, we are all aware that human beings are not completely dominated by them. The world changes and this change, to a large extent, depends on human action and imagination. If sociology is to be useful, it has to contribute to an understanding of change – and it has to change itself. It is on the move and has to be on the move because the world, the societies, collective and individual actors are on the move. "Sociology on the move" means that our discipline contributes to an understanding of our world by defining new objects of research, devising new approaches and reevaluating its rich heritage. It implies a new openness with regard to other disciplines and to normative questions.

The aim of this paper is to encourage dialogue between different voices of the individual/society and theory/practice disputes by presenting a new perspective, neurosociology. Instead of arguing for the significance of one or the other voice that already exists *within* sociology, the "imaginative voice" of empowerment-oriented neurosociology can contribute to the solution of these longstanding challenges and hopes to lead to dialogue with the "critical voices". The necessary building blocks of such a neurosociology and its application through a formula and a model will be presented. This paper proceeds in the following order: (1) what is sociology's and neuroscience's main concerns, (2) what is, or what could be considered as neurosociology, (3) the four necessary cornerstones of an empowerment-oriented neurosociology, and (4) conclusion.

1) Sociology's and neuroscience's main concerns

Tom Bottomore and Robert Nisbet (eds 1978), two sociologists at opposite ends of the sociological spectrum (Bottomore presents "Marxism and sociology" and Nisbet presents "conservatism"), in their joint statement about what produced sociology, refer to the importance of the concept of 'society' that was elaborated in analyses of social structures, social systems, and social institutions. Even Peter Berger (1963) writes that society as a term refers to the object par excellence of the discipline of sociology.

The task of neuroscience is to explain behavior in terms of the activities of the brain. The ultimate challenge is to comprehend the biological basis of consciousness and the mental processes. What we commonly call the mind is regarded as a set of operations carried out by the brain. The actions of the brain underlie not only motor behaviors such as walking or eating, but complex cognitive actions that we believe are quintessentially human, speaking and creating works of art. Neuroscientists want to understand how the brain marshals its millions of individual nerve cells to produce behavior, and how these cells are influenced by the environment, which includes the actions of other people. (Kandel, 2000, p. 5). Neuroscience spans a wide range of research endeavors from the molecular biology of nerve cells to the biological basis of normal and disordered behavior, emotion, and cognition. Neuroscience is multidisciplinary: neuroanatomists study brain's shape, its cellular structure, and its circuitry; neurochemists study the brain's chemical composition, its lipids and proteins; neurophysiologists study the brain's bioelectric properties; and neuropsychologists investigate the organization and neural substrates of behavior and cognition (Bloom, 2008, p. 3).

One branch of neuroscience is social neuroscience that addresses fundamental questions about the mind and its dynamic interaction with the biological systems of the brain and the social world in which it resides. This field studies the relationship between neural and social processes (Cacioppo and Bernston, 2002).

2. Neurosociology as a growing fetus in the uterus

We should not confuse social neuroscience with neurosociology. As mentioned before, the first term is part of neuroscience and is recognized by this community. What is (or since its existence is not widely recognized, perhaps it is better to ask, what could be regarded as) neurosociology? We can hardly shape one single comprehensive definition of this concept. We do not have one single way of defining sociology, why should we have a single definition for neurosociology. To use a metaphor, sociology can be regarded as the mother discipline and neurosociology can be regarded as a growing fetus in the mother's uterus; it is not even a newborn infant. So we should be humble, open-minded and creative in what could be considered to be neurosociology.

It is this author's opinion that neuroscience's contribution to sociology should make a difference and add something new to this discipline within the social sciences. Since sociologists usually focus on society, social structures, or institutions, neuroscience can assist sociology in building a strong knowledge base by focusing on individuals. Society would be studied as individuals—owners of brains—in interaction with their (physical and social) environments. People can be considered as unique individuals with wishes, dreams, thoughts, emotions, strengths, problems etc., and we may attempt to look for social developments and changes in a manner where individuals play the vital role in these events. Such changes in political language are understood to be democratic and bottom-up changes. However, we can contribute to this widely known perspective of change so that it can acquire scientific status as well. This is in line with American pragmatists' desires to use science as the medium of change for meliorism. *Neurosociology's cardinal goal would be to contribute to the theory and practice of a balanced development between individual interests and the character of our changing civilization in a winner-winner manner.* This brings us close to a sociological tradition that includes great social thinkers such as George Simmel and the founding fathers of symbolic interactionism, known also as early interactionists who include pragmatists such as William James, John Dewey, G. H. Mead, and J. H. Tufts. The first thing that we learn about these social philosophers is that they are concerned about "individuality" (Simmel) or the "agency" of individuals (pragmatists). For example, Simmel wrote:

> The deepest problems of modern life flow from the attempt of the individual to maintain the independence and individuality of his existence against the sovereign powers of society, against the weight of the historical heritage and the external culture and technique of life. The antagonism represents the most modern form of the conflict which primitive man must carry on with nature for his own bodily existence. The eighteenth century may have called for liberation from all the ties which grew up historically in politics, in religion, in morality and in economics in order to permit the original natural virtue of man, which is equal in everyone, to develop without inhibition; the nineteenth century may have sought to promote, in addition to man's freedom, his individuality (which is connected with the division of labor) and his achievements which make him unique and

> indispensable but which at the same time make him so much the more dependent on the complementary activity of others; Nietzsche may have seen the relentless struggle of the individual as the prerequisite for his full development, while socialism found the same thing in the suppression of all competition - but in each of these the same fundamental motive was at work, namely the resistance of the individual to being leveled, swallowed up in the social-technological mechanism.

John Dewey (1897/1997, pp. 297-298) had the same worries about individuals when he raised the following questions:

> [T]he problem of modern life is that of reconstruction, reform, reorganization. The entire content of experience needs to be passed through the alembic of individual agency and realization. The individual is to be the bearer of civilization; but this involves a remaking of the civilization which he bears. Thus we have the dual question: How can individual become the organ of corporate action? How can he make over the truth authoritatively embodied in institutions of church and state into frank, healthy, and direct expression of the simple act of free living? On the other hand, how can civilization preserve its own integral value and import when subordinated to the agency of the individual instead of exercising supreme sway over him?

As we see, there has been a strong interest in individuality and agency among classical social thinkers and neuroscience can support this sociological interest. It is safe to suggest that on an individual/society continuum, neuroscientific studies are closer to the individual and sociological studies are closer to society.

Another difference between neuroscience and sociology concerns the types of practices they are engaged in. While neuroscience's development and strengths depend strongly on laboratory research and clinical practice, sociology's development and strengths depend, to a large extent, on theoretical practice. Theoretical papers, not based on empirical research, are widely published in many sociological journals. In neuroscience, "review articles" exist but they do not have the prestige that research-based original articles have. With the help of neuroscience, neurosociology can become more action-oriented and interventionist. Neuroscientist Paul Zak imagines that neuroscientific principles will be used to design better public policies; to create better human environments in homes, schools, and workplaces; and also to help find more ways to naturally increase beneficial, pro-social behavior while reducing destructive antisocial tendencies (in Lynch, 2009, p. 99). In other words, neuroscientific discoveries need not be used only in laboratory and clinical environments but even in larger society. In sum, a strong alternative for neurosociology could be an action-oriented sociology that is concerned with people's individuality and agency and aims to empower people.

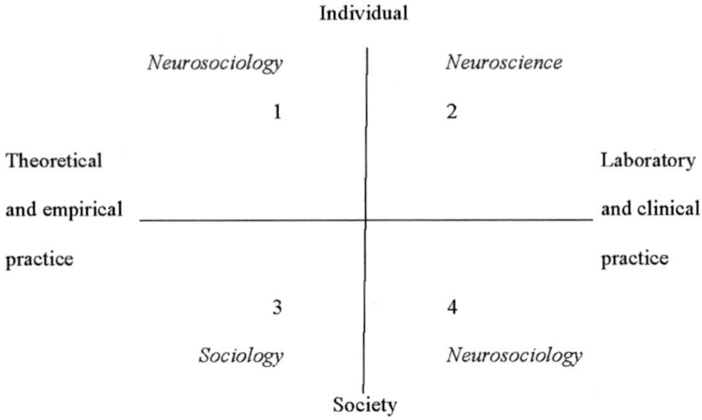

Figure 1 shows a two-dimensional matrix framework where sociology, neuroscience and neurosociology *mainly* exist. On the vertical axis, individual is depicted at one end, and society at the other. The horizontal axis ranges from theoretical & empirical practices to laboratory and clinical practices.

This classification should be understood with some reservation, for example, sociology, or rather microsociology, can exist in quadrant 1 as well. There are even "theoretical neuroscientists" who may be placed in more than one of these four quadrants.

Neuropsychologist Jerome Kagan (2007) writes that scientists who want to choose among alternative conceptions of a phenomenon usually select one of two strategies: (a) to trust the imagination and through disciplined thought detect the initial form of the answers or (b) to begin with a reliable phenomenon whose causes, consequences, and contextual constraints can be probed. In terms of the second option, only after relevant evidence has been gathered can scientists invent concepts that summarize what has been observed. The choice then is between an attractive idea and a plain but reliable fact.

"Science is the pursuit of knowledge at the frontier; along the boundaries between the known and the unknown." (Lynch & Granger, 2008, p. 180) It is possible to be inspired by the two aforementioned strategies and *on the basis of a plain but reliable neuroscientific fact, construct attractive but preliminary propositions and test them in practice in a large social context.* Testing such propositions would be an important strategy for neurosociology. A child (read a child's brain) may learn something in a controlled laboratory environment, but can this child learn the same thing outside a laboratory environment, for example in a playgroup of several children? Here is where neurosociology can contribute greatly. Toward this end, we can be helped by Lyng and Franks' (2002) typology of action. Accordingly, action can move across five levels; (1) on the lowest level of action, an individual thinks about something and uses her/his cognitive capacities and the power of imagination to construct an idea, (2) at the second level, the individual writes down these ideas and develops her/his thoughts and imagination through the help of writing, (3) at the next level, the individual talks to someone about her/his thought and shares her/his imagination, (4) at this level the individual "brings" her/his thoughts into an environment and tries to find out what happens to them, (5) finally, at this stage, the individual decides to bring her/his limited test into the larger social context and put it into "praxis".

Through the help of a neuroscientific fact, a neurosociologist constructs a proposition that has significance for all individuals and she/he develops this idea further by writing it down (level 1 & 2 of the typology of action, quadrant 1 of Figure 1). To move further, the constructed proposition will be discussed with one or more neuroscientists and after getting feedback, it will be put to practice in a few cases, usually known as a pilot study (level 3 &4 of typology of action and quadrant 2). If the result is promising, the proposition will be put to test at a larger level (level 5 of typology and quadrant 3). If the new attempt is satisfactory, further moves are made to form a theory from the primary proposition (I actually add a level 6 to the typology of action, quadrant 4). As we see, neurosociology moves in all quadrants and through this circular movement, it gains inspiration from both neuroscience and sociology. Simultaneously, through this movement, neurosociology can affect neuroscience and make it more social while affecting sociology and making it more individual-based. Although neurosociology moves in all four quadrants, becoming empowered by neuroscience and sociology and creating a bridge between these two disciplines, its ideal places are quadrants 1 and 3. It is in these two quadrants that neurosociology can have its own identity, even if this identity critically depends on both sociology and neuroscience. This idea is neither strange nor completely imaginary (I will give some concrete examples in section 3.4). This is happening in a new area that is called "neuroeducation". There is a newly formed society called Mind, Brain and Education Society with a journal of the same name. Their aim is to bring the basic facts of neuroscience into schools and make education more brain-sensitive. Sociologists can do the same thing and integrate neuroscience with many branches of sociology such as sociology of childhood, sociology of youth (for an example see Moula, Timpka and Puddephatt, 2009), sociology of aging, sociology of education, sociology of emotions, sociology of (mental) health and illnesses, sociology of science, clinical sociology, etc. The table below shows how the integration of neuroscience and sociology can be useful for many branches of sociology.

Neuroscience	**Sociology**
If one were to run a "disease of the decade" contest, ADD and ADHD would be among the plausible contenders (Goldberg, 2009, p. 200). While the underlying causes of ADHD have been debated in the past, recent advances in neuro-imaging, as well as positive responses to medication, have provided strong support for a neural basis for this disorder (Anderson, 2008).	According to Goldberg (ibid), parents actively seek the ADD diagnosis for their children to explain their school failures, and adults seek it for themselves to explain their life failures. He means that ADD and ADHD have now become a social phenomenon. He encourages sociological considerations that can explain how a neurological diagnosis becomes a social phenomenon.
There is strong evidence that the prefrontal cortex continues to grow even in the third decade of our life. This part of the brain is the neurobiological site of reasoning, thinking about possible alterative courses of action and their consequences before acting. There is a very important discussion about how much adolescents can think rationally before acting (Fuster, 2008; Goldberg, 2001, 2009).	The sociology of youth can become active in important issues such as the plausibility of giving a driving license to adolescents (in some states in USA at 16- years-old), or the early marriage of teenagers in some countries under 18-years-old (see section 3.4).

Education and cognitive training seems to be a protective factor against mental decline and against dementia (Goldberg, 2005). School Sisters of Notre Dame, are the much-studied and reported nuns from Mankato, Minnesota. Remarkable for their longevity, they were also known for the absence of debilitating dementia among them. The phenomenon was unanimously attributed to the lifelong habit of being cognitively active. The nuns constantly challenged their minds with puzzles, card games, debates of current policy issues, and other mental activities. Furthermore the nuns with college degrees, who taught and engaged in other mind-challenging activities on a systematic basis, on average lived longer than their less well-educated counterparts. (Goldberg, 2009)	The population of senior citizens is increasingly growing and the welfare of this population is a main concern in many societies. Here important questions for sociologists are raised: If we accept that people with lower education levels are more vulnerable to dementia, how should society respond to this? Should society facilitate and encourage education even at university level for almost all people? That education is an important factor for social mobility is known; can we now consider education as a factor in mental health?
Emotions provide a natural means for the brain and mind to evaluate the environment within and around organisms, as well as to respond accordingly and adaptively. We can modulate our emotional response. In effect, one of the key purposes of our educational development is to interpose a non-automatic evaluative step between causative objects and emotional responses. By doing so, we attempt to shape our natural emotional responses and bring them in line with the requirements of a given culture (Damasio, 2003).	The sociology of emotion has engaged a growing group within the community of sociologists. Some of these sociologists show real interest in embracing neuro-scientific knowledge about emotions (see Franks, 1999, 2003).
Some distinguished neuroscientists are referring to the emergence of the "new science of mind" (Kandel, 2005 and 2007; Gazanniga, 2008).	Sociologists interested in the concept of mind can connect neuro-scientific knowledge of brain/mind to classical sociological works such as G. H. Mead's book *Mind, self and society* (1934).
At present, neuroscience is probably the most growing (umbrella) science. In 1971, only 1100 scientists convened at the first annual meeting of the Society for Neuroscience. In 2006, 25,785 scientists	Philosopher of science Godfrey-Smith (2003) writes that the sociology of science is often divided into two periods; Merton's time and more recent periods. The

participated at the society's 36th annual meeting in which 14,268 research presentations were made (Bloom, 2008). Some neuroscientists and experts engaged in "neuro-technology" believe that a kind of "neuro-revolution" is happening and after the era of industrialization and the era of the information society, we are now entering the "neuro-society" (see Lynch, 2009).	first wanted to describe the social structure and social placement of science as a whole. The newer approach has tried to use sociological methods to explain why scientists believe what they do, and how scientific thinking and practice change over time. Here, important questions for the sociology of science are raised: Why is neuroscience growing at such a rate of speed? What real societal needs lie behind this growth? How can this science change our societies? Do beliefs such as that which states we are entering the era of "neuro-society" have any real basis? Does sociology as the science of society want to integrate neuroscience or resist it?
Every mental state is a brain state and every mental disorder is a disorder of brain function (Kandel, 2006). The anxiety states and the various forms of depression are disorders of emotion, whereas schizophrenia is a disorder of thought. Emotion and thinking are complex mental processes mediated by complex neural circuitry. Treatments work by altering the structure and function of the brain. "Insofar as psychotherapy or counseling is effective and produces long-term change in behavior, it presumably does so through learning, by producing changes in gene expression that alter the strength of synaptic connections and structural changes that alter the anatomical pattern of interconnections between nerve cells of the brain" (Kandel, quoted in Brendel, 2006). The last few decades have seen an incredible advance in neuroscience applicable to neuropsychiatry. Studies in genetics and molecular biology have revealed mutations that cause major neuropsychiatric disturbances including	With the rise of the 21st century, there is an increasing awareness of the crucial role of mental health and illness in many societies. Currently, the contribution of the social sciences to the actual discussion of the aetiology of mental illness and the further development of interventions for mental health promotion and mental health is rather small and it seems that the biological perspective has become the leading scientific paradigm. Nevertheless, in spite of the tremendous success of the biological sciences in analyzing the functioning and malfunctioning of the brain as well as the aetiology of mental diseases, the evidence for the relevance of social factors in mental disorder prevention and mental health promotion is still

familial Alzheimer's disease, familial Parkinson's disease and many other developmental disorders (Cummings & Mega, 2003). Psychotherapy emphasizes the importance of subjective human experience and the power of relation-ships to transform the growing mind. Neuroscience focuses on quantifiable, objective data and the scientific method to create models of mind and brain. Embracing the view that subjective experience is real as the quantifiable data of science, is a first step in finding a convergence of knowledge across the two disciplines (Siegel, 2002).	growing. However, the rising insight into the complexity of the interaction between biological and social processes underlines the need for interdisciplinary approaches and cooperation between social and natural sciences. (2010, Research Committee on Sociology of Mental Health and Illness RC49, XVII World Congress of Sociology, ISA homepage).

Table 1 shows how neuroscience can contribute to many branches within sociology. This is a limited table that could be developed widely in the areas mentioned as well as in many more areas not mentioned here.

Neuroscientists are increasingly making their discipline social. Paul Zak, a leading researcher of oxytocin in humans, believes that we need to increase our understanding of the brain mechanism associated with pro-social behavior. He sees the potential that this kind of knowledge will lead us forward as a society, helping us evolve a world that naturally calls forth our best characteristics (in Lynch, 2009, p. 99). Damasio (2003) writes that the beneficial role of culture depends, in large measures, on the accuracy of the scientific picture of human beings that the culture uses to forge its future path. This is where modern neurobiology integrated within the traditional fabric of the social sciences may come to make a difference. Kandel (2006, p. 425) writes, "I like the idea of applying molecular biology to link my area, the molecular biology of mind, to Denise's [Kandel's wife who was doctoral student of Robert Merton] area, sociology, and thus develop a realistic molecular sociobiology." Some other neuroscientists, for example, Steven Quartz, a social cognitive neuroscientist and Terence Sejnowski, a theoretical brain scientist (2003, p. 255) talk about "a nascent neurosociology" and refer to some of the characteristics of such a sub-discipline:

> Ultimately, the human search for happiness depends on unraveling the dynamic of brain development and the influence that social life has upon it. This enterprise constitutes a nascent neurosociology, an approach to understanding the human condition and possible avenues for its betterment that recognizes the power of both our biology and our social context to shape us.

Neuroscience demonstrates that the brain is social and neuroscientists have shown genuine interest in understanding how social contexts affect and form our brains. Cacioppo and Berntson (2004) write that the past decade has seen the emergence of the interdisciplinary field of social neuroscience, which involves collaboration among cognitive scientists, social psychologists, neuroscientists, anthropologists, geneticists, biologists, neurologists, endocrinologists, and many others in related disciplines. The amazing point is that in the list of collaborations among scholars for the creation of *social* neuroscience, sociologists are absent!

The questions are: what do we sociologists want to do and do we want to embrace neuroscience and understand the biological root of our thoughts, emotions and actions without the fear of "being eaten by biologists"? Do we, through dialogue and cooperation with other disciplines, want to play the main role in constructing neurosociology or do we want to wait until non-sociologists do it for us? Below, in a preliminary attempt, I try to describe the characteristics of such a neurosociology.

3. Primary steps toward constructing an empowerment-oriented neurosociology

> One should not be afraid to try new things, such as moving from one field to another or working at the boundaries of different disciplines, for it is at the borders that some of the most interesting problems reside (Kandel, 2006, p. 427).

In this paper I cannot, in *great* detail, present what I consider to be essential elements of an empowerment-oriented neurosociology. Nor can I focus on just one building block and present a partial picture. Thus, in order to show an understandable picture of *how such a sociology could look*, I present, in *some* detail, what I consider to be the necessary cornerstones of an empowerment-oriented neurosociology. With inspiration from pragmatists, especially Tufts (1917), and the emphasis of neuroscience on the individuality and uniqueness of each human being, I present a preliminary definition of empowerment-oriented neurosociology. Empowerment, as a special kind of change, is the intention and the process within which individuals learn to discover and develop some of their biopsychosocial capacities in order to make decisions, solve problems and regulate their lives according to defined goals. This description of empowerment can facilitate the integration of neuroscience with sociology (for more about empowerment and neuroscience, see Moula, 2009). Empowerment-oriented neurosociology is considered with both theoretical and practical issues concerning the interaction of the individual with the environment. Four necessary and complementary elements are part of this consideration:

- *Individuals as owners of brains.* Human beings are part of nature and its historical evolution. Individuals are embodied and the brain is the leading organ of the body.
- *Brains are social.* Brains are the necessary, but not sufficient, condition for human beings to become social. For a brain to become functional, it needs other brains; we are wired to connect (Goleman, 2006).
- *The human prefrontal cortex gives humans exceptional capacity for reasoning and intelligent action in daily life.* With the help of knowledge of the functions of the prefrontal cortex, we can recognize and facilitate each human being's biopsychosocial capacity to actively lead her/his own life.
- *Individuals as problem-solvers and decision-makers.* Life can be considered as the process of judgment and choice and how, through mass-education, we can empower individuals in their interactions with their environments.

Below, I present, in some detail, the four elements of empowerment-oriented neuro-sociology.

3.1 Sociology can study individuals as owners of brains: Societies do not have nerve systems but human beings do.

One of the most ancient ideas about human beings comes from Protagoras' statement that "man is the measure of all things." Despite remarkable pragmatist insights provided by other Greek scholars, no one seems to have been more thoroughly attentive to the humanly known and engaged world than Aristotle. He spent considerable time comparing and differentiating people and other animals along dimensions of capacity and behavior. Aristotle (considered by Peirce as "the greatest intellect that human history has to show") was highly sensitive to matters of human agency, both in terms of invoking reflective choice and actively, mindedly, influencing and resisting one another (See Prus, 2003, pp. 26-30). Aristotle was a great biologist of his time. He, however, clung to the belief that the brain was a radiator for cooling blood that was overheated by the seething heart (Bear, Connors & Paradiso, 2001, p. 4). To those of our time, this is a very strange idea. In spite of his idea about the brain's function, Aristotle believed that part of the "soul" deals with what he called "excellent reasoning" (Aristotle, 1998; Kraute, 2002). The Greek word used by Plato and Aristotle is *psuche*, the root of the English words psyche, psychology etc. It is usually translated into English as soul. But the Greek has no particularly religious, spiritual, or even specifically mental connotations. For Aristotle, psuche is simply what animates the matter of any living organism: all living things have a psuche (Sayers, 1999/2006).

Pragmatists were deeply influenced by Darwin and regarded human beings as part of nature and evolution. Living in a period that paralleled the rise of scientific research about the human central nervous system, these philosophers had a scientific approach to the role of the brain in our lives. William James, one of the founders of pragmatism, is one of the best examples of a modern thinker with a biopsychosocial perspective. As an educated physician, he started to teach physiology at Harvard University. Later, he taught psychology and at the end of his career, he taught philosophy. William James' (1890) two volumes of *Principles of Psychology,* usually referred to as his "masterpiece" (Edelman, 1999), are a combination of philosophy, psychology and what may be called an early neuroscience. "The fact that the brain is the one immediately bodily condition of the mental operation is indeed so universally admitted nowadays," wrote William James, that "the psychologist is forced to be something of a nerve-physiologist" (ibid, p. 18). More than a hundred years later, we can safely say that his prediction is quite correct! Psychology is rapidly becoming either neuropsychology or is integrating with cognitive science. Gazzaniga (2000. pp. xi-xii) wrote that his college, Dartmouth, is constructing a magnificent new building for psychology. Its four stories are allocated to the following: the basement is all neuroscience, the first floor is devoted to classrooms and administration, the second floor houses social psychology, the third floor is cognitive science, and the fourth, cognitive neuroscience. And he asks: "Why is it called the psychology building?"

Even G. H. Mead, usually considered to be one of the founders of the sociological perspective known as symbolic interactionism, showed considerable interest in brain science. In *Mind, Self and Society*, Mead (1934) writes that while minds and selves are essentially social products – products or phenomena of the social side of human experience – the physiological mechanism underlying experience is far from irrelevant, indeed is indispensable, to genesis and existence; for individual experience and behavior is, of course, physiologically basic to social experience and behavior: the processes and mechanisms of the latter (including those which are essential to the origin and existence of minds and selves) are dependent physiologically upon the processes and mechanisms of the former, and upon the social functioning of these. Indeed, Mead (1934, p. 348) always emphasized the importance of understanding the development of mind and self, as well as the construction of action, from the standpoint of the "biologic individual", in that the human organism lives within a matrix of 'impulses sensitizing us to stimuli'. According to Mead, people experience relatively automated physiological responses to their environment in terms of balance, food, sex, attack and defense, escape from danger, and so on. It is from this confluence of impulse-response patterns, both at the individual and social interactional level, that the symbolic, reflective, and socialized aspects of human consciousness must be built. Thus, social experience and symbolic interchange will never free people from a direct connection of the physical and social environment to the 'physiological mechanism of the human central nervous system' (Mead 2001, p. 98). Mead also pointed out that the practical use of technical artifacts employed to extend our sense organs, such as the telescope, still require "an advanced brain"; that is, the person needs a "particular nervous system" or "a whole series of neurons" (1934, p. 5) in order to use these artifacts in any meaningful way. So we are always constrained by our brain's makeup despite products of modern technology. Morris (1934, pp. xv–xvi) points out that for Mead "psychology and sociology are united upon a biological basis". He adds that 'how the human mind and self arises in the process of conduct is answered by Mead in biosocial terms' (ibid, p. xv).

Despite the fact that James' pragmatism also inspires the work of many sociologists, most do not show much interest in neuroscience. The issue is not that sociologists lack an interest in brain science *per se*, but also that there is a kind of skepticism about biological science as an explanation of human behavior generally. Many sociologists believe that biology is synonymous with biological determinism and a resulting reductionist position. This has probably changed to some degree in the last decade or so, as some sociologists are more willing to engage in a dialogue with evolutionary biology, genetics, biology, psychology, and brain research generally. For example, Turner (2000) has collected evidence from evolutionary biology to generate a sociological theory of human emotion that assumes certain hardwired characteristics as opposed to a purely constructionist approach. Dingwall et al. (2003) have tried to consider ways in which genetics research and social science research might complement each other in more integrative research designs. Franks (2003) has taken neuroscientific research seriously to enrich and extend neo-pragmatist social theory. Bergesen (2004) has considered evidence from psychological studies of early childhood development to argue against the "blank-slate" view of socialization and language acquisition often assumed in symbolic interactionist designs (see Puddephatt and Segaert 2005). Moula, Timpka and Puddephatt (2009) have connected pragmatism to neuroscience in studying adult-adolescent interaction. While these research directions are encouraging, the majority of sociologists are still rather hesitant to consider biological arguments for human behavior. So, more than a hundred years ago American pragmatists understood human beings as biopsychosocial beings and showed great interest in brain science. However, after a century, sociologists, even symbolic interactionists who consider themselves to be followers of these pragmatists, still show a very weak interest in neuroscience.

3.2 We are born to connect: People transform themselves and their social world as they engage in communication

We are preprogrammed to reach out (De wall, 2009, p. 43).

Eric Kandel (2005) maintains that while both the brain and human behavior are shaped by genetic and developmental factors, there is also a crucial role played by the social environment that shapes these as well. According to Ratey (2001), the brain is not a computer that operates according to a pre-determined genetic program. Neither is the brain passive grey matter totally under the control of the environment. Rather, the mind and self emerge and develop through the ongoing interaction between the brain and the physical and social environment (LeDoux 2002). Our experiences and choices are important factors that change the structure of the brain and shape our behavior (Goldberg 2005; Ratey 2001; Schwartz and Begley 2002). By its very nature, the brain is 'social'; either the neurons succeed in creating contact with other neurons or they die (Freeman 2000; Ratey 2001). Siegel (1999) explains that most neuroscience rejects biological determinism, and he has focused his research on how relationships are able to change the brain's structure. Here, I suggest that it is vital to communicate with sociologists that human biology is the result of years of interaction with environment. It is in these processes that human beings have changed nature (for the better or for the worse), have created civilizations and have become who we are. The point is to see humans as biopsychosocial beings. Mesulam (2002, p. 26), an expert on the prefrontal cortex, writes:

> The phylogenetic emergence of the prefrontal cortex may have introduced the capacity for choice, change, and reflection but has not specified the nature of the choices, the content of the reflections, or the direction of the changes. The mental faculties promoted by the prefrontal cortex are thus likely to lead to the greatest feats of culture and civilization as they are to Hiroshima and Auschwitz.

The three concepts of *choice*, *change* and *reflection* used in the quotation above are crucial for human beings in order to regulate their lives in interaction with nature and the social environment. As Mesulam indicates, the nature of the choices, the content of the reflections, or the directions of the changes depend mainly on the quality and quantity of interaction among humans as social beings. "Symbolic interactionism, the way I have come to understand and use it" writes David Maines "is merely the concerted application of pragmatist principles to sociological inquiry" (2001, p. xii). With reference to Stone and Farberman, Maines (ibid., pp. xii-xiii) concludes that interactionism is much about the "processes through which people transform themselves and their social world as they engage in communication." He adds four elements to this characterization:

1. *People transform themselves*: people are "selved beings" who reflexively form their conduct and thus are capable of adjusting their lines of action and creating new ones.
2. *People transform their social worlds*: human action takes place in contexts of situations and social worlds. People can modify the social matrices in which they act, and thus people are agents of change.
3. *People engage in social dialogue*: communication is generic and is at the heart of both stability and change.
4. *People respond to and deal with their transformations*: people construct situations and societies, they establish social structures and cultures. These are the consequences of human action, and once formed they reflexively function to direct and channel conduct.

These points clearly show how symbolic interactionists emphasize the centrality of human beings' actions and communications in creating social worlds. "So much has been written in recent years on the social nature of man that it seems unnecessary to elaborate the obvious. Protest has been raised against the exaggeration of the social." These lines were written by Tufts almost a century ago (1917, p. 361). The problematic point has been that scientists from the "biological side" and sociologists from the "social side" have not tried to cooperate or convince each other. However, biologists and neuroscientists recently have accepted that "our biology is social in such a fundamental and thorough manner that to speak of a relation between the two suggests an unwarranted distinction." (Wexler, 2006, p. 13) Wexler (ibid) with reference to Lowontin, Rose and Kamin raises the issue that humanity cannot be cut adrift from its own biology, but neither is it enchained by it. "Our bodies and minds are made for social life," writes known primatologist France De Waal (2009, p. 10) who adds that "we become hopelessly depressed in its absence. This is why next to death, solitary confinement is our worst punishment." Gazzaniga (2008, p. 112) writes that as we move on, "realizing our social nature is deeply rooted in our biology not simply in our cognitive theories about ourselves, we begin to see how the rest of our human equipment helps to guide through the social maze." The point is to move toward an integrated understanding of the relationship between the biological and the social in a way that sociologists become an important part of this process.

3.3 Frontal lobes and executive functions: Each human being has the neurobiological capacity to become the CEO of her/his own life

The main task of the brain is to help maintain the whole body in an optimal state relative to the environment, in order to maximize the chances of survival. In this process, frontal lobes, especially a part of them, the prefrontal cortex, has the capacity of playing the role of the leader. The human frontal lobes comprise a large region of the brain, representing 30% of the cortical surface (Miller, 2008). This part of our brain mediates those behaviors that most distinguish man from animals. Even higher, non-human primates lack the empathy, regret, sarcasm, social awareness, planning, and judgment characteristic of human behavior. These frontally mediated behaviors define the highest level of human culture and achievement (Cummings & Miller, 2008).

What controls your thought? How do you decide what to pay attention to? How do you act appropriately while dining in a restaurant or listening to a lecture? How do you plan and execute errands? How do you manage to pursue long-term goals, like obtaining a college degree, in the face of many distractions that can "knock you off task"? In short, how does the brain manage to orchestrate the activity of millions of neurons to produce behavior that is willful, coordinated, and extends over time? This is called cognitive control, the ability of your thoughts and actions to rise above mere reactions to the immediate environment and to be proactive: to anticipate possible futures and coordinate and direct thought and action to them. Cognitive control necessarily involves the coordination of many different brain processes, however the prefrontal cortex plays a central role (Miller & Wallis, 2008).

Frontal lobes are the latest achievement in the evolution of the nervous system; it is only in human beings that they reach such great development (Goldberg, 2001, 2009; Fuster, 2003). The expansion of the human prefrontal cortex to one-third of the total cortical surface, compared with only one-tenth in the gorilla, appears to be a seminal reason for the display of the psychological properties that differentiates humans from all other primates. These eight distinctive human properties are: (1) an expanded working memory; (2) the ability to maintain a representation of a goal, despite distraction, due to the inhibition of thoughts and responses irrelevant to the goal; (3) the ability to retrieve representations of events from the distant past; (4) the ability to generate representations of events that might occur in the distant future; (5) conscious awareness of the self's feelings, thoughts and properties; (6) creation of the concept of prohibited acts, understanding the semantic categories "good" and "bad" and capacities or shame over violations of a standard; (7) the seeking of new experiences that can be understood or coped with effectively; (8) the ability to invent relations of similarity and difference among varied classes of representations. These eight competencies require participation of the prefrontal cortex and its connections to other brain sites (Kagan & Baird, 2004, p. 100).

The entirety of the frontal cortex, including its prefrontal region, is the "action cortex" in the broadest terms. It is the cortex devoted to action of one kind or another, whether skeletal movement, ocular movement, the expression of emotion, speech, or visceral control; it can even be the kind of internal, mental, action that we call reasoning. The frontal cortex is the "doer" cortex, much as the posterior cortex is the "sensor" cortex. Most certainly, however, the frontal cortex does nothing by itself; all it does is in cooperation with other cortices, with subcortical structures, and with certain sectors of the sensory and motor apparatus and the automatic system (Fuster, 2008).

Like a large corporation, a large orchestra, or a large army, the brain consists of distinct components serving distinct functions. And like these large-scale human organizations, the brain has its **chief executive officer**, its conductor, its general -- the frontal lobes. To be precise, this role is mediated by one part of the frontal lobes, the prefrontal cortex. It is common shorthand, however, to use the term frontal lobes (Goldberg, 2001). Frontal lobes are the focal point for the integration of information from the environment, the internal milieu of the body, and the emotional state of the individual. Organized volitional activity requires an environmental assessment, the integration of perception with past historical information and one's emotional state, the generation of action or response contingencies, the choice of an action plan that both accomplishes one's goal and anticipates environmental consequences, the implementation of the plan through appropriate motor activity, monitoring of the action, and adjustment based on feedback. The accomplishment of this complex series of activities is dependent primarily on frontal lobe function (Cummings & Mega, 2003).

"Despite the building evidence base supporting a key role for the frontal lobes, and prefrontal cortex in particular, in mediating executive function, each of these research endeavors has also documented findings consistent with the argument that executive functions and frontal lobes are not synonymous" (Anderson, Jacobs & Anderson, 2007, p. xxix). Executive control, a broad term to describe the neuropsychological functions of the frontal lobes, incorporates many distinctive cognitive processes. Components of executive control include: working memory, generation, inhibition, set shifting, concept formation, temporal sequencing, insight, interpersonal perspective taking (theory of mind), as well as social and real-world executive performance (Miller, 2007). However, Fuster (2008, p. 178), a distinguished expert on the prefrontal cortex, wrote that "executive function, the principal function of the prefrontal cortex, is the ability temporally to organize purposive behavior, language, and reasoning." Metzer (2007) indicated that most definitions of executive functions include many of the following elements:

- Goal setting and planning
- Organization of behavior over time
- Flexibility
- Attention and memory systems that guide these processes (e.g., working memory
- Self-regulatory processes such as self-monitoring

In sum, human beings' brains, through interaction with the natural and social environment, have acquired a unique position among animal central nerve systems. The large and very complicated human prefrontal cortex has given us *the capacity* to reason, deliberate and choose among diverse courses of action in a highly complex society. *It is important to note how social environments play their parts in developing these individuals' capacities so that each human being can become the chief executive officer, that is, the leader, of her/his own life.*

3.4 Practicing neurosociology: People can learn to make intelligent decisions and regulate their lives according to certain goals

> Theories thus become instruments, not answers to enigmas… We do not lie back on them, we move forward… Pragmatism unstiffens all our theories, limbers them up and sets each one at work. (William James, 1907, p. 21)

Hewitt (1988, p. 276) writes, "If sociology and social psychology represent potential contributions to the advancement of human intelligence – and hopefully to the betterment of the human condition – they do so only if exercised." Richardson (1994, p. 481) writes the "outstanding success of feminist scholarship across disciplines arises from its explicit link to the feminist movement, a continuity of purpose between research and activism, namely the empowerment of women." In sections one through three of this article, I lay the possible scientific basis of an empowerment-oriented neurosociology. In this section, I suggest that this sociology can become more action-oriented and present a formula and model that can be used in mass education with the aim of empowering people to make intelligent decisions in order to solve problems and regulate their lives according to their goals. "The elegance and beauty of human life is augmented within a vision of decision making, and human spirit is elevated. The idea of decision making gives meaning to purpose, to self, to the complexities of social life." (March, 1994, pp.271-272) Concepts of intelligence and decision-making are discussed widely in philosophy, psychology and neuroscience. Our world has become very complex and uncertainty is a crucial factor that we should take into account. But this should not lead to pessimism. Pragmatists believe that human beings, through intelligence, can confront uncertainty and plan for a better future. With reference to Dewey, Bhattacharryya (1969) writes that intelligence is a new quality in the natural world which has actualized in the lengthy processes of development. This pragmatist understanding fits well with how the brain has developed as well as with the primary difference between human beings and other animals. The pragmatists' concept of intelligence and the functions that they attribute to this concept fit well with the functions that neuroscientists attribute to the prefrontal cortex (see Fuster, 2003, 2008). In a fine collection written by eight pragmatist social thinkers, *Creative Intelligence: Essays in the Pragmatic Attitude*, Dewy (1917, p. 63) indicates that:

> As a matter of fact, the pragmatic theory of intelligence means that the function of mind is to project new and more complex ends—to free experience from routine and from caprice. Not the use of thought to accomplish purposes already given either in the mechanism of the body or in that of the existent state of society, but the use of intelligence to liberate and liberalize action, is the pragmatic lesson.

Bhattacharyya (1969, p. 193), with reference to Dewey, writes that intelligence has given man "not only the ability to think scientifically, but other capacities including sensitiveness and sympathy, sense of responsibility and faithfulness, tolerance and conscience." According to Dewey, meliorism is the belief that the specific conditions which exist at one moment, be they comparatively bad or comparatively good, in any event may be bettered. Meliorism encourages intelligent action and arouses confidence and hopefulness. Progress is not inevitable, it is up to men as individuals to bring it about. Change is going to occur anyway, and the problem is to control change in a given direction (in Campbell, 1995, p. 261). "The pursuance of future ends and the choice of means for their attainment are thus mark and criterion of the presence of mentality in a phenomenon. We all use this test to discriminate between an intelligent and a mechanistic performance." (James, 1890/1981, p. 21)

As we have seen, the interaction of the individual with her/his environment is central in neurosociology. The equation below shows how different factors in the interaction between an individual (A1and A2) with her/his environment (B1 and B2) determines the probability of achieving a goal (C).

Figure 2 shows the interaction of individual and environment for achieving a goal. The purpose is not to put life into a mathematical formula, but to increase people's consciousness about factors that influence the process of our decision-making.

The main assumption is that (A1 & A2) + (B1 & B2) = C. This means that the increase of conscious thinking, reasoning and planning can: decrease the unconscious influence of emotions and habits, increase awareness of and capability in using environmental resources, decrease the influence of barriers and, eventually increase the probability of achieving a defined goal in many life situations. This formula raises an individual's consciousness about the factors that affect the process of decision-making.

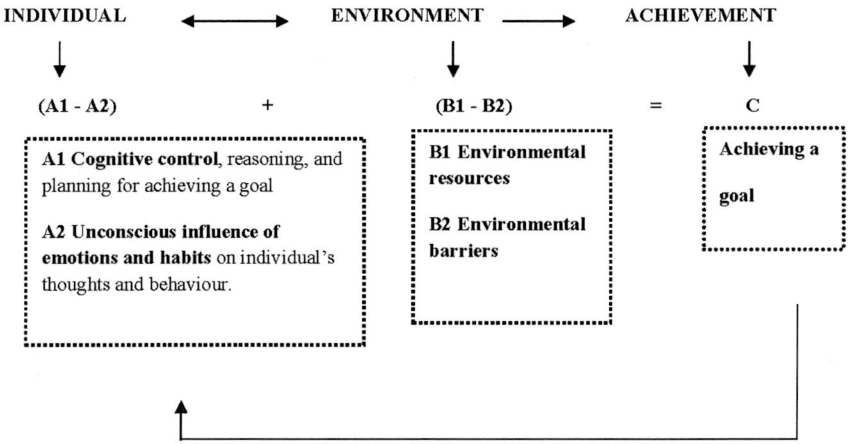

Figure 2 Moula's person-in-environment formula

This formula can be completed with a model, RAHYAB (Moula empowerment model) that has been in use in Sweden and Iran since 1997. In 1997, through a pilot project and with the help of local radio programs and a helpline telephone, I worked as a family counselor with Iranian families in Stockholm. After a year, the project became popular among Iranians and with the support of BRIS (a national Swedish children's organization) I developed the program to reach Iranians in all of Sweden. Between 1998 and 2000, eighty-three live radio programs were broadcast in several cities in Sweden and about 600 Iranians had conversations with me on the helpline telephone. In these radio programs and helpline conversations, people learned RAHYAB (a Persian word which means "searching for a way") in order to use it in daily decisions they made. Subsequent to the popularity of these projects among Iranians in Sweden, I regularly travel to Iran where I have introduced this model for empowerment-oriented interventionist work. Since the year 2000, several thousand students, social workers, psychologists and psychiatrists in Iran and Sweden have learned and are using this model in their daily practice (for a comprehensive report on these projects, see Moula, 2005/2010 and 2009). I present the model below:

RAHYAB (Moula Empowerment Model)

Capacities to develop	**Steps taken to engage in empowerment-oriented problem solving**
To organize thoughts and emotions through story-telling or story-writing	**1. Define your situation and identify a problem**. You may talk to a friend or a family member but *you* are the expert of your own situation. You have lived your life and have experienced its "ups and downs". You may pick up a piece of paper and a pen and describe the problem that mostly engages you. Simultaneously find out how your *thoughts* and *emotions* are involved in the situation you try to define. It is also important that you consider your relationships with significant others (family members, friends, etc.) who play an important role in your life and especially in the situation that you want to define.
To identify desires, strengths and necessary changes	**2. Identify (a) your desirable situation, (b) your strengths and (c) what is it in the problematic situation that you want to change.** Consider how your emotions/feelings are related to your desires. What do you feel when you think about the changes that you wish to make? Consider your relationships with significant others. How do these people affect your desires? When you think about necessary changes, what role can these people play in bringing about these changes? You should also think about how these changes can affect the lives of other people.
To mindfully consider various alternatives	**3. Imagine and map several possible alternatives**. Think systematically to find out which alternative may help you move from the problematic situation toward the desirable situation. This includes thinking about the (a) the possible consequences of each alternative, (b) resources that you can identify and use to realize the chosen alternative and (c) barriers that can impede against each alternative (it is hardly possible to manage this step of the model without writing your thoughts on paper). Consider: • Your emotions/feelings in regard to these various alternatives. • How each alternative can affect the lives of others, and which roles others can play in realizing each alternative. Other individuals can be a resource for realizing an alternative or they may be a barrier. How your values are attached to different alternatives. It is crucial that you are aware how your values influence the choice or rejection of any alternative.

To choose the best possible alternative, define a goal and plan for action	**4. Choose the best possible alternative and according to that define your goal and then make a plan to realize it.** Your goal relates to and gains strength from the desire that you identified in step 2. Now, after considering several alternatives (step 3), you can formulate your goal in line with the possibilities of realizing each of the alternatives.
To evaluate and learn from the situation	**5. Evaluate the situation** and learn from this for future actions

Table 2 The most important aspect of this model is its circular development in theory-practice-theory since 1997. The model is based on pragmatist philosophy and neuroscience. (Here, I refer only to a few sources: Dewey, 1910, 1916, 1922, 1948; Damasio, 2003. For more on the scientific and philosophical base of the model see Moula, 2005/2010, 2009).

This model empowers individuals and helps them move *toward* using their optimal cognitive capacity for reasoning, decision-making and problem solving. People learn and use several concepts in a systematic way in their decision-making processes. These concepts help individuals to (1) describe the situation and define the problem, (2) imagine a desirable change, (3) think about different possible and reasonable alternatives, the consequences of each, the resources available for the realization of alternatives, and barriers impeding alternatives (4) choose the best possible alternative, formulate a goal, and plan for action, (5) evaluate the new situation and learn from the consequences of actions.

This limited project, which started in 1997 to help a number of Iranian families in Stockholm, has developed along several dimensions:

1. In the domain of practice, from a local project in Stockholm to all of Sweden. Later, from the year 2000, I traveled to Iran and taught it to social workers.
2. In the domain of education, with the help of UNICEF, I have educated several groups of educators from the year 2006 including social workers, psychologists and psychiatrists. These practitioners have learned RAHYAB and are teaching it to others. With the support of UNICEF, Iran's Central Social Welfare Agency, and a group of educators in the Ministry of Health, many practitioners are learning and using RAHYAB in many cities in Iran.

3. In the domain of research, I have applied RAHYAB, in participatory action research, to find out how people can learn and use this model in their daily lives. This has been accomplished through three projects in Sweden, and my two doctoral students through, up to now, 6 completed projects in Iran. (A number of research articles are coming out in 2010-2012 about these projects).

In section two, I describe that neurosociology can take a fact from neuroscience and construct a proposition (quadrant 1), then, put this proposition into limited practice and subsequently put it into praxis (quadrants 2 and 3). There is a heated debate in neuroscience about the fact that the prefrontal cortex (the neurobiological site of reasoning, deliberating, planning, choosing and other highly cognitive functions) continues to develop even in the third decade of our lives (Goldberg, 2009; Anderson, Jacobs & Anderson, 2008). This fact implies enormous consequences for all societies and opens the door to very important discussions that sociologists cannot stay outside of. If adolescents are not mentally mature, how can they get driving licenses at 18? (In some states in the USA, even a 16-years-old can get a driving license). How can they marry at 18? (In some countries, they can marry at ages even younger than 18)

There are three options in dealing with this important societal issue. First, continue on the basis of laws to accept 18-year-old adolescents as adults ready for driving, marriage and other affairs. Second, question this practice and indicate that the adolescent brain has not reached maturity and suggest that efforts to teach adolescents to refrain from risk-taking activities would not be very successful. This view implies that "a more profitable strategy might focus on limiting opportunities for immature judgment to have harmful consequences" (Steinberg 2004, 57). For example, rather than advising teenagers about the dangers of adventurous driving, raise the driving age. Third, using other supportive facts from neuroscience such as brain plasticity, educate and train developing adolescents' cognitive capacities. Masten (2004) emphasizes the importance of supportive contexts in providing a frame for the adolescent in the process of developing regulatory competence, particularly during the early part of adolescence when rapid biological development and new interactive contexts challenge regulatory skills that are emergent, but still highly immature. In light of these realities, my doctoral students and I are in the process of conducting several projects in Iran and Sweden. (The resulting articles are set for 2010-2011.)

The purpose of this section is to show how neuroscientific findings can be put into practice in society. We can never know if it is possible to empower individuals to make intelligent decisions – that is, decisions based on achieving a defined goal – unless we test this notion of empowerment in practice. That the prefrontal cortex has the necessary biological capacity for intelligent decision-making is clear, but how this capacity will develop and will be used depends upon social contexts. I conclude this section with a very famous quote from William James (1890/1981, p. 130), "The physiological study of mental conditions is thus the most powerful ally of hortatory ethics… Could the young but realize how soon they will become mere walking bundles of habits, they would give more heed to their conduct while in the plastic state." Interventionist programs based on education of the young are crucial in neurosociology.

4. Conclusions

> When a theory of knowledge forgets that its value rests in solving the problems out of which it has arisen; when it forgets that it has to work out the conditions under which the individual may freely direct himself without loss of the historic value of civilization—when it forgets these things it begins to cumber the ground (Dewey, 1897/1997, pp. 298-299).

It is time to recognize that everything a human being does is simultaneously biological (including neurobiological), psychological and social, without exception. Without the leading organ of the body—the brain—there is no thought, no emotion/feeling and no behavior. Resistance to neuroscience, once based on the fear of "biological determinism", does not have a scientific basis and has now become a scarecrow that is more odd than frightening. Every human act is performed by a unique human being, a unique mind, and has therefore a psychological base. Finally, everything we do has a social basis, no exception. The brain's normal development occurs only when the owner of a brain is engaged in interaction with other individuals. No functioning human brain exists that is not social. If a researcher "separates" these three dimensions (biological, psychological and social) from each other, it is only for an analytical purpose that cannot occur in reality. The separation of the disciplines of biology, psychology and sociology may have some advantages; for example, each can go in-depth and study various dimensions. But this separation also has its disadvantages, it becomes difficult to study the whole—humans as biopsychosocial beings. The question is what can neurosociology do about this problem? What kind of new knowledge can be added by this nascent neurosociology? Below, I try to answer these questions.

From the individual to society. Neurosociological knowledge can provide us a new answer to the individual/society dispute within sociology. Suddenly, the issue regarding whether or not sociology's concern should be to focus on society or the individual becomes unnecessary. Now, we can gain knowledge from two disciplines; one begins with individuals and the other is mainly concerned with society. A good example here is the acceptance of what is usually called "a psychosocial approach" where researchers and practitioners try to use a mix of knowledge from both sociology and psychology. However, today, psychology is becoming neuropsychology and a psychosocial approach is becoming a biopsychosocial approach. Neuroscience provides knowledge of the uniqueness of each individual; how each individual (read each individual's brain) learns and changes. Sociology may contribute to spreading this knowledge base – usually achieved through laboratory and clinical activities – to various social contexts.

Knowledge of and knowledge for. Neurosociological knowledge would not be only knowledge *of* the people but knowledge *for* the people. Through this knowledge, people could realize their desires and eventually change their societies in ways that maintain civilization. Pragmatists' interests in individuality and agency do not mean raising selfish children. Self and society are considered to be two sides of a single coin (Dewey, 1897/1997, 1916, 1922; Mead, 1934).

Mass education as the medium of societal change. With the help of the neuroscience of education and the pragmatist philosophy of education, mass education becomes the main source of changing existing societies. Mass education would realize the desires of the father of progressive pedagogy, John Dewey (1916, 1948), that is to say, a democratic, humanistic and effective way of creating better societies that are constructed from the bottom up.

Concrete methods and tools for practice. A neurosociological formula about the individual's interaction with their environment and an empowerment model that provides concrete methods and tools to put this neurosociological knowledge into practice, finds its weakness and strengths, and moves forward by continuous reconstruction.

To conclude, neurosociology can study individuals' interactions with their environments. Empowerment will take place by a kind of social intervention based on the biopsychosocial knowledge of human beings. With the help of the knowledge about our brains, we can cognitively empower individuals toward intelligent interaction with their environments. Empowerment-oriented neurosociology will grow through continuous development of four interdependent knowledge bases:

1. *Individuals as brain-owners*. It is not reasonable to neglect neuroscience while studying societies that are formed by individuals as brain owners. Such neglect consciously throws away vital and relevant knowledge of self, mind, emotion, cognition, behavior, etc. Neurosociology can contribute to sociology in a way that causes this branch of the social sciences to become more individual-based and brain-sensitive.
2. *Brains are social*. Brains of newborn infants become functional through interaction with physical and social environments. Neurosociology actively integrates neuroscientific knowledge with the sociological knowledge of individuals' social interactions. The resulting neurosociological knowledge will be used in diverse social environments like families, schools, working places, etc.
3. *Individuals as owners of PFCs and as CEOs of their lives*. An important and concrete integration of points one and two is the study of how we can empower individuals—as owners of a prefrontal cortex (PFC)—so that they become the CEOs (chief executive officers) of their daily lives.
4. *Neurosociology in social praxis*. Basic neuroscientific facts should be put into practice in diverse social spaces outside laboratories and clinics. A concrete example comes from the domain of the neurosociology of adolescence where individuals, through education, can learn to interact intelligently with their environments. In particular, individuals learn intelligent decision making in their daily lives (Moula's school project in Sweden).

To return to our metaphor, there are some signs that the fetus in the uterus will grow healthy and will be born as an infant—a new branch within sociology will emerge. My hope is that this emergence takes place through dialogue within the community of sociologists and even through cooperation with some neuroscientists, neuropsychologists, philosophers and educational researchers. The creation of a session for neurosociology at the International Sociological Association's research committee number

54, is one such promising sign. However, as Mitchell (2009) writes, pursuing ambitious goals by going beyond mainstream science into ill-defined and uncharted territory does not always lead to success. A researcher should be ready to take risks and learn from possible failures for further well-defined attempts.

References

Anderson, V. (2008). "Attention deficits and the frontal lobes". In Anderson, V. Jacobs, R. & Anderson, P. (2008)(eds.). *Executive functions and the frontal lobes (pp. 317-344).* New York: Taylor & Francis.

Anderson, V. Jacobs, R. & Anderson, P. (2008)(eds.). *Executive functions and the frontal lobes.* New York: Taylor & Francis.

Aristotle (1998). *Nicomachean Ethics*. New York: Dover Publications.

Berger, P. & Kellner, H. (1981). *Sociology reinterpreted. An essay on method and vocation*. England: Penguin books.

Caine R. N., et al. (2005). *12 Brain/Mind learning principles in action.* Thousand Oaks, California: Corwin Press.

Campbell, J. (1995). *Understanding John Dewey*. Chicago: Open Court.

Cummings, J. L., & Mega M. S. (2003). *Neuropsychiatry and behavioral neuroscience.* Oxford: Oxford University Press.

Curren, R. (2006)(ed.). *A companion to the philosophy of education.* MA, USA: Blackwell Publishing.

Damasio, A. (2003). *Looking for Spinoza: Joy, sorrow, and the feeling brain*. New York: Harvest Book.

De Wall, F. (2009). *The age of empathy*. Harmony books: New York.

Dewey, J. (1910/1997). *How we think*. New York: Dover Publication.

—. (1916/1944). *Democracy and education: An introduction to the philosophy of education.* New York: The Free Press.

—. (1922/2002). *Human nature and conduct*. New York: Dover Publication.

—. (1929/1958). Experience and education. New York: Dover Publication.

—. (1938/1998). *Experience and education (The 60th anniversary edition).* West Lafayette, Indiana: Kappa Delta PI.

—. (1897/1997). The significance of the problem of knowledge. In John Dewey, *The influence of Darwin on philosophy and other essays* (pp. 271-304).New York: Prometheus books.

Edelman, G. M. & Tononi, G. (2000). *A universe of consciousness: How matters becomes imagination.* New York: Basic Books.

Franks, D. (2003). Mutual interest, different lens: Current neuroscience and symbolic interaktionism. *Symbolic Interaction 2* (4), 613-630.

—. (1999). Some convergence and divergence between neuroscience and symbolic interaction. In D. D. Franks & T. S. Smith (eds.), *Mind, brain and society: Toward a neuroscience of emotion. Social Perspectives on Emotion, vol. 5* (pp. 157-182). Stamford CT: JAI Press.

—. (2003). Emotions. The importance of emotions to symbolic interaction. In Reynolds and Herman-Kinney (eds.) *Handbook of symbolic interactionism* (pp. 787-809). New York: Altamira.

Fuster, J. (2003) *Cortex and mind: Unifying cognition.* New York: Oxford University Press.

—. (2008). *Prefrontal cortex.* (4th ed.) Amsterdam: Elsevier, Academic Press.

Goldberg, E (2001). *The Executive brain: Frontal lobes and the civilized mind.* New York: Oxford University Press.

—. (2005). *The wisdom paradox: How your mind can grow stronger as your brain grows older.* New York: Gotham Book.

—. (2009). *The new executive brain. Frontal lobes in a complex world.* New York: Oxford university press.

James, W. (1907/1995): *Pragmatism.* New York: Dover Publications.

—. (1890/1981). *The Principles of psychology.* Cambridge, Massachusetts: Harvard University Press.

—. (1899/2007). *Talks to teachers and students on some of life's ideals.* Charleston, SC: BiblioBazaar.

Kagan, J. & Baird, A. (2004). Brain and behavioral development during childhood. In M. Gazzaniga (Ed.), *The cognitive neuroscience III* (pp. 93-103) Massachusetts: MIT Press.

Kandel, R. E. 2005. *Psychiatry, psychoanalysis, and the new biology of mind.* Washington: American psychiatric publishing, Inc.

—. 2006. *In search of memory: The emergence of a new science of mind.* New York: W.W. Norton & company.

Kandel, E. Schwartz. J. & Jessell, T. (2000). *Principles of neural science.* (4th ed.). New York: McGraw Hill.

Kraut, R. (2002). *Aristotle: Political philosophy.* New York: Oxford University Press.

LeDoux, J. (2002). *Synaptic self: How our brains become who we are.* New York: Viking.

Lynch, G. & Granger, R. (2008). Big brain: The origins and future of human intelligence. New York: Palgrave.

Lynch, Z. (2009). *The neuro revolution. How brain science is changing our world.* New York: St. Martin's press.

Lyng & Franks (2002). *Sociology and the real world*. Lanham, MD: Rowman & Littlefield Publishers

Masten, S. (2004). 'Regulatory Processes, Risk, and Resilience in Adolescent Development.' *Annual Meeting of the New York Academy of Science* 1021: 310–9.

Mesulam. M. (2002). The human frontal lobes: Transcending the default mode through contingent encoding. In D. T. Stuss & R. T. Knight (Eds.). *Principles of frontal lobe function*. Oxford: Oxford University Press.

Malten, A. (1981). *Vad Är Kunskap*? (Swedish book about *What is knowledge?*) Malmö: Liber Läromedel.

Mead, G. H. (1934). *Mind, self, & society, from the standpoint of a social behaviorist*. Chicago: University of Chicago Press.

Meltzer, L. (Ed.). (2007). *Executive function in education: From theory to practice*. New York: The Guilford Press.

Miller, B. L. (2007) The human Frontal lobes. In Miller, B. & Cummings *The human frontal lobes: Functions and disorders (pp. 3-11)*. New York: The Guildford Press.

Mitchell, M. (2009). *Complexity: A guided tour*. New York: Oxford University Press.

Moran, S. & Gardner, H. (2007). Hill. Skill, and will: Executive function from a multiple-intelligence perspective. In Meltzer L. (ed.) *Executive function in education: From theory to practice* (pp. 19-38). New York: The Guilford Press.

Moula, A. (2003). *Kunskapsträdet*: *Erfarenheter från en forskningscirkel om lösningsfokuserad arbetsmodell.* Skriftserie 2003:3, Kalmar: Fokus, Kalmar län.

—. (2005). *Population based empowerment practice in immigrant communities.* Doctoral dissertation. Linkoping, Sweden: Linköpings universitet.

—. (2009). *Empowermentoriented socialt arbete*. Lund: Studentlitteratue.

—. (2010) *Population based empowerment practice in immigrant communities. Saarbrucken: Lambert academic publisher (published doctoral dissertation with minor changes).*

Moula, A. Timpka, T. Puddephatt, A. (2009). Adult–Adolescent interaction and adolescents' Brain Development: Integrating pragmatism/ interactionism and neuroscience to develop a platform for research on adolescents' life regulation. *Sociology Compass,* 3/1, 118-136.

Sacks, O. (2001). Foreword. In: Goldberg, E. *The executive brain: Frontal lobes and the civilized mind* (pp. vii-xiv). Oxford: Oxford University Press.

Samuels, B. M. (2009). Can the differences between education and neuroscience be overcome by mind, brain and education? *Mind, Brain and Education* 3/1, 45-55.

Squire, L. et. (2008). The prefrontal cortex and executive brain functions. In Fundamental neuroscience (3th ed.) (pp. 1199- 1222). Amesterdam: Elsevier.

Steinberg, L. (2004). 'Risk Taking in Adolescence: What Changes, and Why?' *Annual Meeting of the New York Academy of Science* **1021**: 51–58.

CHAPTER THIRTEEN

MEMORY AND POLITICAL CULTURES IN EUROPE WITH SPECIAL REFERENCE TO ITALY

UMBERTO MELOTTI

Abstract

Memory is the necessary basis of culture, in particular political culture: in our definition, the way each country conceives state, people and nation, the relationship between ethnicity, nationality and citizenship, and the rights and duties which derive. This is quite evident in both the main European political cultures: the Romantic one, which emphasizes the "objective" elements of the nation (foremost language and the ethnic characteristics), as found in Germany, and the one going back to the spirit of the Enlightenment, which emphasizes the civic "subjective" elements, as found in France. The role of memory, however, is very important also in other countries, such as Israel, whose political culture is connected both to Bible stories and to the Shoah, which is remembered annually on the "Day of Memory".The case of Italy clearly confirms the close relationships between memory and political culture. Suffice it to recall the verses where, at the beginnings of the Risorgimento, humus of the Italian nation-state, Alessandro Manzoni, one of the most important Italian poets, called for a nation "one of arms, language, altars, memory, blood and hearts".

The great Italian poet Dante Alighieri pointed it out clearly: "Non fa scïenza / sanza lo ritenere avere inteso": [There is no knowledge without retaining what has been understood] (*Paradise*, V, 40-42).

Memory, perhaps the most important function of the human mind, is the necessary basis of culture and, therefore, of political culture: its part concerning politics and policies. This becomes even clearer if we remember the crucial role that in modern political cultures is played by the idea of nation, which, to a large extent, depends on memory.

However, there is little agreement about the respective importance in nation-making of the ethnic and the civic factors and about the balance between its "subjective" elements, like memory and will, and its "objective" ones, like territory and language. And even less agreement exists about the real nature of ethnicity and nationality and their role in individual and collective identity, with respect to other main identity factors, such as race, gender, class and social status (see Hutchinson and Smith, 1994).

In Europe this long-standing debate has been recently renewed and revitalized by the synergic effects of three epoch-making processes: 1) imposing immigration from other countries both inside and outside the continent; 2) the transformation of the previous European communities with mainly economic competence into the present European Union with even important political competence, and its subsequent remarkable enlargements; 3) the development of economic and cultural globalisation, with the resulting partial homologation of uses and customs (see Melotti, 2000a, 2008a, 2008b, 2010).

In this chapter we'll discuss some of these points by reviewing and comparing the idea of nation in some European and non-European countries and then by dwelling upon the Italian case, which, despite its substantial importance, is still little known, not only abroad.

Memory and political culture

To begin with, it may be useful to define clearly what we mean by "political culture". In fact, this expression is often used in a too loose and elusive way to generically refer to orientations, feelings, attitudes and values that appear to be variously connected with political behaviour (see Almond, 1956, and Almond and Verba, 1963)[1]. Somewhat delimiting this broad use, with this expression we'll mean the way in which, in a country, state, people and nation are regarded, the relationships between ethnicity, nationality and citizenship are conceived, and the resulting rights and duties are defined (Melotti, 2000b). Though innovative, this definition is not so far from some late elaborations of one of the pioneers of the study

[1] Gabriel A. Almond and Sidney Verba are the scholars who have introduced the study of political culture. According to Verba (1965: 513), "the political culture of a country consists of the system of empirical beliefs, expressive symbols, and values which define the situation in which political action takes place". This was only "one aspect of politics", but of paramount importance, because "it provides the subjective orientation" to political behaviour. He recognized, however, that "few words in the social science have a greater variety of use".

of political culture, Gabriel A. Almond (1990), who eventually stressed that political culture consists of the set of subjective orientations to politics in a national population or in some of its subpopulations. Anyhow, also in this more precise meaning political culture has cognitive, affective and evaluative aspects: it includes knowledge and beliefs about political reality, feelings with respect to politics and commitment to political values; it depends to a large extent on childhood socialization, school education, media exposure and adult experience with social, economic and political activities; and it affects both political structures and political attitudes. Moreover, political culture is a quite persistent fact in the *longue durée* of history: its changes are generally slow, limited, infrequent and, mostly, have a generational character (see Pasquino, 2010).

Especially where nation is conceived in ethnic-cultural (*völkisch*)[2] terms, as in Germany (the paradigmatic case of *Kulturnation*), memory is the basic factor of political culture. But also where nation is conceived in ethic-political terms, as in France (the paradigmatic case of *Staatsnation*)[3], memory plays an important role. General de Gaulle, for instance, used to underline the age-old process of French nation-building, which, according to him, was started in the Middle Ages by Clovis (the first King of the Franks to unite all the Frankish tribes under one ruler, who was converted to Catholicism and expanded his dominion over almost all of the old Roman province of Gaul, roughly modern France) and was then fostered throughout the centuries by such great personalities as Charlemagne (Charles the Great), Jeanne d'Arc, Louis XIV and Napoleon I. Similarly, other French politicians (such as Napoleon III and Philippe Pétain) and even some well-known and appreciated historians went so far as to present Vercingetorix (the Arverni chieftain who led an unsuccessful resistance to the Roman conquest) as the first French national hero[4], just as in Germany

---

[2] *Völkisch* derives from *Volk* (people), but it does not mean popular (in German *populär*). Rather, it is used to define the alleged original (*ursprungliches*) and autonomous (*selbstständig*) traits of a people. This term was already used in this sense by the philosopher Johann Gottlieb Fichte (1762-1814), one of the fathers of German nationalism (in particular, see Fichte, 1808).

[3] This useful but rather stereotypical distinction between *Kulturnation* and *Staatsnation* was cogently theorised by the German historian Friedrich Meinecke (1907), with reference to the long-standing German and French debate on the idea of nation. To summarize, we could say that *Kulturnation* is the premise of some types of nation-state (as was firstly theorized by Johann Gottfried Herder and Johann Gottlieb Fichte between late 18th and early 19th century) and *Staatsnation* is the result of the presence of a centralized and bureaucratic state in a territory (as was firstly theorized by Emmanuel Sieyès during the French revolution).

[4] Among the historians, see Amédée Thierry, *Histoire des Gaulois depuis le temps*

other scholars did with Arminius (the Cherusci chieftain who defeated the Romans at Teutoburg, on the first September 11 of history). Yet, what has always prevailed in France is the civic view of the nation, which Renan (1823-1892), in his famous 1882 lecture on this subject, could even define as "a daily plebiscite"[5]. This pregnant metaphor entailed some important political consequences. According to him, "a nation never has a real interest in being annexed or holding on to a country despite itself" and "the desire of nations to be together is the only real criterion to take into account".

Not by chance, Renan's statements were expressed during the hot French-German contest over Alsace and Lorraine, two border regions determined to remain French despite their alleged prevailing German background, which had induced Germany to annex them after the Franco-Prussian war (1871), though actually they were a patchwork of ethnicities and dialects, often hardly intelligible to their neighbouring populations, either French or German speaking.

Yet, some years later a prominent German social scientist, Max Weber (1864-1920), despite his ardent nationalist feelings, openly recognized that "the reason for the Alsatians' not feeling themselves as belonging to the German nation *had* to be sought in their memories": "their political destiny *had* taken its course outside the German sphere for too long" and, therefore, "their heroes *were* those of French history". "If the custodian of the Kolmar museum wants to show you which among his treasures he cherishes most, he takes you away from Grünewald's altar to a room filled with tricolours, *pompier* and other helmets and souvenirs [...] from a time that to him is a heroic age". This was consistent with Weber's view of the nation as "a community of sentiment which would adequately manifest itself in a state of its own"; therefore, "a community which normally tends to produce a state of its own" (1994: 25).

---

les plus reculés, 1828; Henri Martin, *Histoire de France populaire*, 1867-1875; Camille Jullian, *Vercingétorix*, 1901; [Philippe Pétain] *De Vercingétorix à Pétain. Le don de soi-même*, Bureau de documentation du chef de l'État, Paris, 1942; Pierre Miquel, *Histoire de France: de Vercingétorix à De Gaulle*, Marabout, Paris, 1996; Georges Bordonove, *Vercingétorix, avec lui commence véritablement l'histoire de la France*, Pygmalion, Paris, 1997. Alse see the documentary film *Vercingétorix: Le roi des guerriers, le héros national, le dernier Gaulois*, by Jérôme Prieur, 2007. For a criticism of such views see André Simon, *Vercingétorix et l'idéologie française*, Imago, Paris, 1989.

[5] Yet, Renan maintained the idea that "a nation is a grand solidarity constituted by the sentiment of sacrifices which one has made and those that one is disposed to make again"; a sentiment which "supposes a past" (1882: 17).

This view was not so far from the idea of nation worked out by Otto Bauer (1882-1938), the leading theorist of Austro-Marxism, who regarded the nation as a "community of fate": a community created by a long historical process with complex economic, social, political and cultural aspects and destined to persist in the future (Bauer, 1907 and 1924)[6].

As for France, the long-standing existence of a centralized nation-state eventually succeeded in imposing a certain cultural homogeneity throughout the country. But recently its great assimilation machinery has shown important signs of grinding to a halt (see Melotti, 2007). Moreover, the traditional tendency to treat regional and local feelings (not to mention immigrants' ethnic identities) as dangerous for social cohesion and national allegiance has become quite obsolete, and even counterproductive, in a world increasingly characterized by strong trends toward some forms of global civil society (see Amselle, 1996; Wieviorka, 1996; Touraine, 1997 and 2010).

In Italy the idea of the nation has long fluctuated between the Romantic view, emphasizing the traditional "objective" elements of ethnic and cultural character (as in the prevailing German political culture) and the civic view, largely inspired by the Enlightenment values, emphasizing the "subjective" factors (as in the prevailing French political culture). In other words, the prevailing Italian position on this subject seems to be a sort of compromise between the two main traditions above-mentioned – both deeply enrooted, though partially "invented", to use Hobsbawm's (1983a) term. But the Italian idea, as we'll see better below, also present original traits, due to the persistent memory of a specific historical past and a unique cultural legacy.

Memory has played an important role also in other countries. A quite notable case is Israel, where the Bible itself has represented, and still represents, an extraordinary source of national memory. In fact, even before the constitution of the Jewish state (1948), the Jewish national movement, tracing back to the mid 19th century (see Melotti, 1965), used the biblical myths to present the "chosen people" as the "first nation" in world history, which had to recover its lost territory (see Walzer, 1985, and Sand, 2008: 377)[7]. In addition, the memory of the *Shoah* (the

---

[6] Otto Bauer was a leader of the Austrian Social Democratic Party and one of the major Marxist theoretician of his time. He criticized mainstream Marxist thought, which neglected or despised the national fact owing to the traditional class analysis and internationalist orientation and integrated his Marxism with a refined analysis of the nation and much interest for the national claims emerging in a great multi-national state like the Habsburg empire.

[7] These two scholars stand among the growing ranks of contemporary academics

destruction of the Jewish people criminally pursued by the Nazis during the Second World War) has been used – and even exploited, according to some (e.g. Finkelstein, 2000) – to justify the *aliyha* (the return or, rather, the "ascent") of the Jewish survivors, their descendants and all legitimate Jews to the "promised land" (which had already been re-promised to the Jews as their "national home" by the British government during the First World War, with the famous Balfour declaration)[8]. Moreover, since its origins, Zionism (the Jewish national movement), especially in the great Askenazy area in central and eastern Europe, had been influenced to a large extent by the German *völkisch* view of the nation (linked to the primordial idea of blood ties) and by the Polish romantic form of nationalism, which, in turn, had been largely affected by Messianic ideas (see Plumiène, 1979)[9]. Anyhow, the *Shoah* is directly remembered – not only in Israel and not only by the Jews – in the so-called "Day of Memory", which is celebrated annually throughout the world[10]. But memory has always played a main role in the Jewish identity. Suffice it to recall the famous Bible verse "Remember what Amalek did to you" (*Deuteronomy* 25: 17), which has become one of the *mitzvot* (precepts) that orthodox Jews must observe daily (Amalek, the tribal chief who, according to *Exodus* 17: 8-10, led a treacherous attack against the faint and weary Hebrews escaping from Egypt, has become the symbol of all enemies the Jews had to face throughout their millenarian history)[11].

---

who are discovering in biblical and rabbinic sources a literature rich with significance for sociological interpretations of modern societies.

[8] This declaration (1917) was the first political recognition of the Zionist aspirations by a great power. Arthur James Balfour was then the foreign minister of the United Kingdom, in war with the Ottoman empire, which incorporated the ancient Palestine, an area including today's Israel and Palestinian territories, as well as parts of Jordan, Lebanon and Syria.

[9] In particular, see its chapter 8 ("The Polish Messianism"), which quotes the explicit references by Polish nationalists to the Bible and its mythical history of the Jewish people.

[10] The annual commemoration of the victims of the Holocaust was even sanctioned on 22 March 2007 by a U.N. resolution, which fixed its date on January 27, the day when in 1945 the surviving prisoners of the Auschwitz *lager* were freed by the Soviet troops. Further, the same year, UNESCO, the U.N. specialized agency for education, science and culture, adopted a resolution to promote Holocaust Remembrance through education.

[11] A Jewish historian, Anna Foa, when she was asked whether memory of past sufferings could entail forgiveness, answered that the issue to discuss was not forgiveness (a Christian idea) but the elaboration of mourning, which was treated by the founder of psychoanalysis, Sigmund Freud, perhaps not by chance of Jewish

Even more generally, according to Elie Wiesel (1990: 14), “to be Jewish means to remember”.

Another important case that deserves some remarks is Greece. In the 19th century Greece was not reborn, as is commonly said; it was created, since between ancient and modern Greece there was no historical continuity. However, in the Greek nation-building memory was of paramount importance. When, after the French revolution, the new national ideals arrived at this remote corner of the Mediterranean area, Greece was under Ottoman rule and before its conquest by the Muslims (after the fall of Constantinople in 1453) it had been under the Roman empire and Byzantium. But even in ancient times there had never been a Greek state. Ancient Greece was a non-harmonious whole of tiny political entities, the *póleis*: independent city states, politically divided and often in open conflict, which were only loosely connected by cultural factors (language, religion, rites and myths), though at times able to unite and to combat together against foreign invaders, such as the Persian “great kings”, in what was perhaps the first “clash of civilization” in Western history. However, at the beginnings of the 19th century, memory of the cultural greatness of ancient Greece influenced both Greek and foreign intellectuals, educated to the cult of the classical world, which had been rediscovered in the Renaissance and had soon become an incomparable example in the resulting Classicism. Later, the diffusion of Romantic political ideals induced many people to regard the contemporary inhabitants of Greece as the legitimate descendants of the revered makers of its ancient civilization, all the more so because they spoke a language much more similar to ancient Greek than the various Romance languages are to Latin. Nevertheless, the modern Greeks were incited to fight for their freedom less by the memory of ancient Greek glory than by the underground work of the Orthodox Church, which had kept high the torch of the Christian faith under Muslim rule. In orthodox milieus the Greeks were even presented as the “new chosen people”, the “new Israel”, the “first-born people of the Gospel”, which other Christians had betrayed under the misleading influence of the Roman Catholic Church or the various national Protestant Churches (see Koliopoulos and Veremis, 2002: 228).

However, the memory of a generally admired cultural heritage influenced the Greek nation-building to a large extent. It was the strong ideological identification with ancient Greece that induced its founders to name the new independent state with the classical term, *Hellas*. Later, the

---

origin (interview by Maurizio Cecchetti, in “Avvenire”, 16 January 2001).

ideological necessity of bridging the gap between classical heritage and Christian Orthodoxy led some Greek scholars to reinstate Byzantium's millenarian empire in the history of the Greek nation – an appropriation or a rehabilitation, depending on the point of view (see Stefanidis, 2007: 12). This was certainly consistent with the irredentist policy of the Greek state (which in 1832 had been established in a territory inhabited by only one quarter of the then Greek population) and with its later expansionistic ideology, the so-called "Great Idea" (*Megali Idea*). But this led to the superimposing of "sentiments" and "consciousness" on language and descent as constitutive elements of the Greek nation.

There is also a mere nationality of memory. A quite surprising case (human beings are often surprising!) is the "Yugoslavian nationality" declared by some minorities of South Slavs, both in the countries of former Yugoslavia and in their substantial Diaspora, after the bloody dissolution of the Socialist Federal Republic in the '90s. Only in Serbia about 80,000 people have claimed to belong to this nationality (the fourth national minority after the Hungarians, the Muslim Bosnians and the Roma) and they regard themselves as the "Kurds of Europe".

In former Yugoslavia, 1943-1991, official designation for those who wanted to define themselves that way was "Yugoslavs" in quotation marks, to distinguish them from the other Yugoslavian citizens (according to the use introduced by the 1971 census). But shortly before the implosion of the country many of those who had declared themselves so reverted to, or adopted, one of the traditional ethnic nationalities officially recognized (Serbs, Croats, Muslim Bosniaks, Slovenes, Albanians, Macedonians, Montenegrins and Hungarians), or accepted one of those recently emerged (including Janjevci, Bunjevci and Šokci), or turned to the old regional identifications, especially in such multi-ethnic historical areas as Istria, Vojvodina and Bosnia. The "Yugoslav" designation, however, continues to be used by thousands of people.

Of course, Yugo-nostalgia is a case in point. This term – comparable to the German term *Ostalgie*, used to indicate the wistful longing for Eastern Germany[12] – defines the psychological, cultural and political attitude of those which miss the former Socialist Federal Republic of Yugoslavia. But the term is used in two distinct ways: as a proud identity claim or as a negative and even derogatory label. In the first way, it defines the

[12] *Ostalgie* – a combination of the words *Ost* (east) and *Nostalgie* (nostalgia) – is the German neologism referring to nostalgia for East Germany. The term (along with the phrase "Soviet chic") is also occasionally used to refer to nostalgia for life in other Eastern European countries that had been ruled by communist parties, most notably Poland and the Soviet Union (see Banchelli, 2006).

emotional attachment to an idealized picture of some aspects of the former Yugoslavian Republic (socialist ideology, economic security, multiculturalism, internationalism, non-alignment, history, customs, traditions etc.), in contrast with the alleged faults of its successor states, which are still burdened by the effects of the cruel ethnic wars between them and the difficulties arising from a complex political and economic transition. In the second way, it is used to blame the anachronistic and unrealistic attitude of the supporters of the previous totalitarian regime. Its manifestations include music activities, art works, films and theater performances and even organized themed tours in the main cities of the former Republic. This Yugo-nostalgia, like the better known *Ostalgie*, is based on the tendency of human beings to not completely eliminate the memory of their past experiences, though contradictorily suppressing their most unpleasant aspects.

In effect, at both the individual and the collective level, memory is quite a selective faculty. It not only serves the need to remember the past; it also serves the need to forget some of its parts. However, whether selectively remembered or selectively forgotten, the elements of the past always shape individual and collective attitudes. In fact, between remembering and forgetting there exists a fundamental dialectic. Moreover, the social context exerts some pressures on memory and, hence, on culture, behaviour and history itself. Especially in collective memory the past is shaped, to a large extent, to serve the present and to prepare the future (Elayyadi, 2004).

Memory acts in this way also as regards political cultures. Renan (1882) made it clear in his famous lecture, the classical text of French nationalism: "Oblivion, and even historical error, is a most essential factor in the creation of a nation"[13]. In this process forgetting is by no means less important than remembering, not to mention the effects of both mistakes and deliberate falsifications on the reconstruction of historical events or their tendentious interpretation and manipulation to adapt them to the prevailing ideologies. The national definition of the present seems to demand justification through a certain idea of the past. Therefore, the relationships between history, memory and nation show a certain

---

[13] However, Renan did not overlook the role of memory. The nation, for him, was "a large-scale solidarity, constituted by the feeling of the sacrifices that one has made in the past and of those that one is prepared to make in the future. It presupposes a past; it is summarized, however, in the present by a tangible fact, namely, consent, the clearly expressed desire to continue a common life". This was the basis on which he based his above-mentioned definition of the nation as "a daily plebiscite": a metaphor that was all too frequently misunderstood.

circularity (see Nora, 1989: 10). By appropriating and re-appropriating the past, memory transforms history into something else and makes it suitable for the ideological purposes and political aims of a society or some of its components.

At least in part, even the writing of history reflects the tensions and conflicts existing in a society. A quite clear example is the historiography of Belgium. In the 19th century many historians maintained that the kingdom proclaimed in 1830 was only an artificial construction, invented for its purposes by European diplomacy (a French minister, Camille Hyacinthe Odilon Barrot, even went so far as to define it as a mere *nationalité de convention*). In contrast, the great Belgian historian Henri Pirenne (1862-1935) tried to show that Belgium was a glorious nation with a proud past and that the Belgian national feeling had appeared many centuries before its political union and independence[14]. But, in the 20th century, after the development of a strong Flemish nationalist movement in its northern part, a new generation of Flemish historians began to reject that idea, which seemed to implicitly support the unity of the Belgian state. Anyhow, from the '70s onwards, both Walloon and Flemish historians appeared much less interested in what the Walloons and the Flemish share (first and foremost the Roman Catholic religion) than in what distinguishes them (language and culture). Therefore, they have begun to write specific histories of the different parts of the country and their prevailing ethnic groups. Thus Belgium has returned to being presented as an artificial construction, easy to divide, according to the increasing separatist claims, which have recently been supported by the results of the latest general elections (June 13, 2010).

Also in Spain the influence of the social and political context on national memory has been quite substantial, especially during the gradual transition to democracy after Franco's death (1975). The civil war and its legacies have represented a potent source of historical division between the left and the right. Memory of the lacerating tragedy of the civil war (1936-1939) between *Nacionales* and *Republicanos* (which caused almost 1 million dead) and the subsequent Franco regime (1939-1975) (which caused another 100,000 dead) persists in spite of the unspoken *pacto del olvido* ("pact of forgetting") aiming at national reconciliation. Even some later events – such as the atrocious train-bombings in Madrid on March 11 2004, which killed 191 people and injured almost 1,900 – have confirmed

---

[14] See Henri Pirenne, *Histoire de Belgique*, Bruxelles, 1899-1932 (seven volumes). This monumental history of Belgium, or, rather, of the Belgian nation, is a real masterpiece. The prominent French historian Marc Bloch wrote that no other country had a work of comparable importance on its history.

Spain's tendency to split deeply over its past (see Balfour and Quiroga, 2007).

The United States is a "nation of immigrants", composed since its origin of many different ethnic groups, sharing residence and citizenship but without a common descent and a common culture. Therefore, it is a country of plural memories and diverse traditions: a real "nation of nationalities", as Horace M. Kallen (1924) aptly called it. However, even the different interpretations of some important events of its history (such as the arrival of the pilgrim fathers, the so-called Revolution, the Civil War and the long conquest of the West) have not threatened its basic ideology: *e pluribus unum* ("One Out of Many"), to say it with the national motto proposed in 1776 by its founding fathers. In subsequent periods, this entailed policies of assimilation, melting pot and salad bowl, not to mention the final "glorious mosaic", in an increasing multicultural perspective, which eventually reversed the initial claim into the new proud statement "Within One, Many". But, especially since the terrorist attacks of September 11 2001, this rather naïve and simplistic form of multiculturalism has met increasing criticisms.[15]

Other significant examples are Russia and the minor post-Soviet states in Eastern Europe. After the collapse of their bureaucratic-collectivist regimes between the late '80s and the early '90s, these countries experienced a profound transformation from their previous totalitarian systems to some forms of more or less democratic kinds of political organization (see Klingemann, Fuchs, Zielonka, 2006). Especially in Russia, Ukraine and Belarus there has been an important re-thinking of the past and its memory has become a fundamental element of their new political cultures. The institutional change itself, which entails new rules, procedures and patterns of behaviour, has been clearly affected by that memory (see Klymenko, 2010).

In effect, a comparative analysis of the transition to democracy in both western and eastern European countries that experienced some totalitarian regimes in the last century permits us to learn a lot about the impact of memory on political cultures and national identity. Their new constitutions seem to float between a past to be absolutely rejected and a future to be imperatively constructed, often leaving aside any realistic consideration of the problems and needs of the present. This, sometimes, does not exclude a partial revaluation of the past (though without open regret), as has

---

[15] For an early criticism of this perspective see Schlesinger, 1992. Among the many books of similar character published after 9/11, see Rubin and Verheul, 2009. It is worth remembering that only a few years before, the sociologist Nathan Glazer (1995) could sarcastically remark "We are all multiculturalists now".

recently occurred in Romania[16], but has occasionally occurred also in Italy, after the Second World War. Such an attitude is not surprising. After some time, the selective nature of the working memory plays even such nasty tricks.

The Italian case

The Italian case deserves particular attention. Also in this country in the last few years a number of titles have appeared, which more or less explicitly connect memory and nation.[17] This is a remarkable novelty, since in Italy political culture has been a theme hardly treated. In fact, in Italy national identity itself has long remained a sort of taboo, due, among other things, to the memory of the Fascist regime, which exploited national feelings for its own aims, with well-known results. Italian scholars have long preferred to investigate the so-called "political subcultures", dwelling upon the distinction between the "red" subculture, prevailing in the areas under the control of the leftist parties of communist and socialist orientation (included the Democratic Party of the Left[18], as the former Communist Party was renamed after the inglorious collapse of the communist regimes in Eastern Europe) and the "white" subculture, prevailing in the areas under the influence of the Roman Catholic Church (even after the dissolution of the Christian Democratic Party, which after 1945, alone or in coalition with minor centre-right or centre-left parties, dominated the Italian political scene for almost half a century). More recently, Italian scholars have focused their attention on the persistent contrasts between the northern, central and southern parts of the country and on the emergent differences between the western and eastern regions in northern Italy. But throughout the country, in the last two decades, disaffection with traditional political parties has considerably increased

---

[16] See Sergio Romano, *La Romania che rivaluta (in parte) il passato*, in "Corriere della Sera", Novembre 27, 2010: 21.

[17] For a conspicuous example see Mario Isnenghi, ed., 1997. This book is clearly inspired by the monumental work edited by the French *académicien* Pierre Nora (1984-92), whose second volume is specifically devoted to the French nation. Nora is a prominent exponent of the *nouvelle histoire* (among his texts translated into English, see Nora, 1989). According to him, memory and history, far from being synonymous, in modern times appear to be in fundamental opposition, since history requires a critical discourse that is antithetical to spontaneous memory (1989: 8-9).

[18] Later re-named Democratic Party, after the fusion with the left of what remained of the Christian Democratic Party.

and the electorate has become much less identified with their obsolete ideologies, boring rites and tricky practices. This, at least in part, explains the diffusion in the northern part of the country of a new movement of federalist and even secessionist orientation, the Northern League, and, almost everywhere, the very rapid success of the pragmatic movement founded and led by an outsider of the old Italian political system, the charismatic entrepreneur Silvio Berlusconi.

As for Italian political culture, the first point to stress, as above-mentioned, is its idea of the nation, which has long fluctuated between the Romantic and the civic views. But this idea also presents original traits, partly due to the long-standing history of the Italian nation. In effect, though the Italian nation-state could be proclaimed only in 1861, culturally speaking, Italy was the most precocious European nation, as was cogently argued (see Cerroni, 1996; Tullio-Altan, 1999: 142). Yet, in the fist half of the19th century some foreign observers could still regard it as a merely "geographical expression", to quote Prince Metternich's words[19]: a reductive, if not derogatory, description, since a nation can exist even without, or before, being constituted in a nation-state. It was not the Risorgimento that made the Italian nation; it was the Italian nation that made the Risorgimento.

However, it was only with the Enlightenment, and more dramatically with the French Revolution and the Napoleonic conquest, that the pre-existing aspirations for an Italian nation-state became clearer and rapidly spread. The French domination in large parts of the peninsula after 1796 introduced the aims of national sovereignty and popular will, but, paradoxically, they were frustrated by the French governments installed in Italy by Napoleon. An emblematic figure in this transitional period between Enlightenment and Romanticism was the poet Ugo Foscolo (1778-1827), an academic, soldier and exile whose fascinating poem, *Dei Sepolcri* (1807), created a patriotic myth around the memories of the nation's greatest men: a sublime effort to seek refuge in the past from the misery of the present and the darkness of the future.[20]

---

[19] It is usually maintained that he pronounced these words during the Congress of Vienna (1815), but in effect he used them only many years later, in a letter to the then British foreign minister Lord Palmerston (August 6, 1847). Prince Klemens von Metternich was a towering figure in the European diplomacy of the first decades of the 19th century. His work was mainly aimed at the restoration of monarchical Europe after the turmoil of the French revolutionary and Napoleonic wars.

[20] Ugo Foscolo, some year before, when he was a young soldier in the French army, had even ingenuously invited a French general, Jean Étienne Championnet,

The situation in Italy worsened further in the subsequent Restoration period, when Austria occupied a large part of the country and harshly repressed the Italian national movement and its periodic desperate outbreaks.

At the beginnings of the Risorgimento (1821), during its first uprising, a great Italian poet and novelist, Alessandro Manzoni (1785-1873), called for a nation "one of arms, language, altars, *memory*, blood and hearts", according to the then prevailing Romantic ideas[21].

Some years later Giuseppe Mazzini (1805-1872) – perhaps the most important apostle of the principle of nationality (one nation, one state), as was recognized by a well-known specialist, the British historian Eric J. Hobsbawm (1992: 31) – invoked a nation characterized not only by its ethnic and cultural traits, but also, and above all, by the voluntarily consent of its members. In his view – of ethic-religious character, but it was clearly a secular, civil religion – such a nation had to be fraternally connected with all other nations in the world, especially those which were still politically divided and oppressed by both foreign and domestic rulers. Therefore, after "Young Italy" (1831), his association for unification, liberation and republican recovery of Italy, together with some German, Polish and Swiss patriots of similar political orientation he founded "Young Europe" (1834), the first supranational democratic association of the continent. However, for him, duties to Humanity took priority over duties to one's country, and this traced a clear-cut distinction between his ideal of the nation and any later degenerated forms of nationalism, leading to prevarications of other nations' rights (see Mazzini, 1972)[22].

---

who was then in northern Ialy, to act for its unity ("Discorso sull'Italia", October 18, 1799; in *Opere*, Milano - Napoli, Ricciardi, 1981, vol. 2). In incisive synthesis: "Proclaim the unity of Italy and all Italians will be with you" (De Sanctis, 1958, vol. 1: 200).

[21] On Manzoni's national sentiment see the pioneering lecture by Francesco De Sanctis, "Il sentimento nazionale italiano e il coro del *Carmagnola*" (1872), in De Sanctis, 1958, vol. 1: 194-213.

[22] Mazzini, however, also underlined the influence of some natural factors, such as the alleged "natural frontiers". He affirmed that God had endowed Italians with "the best defined fatherland in Europe" by establishing around it the Alps, the highest mountains in Europe, on one side, and the Mediterranean sea, on the others (see Mazzini, 1860, chapter "Doveri verso la patria"). This idea may be traced back to Cicero (106-43 B.C.), whose famous words on this subject ("Alpibus Italiam munierat antea natura non sine aliquo divino numine" [Formerly nature had fortified Italy with the Alps, not without any divine intervention], *De Provinciis Consularibus*, XIV, 34) had been often quoted since the fifteen century by humanists and other well-known scholars, such as Niccolò Machiavelli, Francesco

Building on such ideas, Pasquale Stanislao Mancini (1817-1888), a prominent jurist of Turin university, in his 1851 academic inaugural lecture (delivered more than thirty years before Renan's), aptly underlined the limits of the "objective" elements that many scholars still regarded as the natural bases of the nation and, in sharp contrast, clearly stated the importance of a "subjective" factor, the "consciousness of nationality": the sense of belonging to a specific social and cultural unity, even if not yet constituted in a state (as was then the case of Italy).

When the Italian kingdom was formed (1861), its first premier, Camillo Benso di Cavour (1810-1881), presenting the bill conferring the royal title to Victor Emmanuel II, underlined that Italy was a "noble nation", which, after suffering foreign and domestic oppression for three centuries, had become "one of government and institutions" as it was already "one of stock, language, religion, *memories* of its pains and hopes for its complete deliverance".

The existence of such different views among the activists and theorists of the Italian nation-state is not surprising. The Risorgimento was a very long process (it is usually maintained that it lasted from 1796 to 1861, when the Italian kingdom was proclaimed, but actually it did not end then, since some important parts of the country could join it only later: the region of Venice, the province of Mantua and the western part of Friuli in 1866 (after the Austro-Prussian war, which Italy fought against the Austrian empire with a political success only due to the Prussian victory at Sadowa); Rome itself, the living symbol of Italian nationhood, in 1870 (after the military defeat of the secular power of the Catholic Church, supported by the French, after the French debacle at Sedan)[23]; the two north-eastern regions of Trent and Trieste and the eastern part of Friuli in 1918, after the First World War (which Italy mainly fought against their Austrian occupants). In this long period the main Italian thinkers and politicians were inspired by a very complex and even contradictory set of

---

Guicciardini and Baldasar Castiglione, who, in the Renaissance period, rediscovered Greco-Roman patriotism.

[23] Mazzini described Rome as "the dream of *his* youth, the mother of *his* ideas, the religion of *his* soul"; Garibaldi proclaimed that "Rome *was* Italy"; Cavour, shortly before his death, stressed that "Rome alone *had* to be the capital of Italy". Later, the centrality of Rome in Italian imagery was sung by Giovanni Pascoli (1914), who underlined the importance of the memory of its great past. Subsequently, Mussolini himself harped on this rhetorical theme: "Rome is our point of departure and reference; it is our symbol, or if you like, our myth" (quoted by Finer, 1935: 191). In 1932 Mussolini ordered that in all Italian towns a main street should be named after Rome.

values and ideas, which partly re-emerged later, both in the Fascist regime and in the anti-fascist Resistance, and in part still persist today. Moreover, the Italian state, compared to the French, was much less successful in imposing cultural homogeneity throughout the country. Exactly for this reason, when speaking of Italy it is necessary to recall many different municipal memories. But the memory of ancient Rome's greatness has exerted a major influence all over the country, thanks to the important set of mental representations, aesthetic models, juridical traditions and ideological elaborations that have persisted throughout the centuries, after the collapse of its empire. No other city, in fact, has played a similar historical role, not only in Italy but also in Europe and, therefore, in Western civilization (see Giardina and Vauchez, 2000).

As we have mentioned, the Romantic view of the nation re-emerged in the ideological formulations of the Fascist regime, but in a deteriorated and even degenerated form. Fascism eventually reversed the original Romantic view, by theorising that the state creates the nation and not vice versa.[24] In its last years Fascism even accepted the racist theories of its Nazi allies and even their criminal practices. But some forms of both biological and "spiritual" racism (as it was oddly defined) were already present in Italy as a spin-off of the colonial wars in Africa and the subsequent colonial policies (see Gillette, 2002).

This approach aggravated the distrust of foreigners, which had already been fed by the persistent memories of the long sequence of foreign rulers that the country had suffered after the collapse of the Roman Empire. After the barbarian invasions – which in some northern European countries are euphemistically called "migrations of peoples" (*Völkerwanderungen*) but in Italy still evoke the figure of Attila and the destructive raids of the Vandals – there were the dominations by the Arabs and the Normans in southern Italy and later by the Spanish, the French and the Austrians in many parts of the country, not to mention the recurrent Saracen incursions along its coasts and the expeditions and interventions of the German emperors (one of them, Frederick I Barbarossa, was faced and defeated by the Lombard League to the cry "Out with the barbarians!"). Similarly, Machiavelli, seeing Italy suffering "under this foreign inundation", in the famous last chapter of his *Prince* (1513), in contradiction with his previous concern for the state and not for the nation,

[24] See "Fascismo", in *Enciclopedia italiana*, Milano, 1932: 847-884, *sub voce*. This entry, which was signed by Benito Mussolini himself, was mostly written by one of the main theorists of Fascism, the neo-idealist philosopher Giovanni Gentile (see Emilio Gentile, 2006: 181).

exhorted Italian rulers to unite and fight together to liberate Italy from the barbarians.

This long-standing story has deeply marked the Italian collective memory. It is not by chance that the refrain of the so-called *Garibaldi's Hymn* (the patriotic anthem written in 1858 at the request of the most popular Italian hero) harshly thunders "Go out of Italy, go out by now! / Go out of Italy, oh foreigner, go out!". And, similarly, the most important Italian song of the First World War, *La leggenda del Piave* (1918), boldly proclaims "No foreigner will pass!".

The effects of this defensive idea of the nation became quite clear when, after being for a century the first European country of emigration, Italy became a country of immigration (the turning point was reached in 1973). In sharp contrast, the co-existent components of the Italian idea of nation going back to Enlightenment ideas have inspired a sort of cosmopolitan wishful thinking that quite recently (under other influences, mainly coming from Canada and the United States), has nourished a superficial and rhetorical form of multiculturalism (Melotti, 2000b).

In Italy the national sentiment, according to current sociological research by both Italian and foreign scholars (see Segatti, 1999; Haller, 2001), is far weaker than in all other European countries. Among the main historical reasons, we must recall that since its early literary origins – going back to Dante Alighieri (1265-1321)[25], Francesco Petrarch (1304-1374)[26] and Niccolò Machiavelli (1469-1527)[27] – this sentiment has been tempered by the cosmopolitan heritage of the Roman Empire and the Roman Catholic Church. In particular, there was a special difficulty in inserting the first seeds of the emergent national sentiment into a collective memory that had been profoundly nourished by the old myths, ideas and traditions of the Catholic Church.

Moreover, owing to the depressing effects of internal conflicts and foreign rule, this sentiment was directly contradicted by a strong tendency to mind one's own affairs rather than the public interest. This was even

---

[25] In particular see *Purgatory*, VI, 76-78 ("Ah! servile Italy, grief's hostelry! / A ship without a pilot in great tempest! / No Lady thou of Provinces, but brothel!"), and *Paradise* VI, 76-151, where Dante sings the apotheosis of ancient Rome. This induced some to define him as "the first Italian". But Italy was not his main political aim. He, rather, called for an anachronistic restoration of the two "universal" powers of the late Middle Ages: the Empire (in the political and military field) and the Papacy (in the spiritual and moral field).

[26] See *Canzoniere*, 53, and, especially, 128 ("My Italy, although talking does not serve to heal the mortal wounds which I see so thick on your fair body...".

[27] In particular, see *Prince* (1513), chapter 26.

theorized by some important authors of the Renaissance age – such as Leon Battista Alberti (1404-1472) and Francesco Guicciardini (1483-1540) – which in Italy, in spite of a luxuriant flourishing of arts and literature, was a period of deep political crisis. Anyhow, Italians, who lived for centuries under rapacious foreign oppressors and inefficient domestic sovereigns, to survive were obliged to develop private virtues and public vices (see Barzini, 1964).

More recently "civil religion", to use Rousseau's well-known expression, or the "sense of state", to use the old liberal definition, or rather "civicness", as modern sociologists prefer to say, has undergone the attacks of the two subcultures above mentioned: the one characterized by the ecumenical orientation of the Roman Catholic Church and the other marked by the internationalist stance of the leftist parties and movements. It is worth recalling that many European countries (England, Scotland, Switzerland, Sweden, Greece, Russia etc.) have their national Churches and others (such as France) have a long-standing secularized tradition, while in Italy the Roman Catholic Church (spiritual heir of the Roman *imperium*) has almost always played a central role in political life. In addition, in Italy there was the largest communist party in Western Europe and many of its members are still active on the political scene, though under new labels that enabled its main leaders to take part in some leftist governments (in 1996-2001 and 2006-2008) and even to act as premiers in two of them, overcoming a semi-secular *conventio ad excludendum* (the tacit agreement that, during the cold war, excluded them from government, owing to their close relationships with the Soviet Union).

In the same period there was another important development: a widespread reaction against bad government, centralism, corruption and excessive taxation, not to mention the notorious ties between some members of the central government and the criminal organizations originally based in the southern regions but now present almost everywhere. This has fostered the expansion in northern Italy of an autonomist movement, the Northern League, characterized by both anti-national feelings and strong xenophobic overtones.

Besides, Italy is still affected by its very special historical legacy. The Italian national state was formed comparatively late: after all the others in western Europe except for Germany, if we consider the date of its formal proclamation, but even after that country, if we do not forget that some important regions could join the Italian kingdom only later. Moreover, in Italy the formation of the national state was not due to the initiative of the core part of the country, as occurred not only in France, Britain and Germany but also in Spain, Portugal and Greece, all of which offer important examples of the "ethnic origins of nations" (Smith, 1986)[28]. In contrast, in Italy the first steps of unification were taken by a small, peripheral and rather backward kingdom, the then still ethnically and linguistically composite Piedmont, whose ruling class could hardly speak Italian.

Last but not least, the Italian national state was formed without (and even against) the masses, which were largely pro-Bourbon in the southern regions and Catholic and socialist in the central and northern ones, not to mention those of anarchical orientation, much less numerous but widespread all over the country (Italy, with Russia, Spain and, to a lesser extent, France, was among the European countries with the highest number of anarchists). The masses did not play an important part in nation-building, as Gramsci (1949) remarked, despite the idealized picture of the Risorgimento later "invented" for ideological reasons. Their "nationalization" – to use Mosse's term (1974) – i.e. their social, cultural and political integration into the nation-state, with partial ideological identification with it (which usually requires their education or re-education in national memories, traditions, values, myths and symbols), could be pursued only after the establishment of the nation-state. This seems to be the real meaning of the oft-quoted phrase attributed to Massimo d'Azeglio: "We have made Italy, now we must make the Italians")[29]. This task, soon after Italy's unification, was entrusted mainly to the army and the state schools.

---

[28] Of course, even in these states ideology has played an important role in nation-building (see Albertini, 1960), but this does not justify the superficial theories reducing the nations to "imagined communities" (according to Anderson, 1991). Mario Albertini, though still little-known in the English-speaking countries, is one of the main theoreticians of the nation or, rather, of its demystifiers. His ideas, however, need to be both criticized and developed (see Melotti, 1999).

[29] According to some sources, he pronounced this phrase in 1861, during the first meeting of the Italian Parliament (see Latham, 1970). But, probably, he never pronounced these words. Much less rhetorically, in his memories (posthumously published in 1867), he wrote: "Italy has been made, but, unfortunately, Italians are not in train of being made" (a phrase, however, not present in all editions). In 1877 the great historian of Italian literature, Francesco De Sanctis, who had also been a minister of Public Instruction in the two first Italian governments, paraphrased it in this way: "As d'Azeglio said, it is useless to have formed Italy without the Italians". According to other authors, in the form reported in the text the phrase was pronounced only many years later by a minister of Public Instruction, Ferdinando Martini, during the serious crisis of national identity provoked by the military debacle in Adwa (1896), during the first war against Ethiopia: perhaps the greatest European defeat by Africans since Hannibal's time (see Soldani and Turi, 1993: 17).

The nation-building drive in education is clearly exemplified by Edmondo De Amicis' "Cuore" (1886), a hugely popular school-reader inspired by petty bourgeois principles, social feelings and patriotic ideals. Also the massive occupation of urban spaces played a considerable role: squares, streets and monuments were devoted to the main historical exponents of Italian culture (from Dante, Petrarch, Boccaccio, Leonardo da Vinci, Michelangelo, Giordano Bruno, Galileo etc. to the contemporary Manzoni and Verdi) and to the greatest Risorgimento figures (first and foremost Victor Emmanuel II, Giuseppe Garibaldi and Camillo Benso di Cavour, of course, but also Mazzini, who had to live his last days under a false name, as an exile in his country, because of his never disavowed Republican militancy, but after his death, owing to his appreciated unitary vision, was assumed in the syncretistic pantheon of the new state in search of its civil religion).[30] Also the colonial wars in Africa contributed to the nationalisation of the masses. It seemed to many that they marked the re-awakening of the "great proletarian" nation, as at the beginning of the Libyan war the academic poet Giovanni Pascoli (1911) defined Italy, conceding too much to the incipient nationalist rhetoric.[31] The process was later continued during the First World War, and certainly not in a painless way. It was then fostered by the Fascist regime, though in a contradictory and even counterproductive way that eventually increased their pre-existing distrust towards the state. The Resistance to Fascism and Nazism was important, especially in the last phase of the Second World War and mainly in the northern part of the country, but it could hardly change the situation, since, in contrast with what occurred in France and in many other countries, its partisan social and political aims seemed to prevail over the national ones (though not absent in most fighters), and this impression was later accrued by the self-assured attempt of the Communist party to seize its heritage.

However, the 1948 Italian Constitution, which clearly embodies many important political memories consciously or unconsciously selected by its makers, has become a sort of ideological heritage common to most Italians, despite their persistent divisions. This reduced the level of conflicts

[30] Similar practices were carried out in the same period in other European countries, foremost Germany, the other *verspätete nation* (see Hobsbawm, 1983b).

[31] The definition of Italy as a "proletarian nation" was due to an activist of the nationalist right-wing movement, Enrico Corradini (1865-1931), co-founder of the journal "Il Regno" (1903-1905) and of the Italian Nationalist Association (1910), which after the First World War merged with the Fascist party. Corradini had roughly distinguished between "proletarian" and "plutocratic" nations, anticipating a main argument of fascist propaganda.

between the main Italian political organizations: the pro-Western governmental parties led by the Christian Democrats and the leftist opposition parties led by the Communists, then closely tied to (and partly financed by) the Soviet Union. Only recently has collective memory begun to become less sharply divided, but quarrels and factiousness have not decreased. Actually, they have even increased to an intolerable extent.

It is, therefore, not surprising that one of the most influential Italian journalists of the post-war period, Indro Montanelli, who loved his country sincerely and profoundly, could bitterly say: "Italy is no longer my homeland; for me it is only the regret for a homeland".[32]

These words, however, seem to me too pessimistic. Or, at least, I would like them to be so.

References

Albertini, M. (1960). *Lo Stato nazionale*, Milano: Giuffrè (French transl. *L'État national*, Lyon: Fédérop 1978).

Almond, G. A. (1956), "Comparative political systems", in *The Journal of Politics*, 18: 391-409.

—. (1990). "The study of political culture", in *A Discipline Divided: Schools and Sects in Political Science*, Newbury Park, Cal.: Sage (reprinted in Lane Crothers and Charles Lockhart, eds., *Culture and Politics: A Reader*, 5-20. New York: St. Martin's, 2000).

Almond, G. A., and Verba, S. (1963). *The Civic Culture: Political Attitudes and Democracy in Five Nations*, Princeton, N.J.: Princeton University Press.

Amselle, J.-L. (1996). *Vers un multiculturalisme français*, Paris: Aubier.

Anderson, B. (1991), *Imagined Communities*, London: Verso.

Balfour, S., and Quiroga, Alejandro. (2007). *The Reinvention of Spain: Nation and Identity since Democracy*, Oxford: Oxford University Press.

Banchelli, E. (2006). *Taste the East: Linguaggi e forme dell'Ostalgie*, Bergamo: Sestante.

Barzini, L., Jr. (1964). *The Italians*, New York: Atheneum.

Bauer, O. (1907). "Die Nationalitätenfrage und die Sozialdemokratie", Wien, in *Marx-Studien. Blätter zur Theorie und Politik des wissenschaftlichen Sozialismus*, vol. 2.

[32] Quoted by Gian Antonio Stella, "Quell'amor di patria che il potere non ama", in *Corriere della Sera*, November 28: 6.

—. (1924). Preface to *Die Nationalitätenfrage und die Sozialdemokratie*, 2nd edition, Wien: Verlag der Wiener Volksbuchhandlung (English transl. *The Question of Nationalities and Social Democracy*, Minneapolis - London: University of Minnesota Press, 2000).

Cavour, C. B. (conte di). (1861). Speech at the Italian Chambers of Deputies, March 11. In *Discorsi parlamentari del conte Camillo di Cavour, raccolti e pubblicati per ordine della Camera dei deputati*, Firenze: Botta, 1863.

Cazzullo, A. (2010). *Viva l'Italia! Risorgimento e Resistenza*, Milano: Mondadori.

Cerroni, U. (1996). *L'identità civile degli italiani*, Lecce: Manni.

D'Azeglio, M. (1887). *I miei ricordi*, Firenze: Barbèra (2 volumes).

De Felice, R. (1975). *Intervista sul fascismo*, Bari - Roma: Laterza (English transl. *Fascism. An Informal Introduction to its Theory and Practice*, New Brunswick, N.J.: Transaction Books, 1976).

De Sanctis, F. (1958) *Storia della letteratura italiana nel secolo XIX*, edited by Alberto Asor Rosa, Milano: Feltrinelli (4 volumes).

Elayyadi, A. (2004), "Post-colonial immigration in France: history, memory and space", thesis, Miami University at Oxford, Ohio.

Fichte, J. G. (1808). *Rede an die deutsche Nation*, Berlin: Real Schulbuchhandlung (English transl. *Addresses to the German Nation*, edited by George A. Kelly, New York: Harper Torch Books, 1968).

Finer, H. (1935). *Mussolini's Italy*, New York: Holt.

Finkelstein, N. G. (2000). *The Holocaust Industry*, London - New York: Verso.

Gentile, E. (2006). *La Grande Italia. Il mito della nazione nel XX secolo*, Roma - Bari: Laterza.

Giardina, A., and Vauchez, A.. (2000), *Il mito di Roma*, Roma - Bari: Laterza (2nd ed. 2008).

Gillette, A. (2002). *Racist theories in Fascist Italy*, London: Routledge.

Glazer, N. (1995). *We are all multiculturalists now*, Cambridge, Mass.: Harvard University Press.

Gramsci, A. (1949). *Il Risorgimento*, Torino: Einaudi [selection from *Quaderni dal Carcere*, 1929-1935] (English transl. *Prison Notebooks*, New York: Columbia U.P., vol. 1, 1992 - vol. 2, 1996).

Haller, M. (2001). "Identità e orgoglio nazionale in Europa", in Gaspare Nevola, ed., *Una patria per gli italiani?*", Roma: Carocci.

Hobsbawm, E. J. (1983a). "Introduction: Inventing Traditions", in Eric J. Hobsbawm and Terence Ranger, eds., *The Invention of Tradition*, 1-14. Cambridge: Cambridge University Press.

—. (1983b). "Mass-producing traditions: Europe, 1870-1914", in Hobsbawm and Ranger, eds., q.v., 263-307.

—. (1992). *Nations and Nationalism since 1780*, Cambridge: Cambridge University Press (1st ed. 1990).

Hutchinson, J., and Smith, A. D., eds. (1994), *Nationalism*, Oxford: Oxford University Press.

Isnenghi, M., ed. (1997). *I luoghi della memoria*, Roma - Bari: Laterza (three volumes); 2nd ed. 2010.

Kallen, H. M. (1924). *Culture and Democracy in the United States. Study in the Group Psychology of the American People*, New York: Boni and Liveright.

Klymenko, L. (2010). *Memory, political culture and democratization in post-Soviet Ukraine, Russia and Belarus*, Symposium "New Perspective on Contemporary Ukraine", University of Toronto, January 21-23.

Klingemann, H.-D, R, Fuchs, D., Zielonka, Jan, eds.(2006). *Democracy and Political Culture in Eastern Europe*, Abingdon, Oxon - New York: Routledge.

Koliopoulos, J.ohn S., Veremis, T. M. (2002). *Greece. The Modern Sequel from 1831 to the present*, London: Hurst.

Latham, E. (1970). *Famous Sayings and theirs Authors*, Detroit: Gale.

Laven, D. (2006). *Italy. The Idea of the Nation in the Risorgimento and Liberal Eras*, in Timothy Baycroft and Mark Hewitson, ed., *What is a Nation? Europe 1789-1914*, 255-271, Oxford: Oxford University Press.

Machiavelli, N. (1513). *Il Principe* (English transl. *The Prince*, Chicago: University of Chicago Press, 1985).

Mancini, P. S. (1851). *Della nazionalità come fondamento al diritto delle genti*, Torino: Bocca.

Manzoni, A. (1821). *Marzo 1821* (firstly published in *Pochi versi inediti di Alessandro Manzoni*, Milano: Redaelli, 1848).

Mazzini, G. (1860). *Dei doveri dell'uomo*, Napoli: Zomack.

—. 1972. *Scritti politici*, Torino: Utet.

Meinecke, F. (1907). *Weltbürgertum und Nationalstaat: Studien zur Genesis des deutschen Nationalstaates*, München: Oldenburgh (English transl. *Cosmopolitanism and the National State*, Princeton, N.J.: Princeton University Press, 1970).

Melotti, U. (1965). *Premesse del Sionismo nella storia delle dottrine politiche*, followed by *Stato e Torah in Israele*, Milano: Centro Studi Terzo Mondo.

—. (1999). *Etnicità, nazionalità, cittadinanza*, Roma: Seam.

—. (2000). "L'immigrazione e la nazione italiana", in Bianca Maria Pirani, ed., *L'abbaglio dell'Occidente*, pp. 169-188. Roma: Bulzoni.

—. (2000b). *L'abbaglio multiculturale*, Roma: Seam.

—. (2007). *Le "banlieues": immigrazione e conflitti urbani in Europa*, Roma: Meltemi.

—. (2008a). *Migration, political cultures and urban conflicts in Europe*, in 8 2008 b), in Bianca Maria Pirani and Ivan Varga, eds., *The New Boundaries between Bodies and Technologies*, 106-123. Newcastle: Cambridge Scholars Publishing.

—. (2008b). *Migratory policies and political cultures in Europe: Is there something new?* in Dossier "European Governance of Migration", Berlin: Heinrich Böll Stiftung.

—. (2010). *Increasing diversity in a changing Europe*, in Bianca Maria Pirani and Ivan Varga, eds., *Acting Bodies and Social Networks*, 61-85. Lanham, Md.: University Press of America.

Mosse, G.L. (1974). *The Nationalization of the Masses*. New York: Howard Fertig.

Nora, P. ed. (1984-92). *Les lieux de mémoire*, Paris: Gallimard (seven volumes).

—. (1989). "Between memory and history: les lieux de mémoire, in *Representations*, 26: 7-24 (special issue "Memory and Counter-Memory").

Pascoli, G. (1911). "La grande proletaria si è mossa" (speech, Barga, November 26). In *Prose di Giovanni Pascoli*, Milano: Mondadori, 1946, vol. 1, 557-558).

—. (1914). *Patria e umanità*, Bologna: Zanichelli.

Pasquino, G. (2010). *Memory, political culture and institutional change: a comparative perspective* (working paper), New York: Italian Academy for Advanced Studies, Columbia University.

Plumyène, J. (1979). *Les nations romantiques*, Paris: Fayard.

Renan, E. (1882). *Qu'est-ce qu'une nation?* (lecture delivered at the Sorbonne, 11 March 1882), in *Oeuvres Completes*, Paris: Calmann Lévy, 1947-61, I, 887-907 (English partial transl. in John Hutchinson and Anthony D. Smith, eds., Oxford - New York: Oxford University Press, 1994, 17-18).

Rosati, M. (2000). *Il patriottismo italiano*, Bari - Roma, Laterza.

Rubin, Derek and Verheul, J., eds. (2009), *American Multiculturalism after 9/11*, Amsterdam: Amsterdam U.P.

Salvatorelli, L. (1957). *Pensiero e azione del Risorgimento*, Torino: Einaudi.

Sand, S. (2008). *Matay ve'eich humtsa ha'am hayehudi?* [*When and how was the Jewish people invented?*], Tel Aviv: Resling (French transl.

Comment le peuple juif fut inventé - De la Bible au sionisme, Paris: Fayard, 2008; English transl. *The Invention of the Jewish People*, New York: Verso, 2009; Italian transl. *L'invenzione del popolo ebraico*, Milano: Rizzoli, 2010).

Segatti, P. (1999). "Perché è debole la coscienza nazionale degli italiani", in *Il Mulino*, 1: 15-23.

Schlesinger, A. M., Jr. (1992). *The Disuniting of America. Reflection on a Multicultural Society*, New York: Norton (2nd rev. ed. 1998).

Smith, A. D. (1986). *The Ethnic Origins of Nations*, Oxford: Blackwell.

Soldani, S. and Turi, G. (1993). *Fare gli italiani. Scuola e cultura nell'Italia contemporanea*, Bologna: Il Mulino.

Stefanidis, I. D. (2007). *Stirring the Greek Nation: Political Culture, Irredentism and Anti-Americanism in Post-war Greece, 1945-1967*, Aldershot: Ashgate.

Touraine, A. (1997). *Pourrons-nous vivre ensemble? Égaux et différents*, Paris: Fayard (English transl. *Can we live together? Equality and Difference*, Cambridge: Polity Press - Stanford, Cal.: Stanford University Press, 2000).

—. (2010). *Come si difende il multiculturalismo*, in "Reset", 122: 97-104.

Tullio-Altan, C. (1999). *Gli italiani in Europa*, Bologna: Il Mulino.

Verba, S. (1965). "Comparative political culture", in Lucian W. Pye L. and Verba S., eds., *Political Culture and Political Development*, 512-560. Princeton, N.J.: Princeton University Press.

Walzer, M. (1985). *Exodus and Revolution*, New York: Basic Books.

Weber, M. (1922). *Wirtschaft und Gesellschaft*, Tübingen: Möhr (English transl. *Economy and Society*, Berkeley, Cal.: University of California Press).

—. 1994. "The Nation" (transl. from *Wirtschaft und Gesellschaft*), in John Hutchinson and Anthony D. Smith, eds., 21-25. Oxford - New York: Oxford University Press.

White, S. (1979). *Political Culture and Soviet Politics*, London: Macmillan.

Wiesel, E. (1990). *From the Kingdom of Memory. Reminiscences*, New York: Schocken Books.

Wieviorka, M. (1996). *Une société fragmentée? Le multiculturalisme en débat*, Paris: La Découverte.

CHAPTER FOURTEEN

THINKING HANDS, HANDLING THOUGHT: THE HAND AS METHOD

FLORENT GAUDEZ
AND SOPHIE POIROT-DELPECH

Abstract

This chapter aims to show how 'the hand is the method' which, as both a tool and a sign, ties us to the world, to others and to ourselves. By following the hand's 'path' (*methodos*), by empathy, the socio-anthropological perspective draws together Knowledge and Emotion, Mediation and Technique.

Marcel Mauss (speaking of Maurice Halbwachs) stated that 'Man is an animal that thinks with its hands', however 'if the hand tends towards the brain, it is perhaps because the brain is imminently manual' (André Leroi-Gourhan). This perspective enables us to think simultaneously of knowledge as *pathos* and of the artefact within its cognitive and affective dimensions, both of which constitute an exploration and an experimentation of the human and natural worlds. By insisting on the importance of the manual metaphor in durkheimian sociology (the social fact as 'manner'), this article reconsiders the notions of the social fact and of representation, particularly from the starting point of memory theory (Halbwachs) as a means of reorganisation and recognition. Hands are not only necessary for creating, but for conceptualising inventions intuitively; we need them in order to experience resistance and, therefore, in order to interrogate, to question that which will become 'real'. Finally, from a technological perspective, hands are necessary for appropriating and imagining that which, in the contemporary sciences, is of an unrepresentable order (at the nanoscale level, for example). The hand is indeed that which Man does, as well as the condition in which he realises his *Anthropos*; it is also a manifestation: of his intelligence and of his feelings, in other words, an anthropophany.

"Hands are necessary for instituting language,
to mime the action that will become the verb"
—Valéry

"It is not because a man has hands that he is intelligent,
but it is because he is intelligent that he has hands."
—Aristotle

Introduction

The neurosciences do not have the monopoly on discussions regarding the brain, in which the body and emotions are taken into consideration. Furthermore, such discussions cannot be reduced to a cognitivist bias in a simplistic and reductive manner. To ignore the importance of these elements and the relationships between them, therefore, is to allow free reign to the hegemony of a trend.

In proposing a socio-anthropological perspective on the conditions and the possibilities of such an approach, we intend, here and now, to propose a counter-attack.

The socio-anthropologist strives for this position between the Self and otherness, the societal, the symbolic, the particular and the doubt. One particularity of the socio-anthropological position is its recognition of the fact that knowledge transforms, translates, transposes and gives sense to an experience that is never a given in and of itself.

This invites us to imagine the socio-anthropological position as a mode of social knowledge, while taking into consideration the fact that knowledge is, for the socio-anthropologist, subject, object and tool. From this perspective, the total social fact appears, in its essential symbolizing function, as the foundation of a very particular modality of self-knowledge that is primarily constructed through reflexive sociological work on oneself with regard to its connections to ethnology, history, and so forth.[1]

It is for this reason that we believe that it is important to reflect upon a Socio-Anthropology of Knowledge (including the sciences and technology), whose objective is to understand the role of the societal in the invention and production of the human and its *material* and *ideal* institutions, thereby enabling us, as Claude Lévi-Strauss suggests, "to reconquer our most foreign selves, and to allow us to access that which is the most foreign in others as if it were another self" (Levi-Strauss, 1985 : p. xxxi).

[1] For more on the way in which JC Passeron uses the notion of t Total Social Fact as a nodal symbolic operator, (,see *Le raisonnement sociologique. L'espace non-poppérien de l'argumentation*, Paris: Albin Michel, 2006, pp. 83-4 and pp. 500-2).

At a time when a certain "cognitivism" tends to saturate academic fields concerned with explaining of action and production – whether directly or obliquely – these questions aim to (re)affirm the fertility of a socio-anthropological approach.

Ipseity and Pathos

This is where the notion of *Ipseity* intervenes. How is it possible to distinguish between identity and *ipseity*? Drawing on Paul Ricoeur (1984), we can posit that, on one hand, identity is the product of "sameness" (*Idem*), while on the other, *ipseity* is the product of "one's self" (*Ipse*), a distinction that equally allows Ricoeur to underline the autonomy of the subject.

We thus find ourselves confronted with a coupling of semantic oppositions: Duplication/imitation versus reconstruction (of a self). However, if self-knowledge is paradoxical in a way that requires the meditation of others, it is nevertheless insufficient to merely observe (Alter/others) in order to know (Alter/other). It is also essential to think (Alter/others).

And in order to think, it is necessary to suffer. Here, we are using the term "Pathos" (suffering) in the Greek sense of the word: not in the simple and limited vulgar sense (of pain), but in the complex, sophisticated dialectic of pain versus pleasure; of every feeling, every experience. Moreover, this conception of suffering not only relates to the physical, but to the social, as well: thus, to "suffer with". Here, it is necessary to briefly link this concept to that of Spinoza, in which the passions are neither accidental nor a result of the cruelty of men, but, rather, constitute a part of their reality. Therefore, beyond the form of the object of knowledge – whatever it may be – it is the universality of suffering that remains to be explored. Moreover, the knowledge of suffering does not only occur through the meditation of the intellect, but also through that of the body.

Thus, when confronted with the concept of "suffering with", one thinks etymologically of sympathy resulting from the act of participating (intersubjectivity) in the joy or in the sorrow (pain) of others, that is to say, in *Pathos*. However, it is also a feeling, a spontaneous feeling of benevolence towards someone or something. Sympathy is simply the act of living, of feeling, of experiencing something with someone, together, in the same moment. To unite something in the order of Pathos: marrying, mating... (concepts, people, experiences...)

Empathy (Ability and/or Competence)

Empathy encompasses the collective, social and cultural aspects of the relationship to the subject, particularly that which concerns knowledge of the other by the *Social Me* (in other words, the Otherness/*Ipseity* dialectic).

Empathy: "putting myself in someone else's shoes": this is not it!

Empathy is not to put oneself in the place of others! We are never in someone else's place. We already have enough difficulty being in our own place (constructing it, remaining within it, understanding it, in short, learning how to "stay in one's place" in order to better displace ourselves if need be). Yet, we can never displace ourselves (autonomously) if we do not "know" our place. This, moreover, recalls Edmond Rostand's formula: "In life, one can only count on oneself, and even then, not much'.

Hence, empathy is the ability to be capable of moving oneself, of shifting one's own point of view. It is the ability that we develop – to a greater or lesser degree – of representing the suffering of the other for ourselves, and our attempt to reconstruct it; to produce a representation, a construction, of another suffering, of another pathos.

Let us recall Husserl's contribution in this domain:

- (Husserl) elaborated the notion of empathy as the ability to include the environments of others within our own environment,
- This effectively extends one's field of experience to encompass the dimensions of a world.
- But inasmuch as all subjectivity is constitutive of its field of experience, if I am constitutive of my field of experience, the subjectivity of others never ceases to be constitutive of others, just as when I extend my field of experience in order to encompass that of others (of which I try to imagine, through empathy, how their subjectivity is constitutive), my broadened field of experience is not subjective, but intersubjective, because at least part of it is comprised of both my subjectivity and that of others.
- Husserl did not view empathy as the foundation of intersubjectivity. On the contrary, he saw intersubjectivity as the starting point for an empathetic intelligence.
- Empathy is a relationship of equivalency (in the mathematical sense of the term) between the consciousness. Meaning, in other words, that it is a relationship that is simultaneously reflexive, symmetrical and transitive :

- Reflexive: in the form of Cogito (I do not doubt that I doubt), or the mirror stage; o Symmetrical: in the form of an injunction: "Imagine my suffering and I will imagine yours!";
- Transitive: in the form of an injunction: "Imagine their suffering".

In this manner, we ultimately avoid two illusory pitfalls: egocentrism (I imagine that everyone sees, feels and understands the world through my eyes); and *altercentrism* (I imagine that I am "in the place" of others, and that I am in a position (ability, competence) to see, feel and understand the world through everyone else's eyes). Ability/competence (a job, a production and a meditation) consists of consciousness of the above-described limits, and employing these limits to construct a weapon or an art in order to better think, to think of, to think about and to think with Alter (others) while, at the same time, trying to shift my own point of view in multiple ways.

A weapon or an art? A martial art, and even a martial art of the mind. Indeed, just as the archer surpasses the master in his art, the accomplished socio-anthropologist becomes *in fine* his or her own target. Meaning, in other words, that, in passing by the way of others, they arrive at a better understanding of themselves.

Otherness and knowledge

Furthermore, the effect of empathy must be perceived as a cognitive and discursive construction: Thus, the notion of *Epistemes* aims to designate the process during which the epistemic subject is engaged in a cognitive process in which otherness is, just as is the case with *ipseity*, an object, thus showing how the subject negotiates the problem of self/self and self/other.[2] The above formulation of this notion stems from a thought experiment based on a short story by Julio Cortazar (1973). The "Axolotl" is not an epistemological treaty, but rather describes a route of knowledge whereby knowledge of the other passes through self-knowledge and through the recognition of the other in oneself. This ultimately constitutes a method because it is the path that the knowledgeable subject must follow in order to raise his knowledge to the level of its object (Leenhardt, 1986). This is not the place to address this particular issue in more detail, as the

[2] Formulated and introduced by Florent Gaudez during the international colloquium on "Otherness" organised in 2003 at the University of Athens. Actes du Colloque *: Alterité et Société*, Athènes, CD-ROM AISLF (2004).

study has been published elsewhere. However, let us briefly add to this Paul Ricoeur's (1983 : 85) assertion that " between the activity of *telling a story* and the temporal character of the human experience, there exists a correlation that is not purely accidental, but presents a form of transcultural necessity" .

Let us be lucid!

For the researcher, lucidity – which, according to René Char, is "the closest wound to the sun" – also occurs when the nature of the relationship between the observer and the observed object is taken into account. Finally, the difficulty in recognizing the other in oneself, portrayed in "Axolotl" through the animalistic figure, relates back to the socio-anthropologist's quest which, while being a quest for the other, is also a quest for the other in oneself and, by definition, a quest for oneself by means of a reflexive anthropology; the conjunction between the subject and the object being an eternal problem in the human sciences[3].

Therefore, these points of view, *i.e.* these representations, *i.e.* these constructions of reality all vary for socio-anthropologists along three axes: Time, Space, Culture (the *chronos*, the spatial and the actors). These three dimensions can be illustrated by referring to three formulas:

- Spatial: "The earth does not want new continents, but new men" dixit Captain Nemo (a variation of Lévi-Strauss's "I hate voyages and explorers".).
- Chronological: here, we summon the old master André Leroi-Gourhan, "It is not stones I am looking for, but men".
- Actors: Let us call upon our old friend Michel Verret for support: "I like when men meet one another, especially when they come from afar"; however this "afar" is not necessarily spatial or chronological, but can, for example, refer to my distant (next door) neighbour.

At this point, we encounter the general movement of a returning alteration, of an understanding of the here by way of elsewhere, of a possible co-production between the near and the far. This idea of otherness, of *"I' is others*, also refers us back to Kant, who affirmed that "The human being is not a human because he is a relational being". In short, these are the

---

[3] See *Une Socio-Anthropologie du texte littéraire. Approche sociologique du Texte-acteur chez Julio Cortázar*, Paris, L'Harmattan, 1997. (collection "Logiques sociales").

central preoccupations of the socio- anthropologist who attempts to construct this position between the Self and the other, the societal, the symbolic, the particular, the doubt, the uncertainty, even Deleuze's concept of "*uneasiness*".

As Confucius said: "Look at yourself conscientiously, look at everyone else just discreetly."

This is why it is also crucial to insist upon the importance of empathy with the subject as being central to the socio-anthropological approach according to C. Léonardi's formula: "The other, that untamed strangeness that ebbs away from us each time that we believe that we have begun to define it, and who asks that we, too, endeavour to continually recreate otherness within our own methods and in our own categories of thought".[4]

Indeed, as far as the question of method goes, everything is here.[5] *Meta-Hodos*: a pathway associated with the idea of overcoming, of going beyond and, in so doing, of making connections...One could say "the path that connects", *i.e.*, intelligence; one could also describe it as "climbing a trail": investigating, employing a hypothetical process, a process of abduction...The chase, the search, the indirect and indeterminate path. Michel Verret (1999 :12) said that "There is no route, it is necessary to forge a new one (*i.e.* to connect), but by drawing on the emotions of this tortuous travail". *The Method* is the overcoming, the course, the crossing, the short cuts (bridges and doors, as Simmel would say).[6] But there is no transmission without a trajectory. Thought travels along a path, and intelligence links and reformulates incessantly.

From knowledge as Acknowledgement and *re-handling*, to the problem of techniques

The sociology of knowledge and the sociology of techniques traditionally evolve along parallel lines (in the domains of the arts and the sciences in particular).[7] Here, the socio-anthropological position will attempt to cross

[4] On the occasion of the 2007 socio-anthropology meetings at the Sorbonne, in Caroline Moricot (Dir.), *Multiples du social. Regards socioanthropologiques,* Paris: Harmattan, 2010.

[5] method(n.) : μέθοδος, μέσο, μεθοδικότητα, σύστημα, τεχνική, μέθοδος.

[6] *Brüke und Tür. Essays des Phiosophischen zur Geschichte, Religion, Kunst und Gesellschaft.* (1903). In French: 'Pont et porte. Essais philosophiques sur l'histoire, la religion, l'art et la société' in *La tragédie de la culture et autres essais,* Rivages, 1988.

[7] See *Sociologie dês arts, sociologie dês sciences,* Paris : L'Harmattan, 2007. Gaudez, F. (dir. ; 22 authors), 286p. Collection 'Logiques Sociales' in the series

these parallel paths of thought. Let us remember that, if the French sociological tradition has placed the elaboration of a sociology of knowledge at the center of its preoccupations (thereby examining the possibility of a sociology of science and a sociology of art), it has not incorporated the question of techniques into this reflection. However, scholars working in Germany during this period addressed the problem that the development of scientific knowledge posed for social cohesion from the perspective of the development of techno-scientific rationality. For Durkheimians, the question of technique stems from their particular manner of perceiving "social morphology" in a broad sense, or the expression or incarnation of a society in its material forms. That the social needs to be inscribed in its material form, or in its materiality, in order to exist is a vein that is not sufficiently explored in sociology.[8] The hand is nevertheless very present in Durkheim's metaphors: in the gesture of a potter when the social fact is a *mould*, or when Durkheim inscribes the social in a plurality of "manners" (of thinking, of seeing and of feeling).[9] Social obligation, shaped by man's hand, is "malleable", and if determinism exists, it does not possess the implacability of physical laws. Collective representations are produced and are the product of a "handling", of a "fashioning" of and by society. The hand guides thought, or leads to thought. It is with the hands (in other words, through giving, exchange, communication) that the social bond is woven.[10] They are, along with the face, the organ and the symbol of the symbolic.

This theme is central to the sociological work of Maurice Halbwachs, particularly as regards his theses on memory. The memory reconstructs the past by apprehending it anew within the framework of the present. In this respect, all knowledge is a recognition but, at the same time, all processes of recognition reinitiate the original or unique quality of knowledge. All

---

'Sociologie des Arts'.

[8] For more on this subject, see the work of and about Jean-Pierre Warnier: *Construire la culture matérielle. L'homme qui pensait avec ses doigts,* Paris: PUF, 1999. ; Sophie Poirot-Delpech, « Préface à » : L.Raineau, *L'utopie de la monnaie immatérielle, Paris, PUF (coll. Sociologie d'aujourd'hui), 2004.* More recently, let us draw your attention to Richard Sennett's important work: *Ce que sait La main. La culture de l'artisanat.* Paris: Albin Michel, 2010.

[9] For example: "Yet, we have seen that social facts have this characteristic. Far from being a product of the will, they determine it from without; they are like moulds into which our actions are inevitably shaped" (Durkheim, 1927 : 29).

[10] For more information on this subject, S.Poirot-Delpech, *Techniques de pointe, lien social et socio-anthropologie, in* F.Gaudez (dir) : *Comment être socio-anthropologue aujourd'hui? Autour de Pierre Bouvier ,* à paraître, L'Harmattan, 2011.

recognition is an act of creation, but also a reshaping which, in turn, is a necessary condition for adapting the present in a perpetually changing world. For Halbwachs, this process involves a *re-handling* of the vestiges of the past, whether material (places, in particular) or systems of representation.[11]

Let us pause briefly on this notion of *rehandling*. Mauss said that the man that Halbwachs studied was an animal who thought with his hands.[12] Memory (individual and collective), as a construction (or reconstruction) of the past in the present, reworks "traces", "remains", "vestiges"; it reshapes the past according to the present. Society and, more broadly, the "world", are perpetually changing, a phenomenon that would surely be unliveable if we did not produce ideal and material forms that would allow for sustainability (what Hannah Arendt (1958) calls the sustainability of the common world). Humans inscribe themselves in material forms (which, even when they are classified as immaterial, always refer to a materiality), and technique is among these forms. These *rehandlings* symbolically reinvest material structures and representations.

Yet, while technical activity exists on the margin between materiality and representation, it is simultaneously a part of both of these domains. For Halbwachs, representations are not only images (reflections) of material structures and of the social world, but also encompass significations and creation. Techniques combine "discourse" (interpretation), thought and action, nature and culture, matter and mind.[13] Thus (and this is surely paradoxical in a world that is largely computerized) it becomes necessary to move beyond binary logic and established dualisms in order to consider the question of techniques. Techniques are between. They form a link whose particularity is to simultaneously exist as materiality, representation and action (whilst still maintaining a certain "autonomy").

Collective memory, for example, reworks the image of the only telephone in our childhood home around that of a cellular telephone. Techniques of the past are reworked. We think of them as the same object (or as a continuation of that object) despite the fact that they are very different, both technologically and in terms of the ways in which they are

---

[11] For more on this notion in Halbwach's work, see the last edition of *Topographie légendaire des évangiles en terre sainte*, Part One, Chapter Five, « Portée du lexique halbwaschsien de la mémoire (Éric Brian), Paris: PUF, 2008.

[12] This theme runs through all of the works of Halbwachs listed in the bibliography of this article.

[13] For more information on the centrality of this problem in French sociological thought, see Frédéric Worms, *La philosophie en France au XXe siècle*, Folio Essais Gallimard, p. 96-103.

used. Therefore, the history of techniques, or at least a certain history of techniques, counteracts the historicity that is produced through forgetting.

When a technical object is in working order, time is thus shaped in a continuous manner. The time of progress is possibly that of collective memory, or a manner of giving sense to the present and to the future by reworking the past, and by subordinating the past to the present. However, when confronted with a dysfunction, whether technical (and, more often than not, a socio-technical) or institutional, or a crisis (a catastrophe in the aeronautical world, for example), the feeling of uncertainty that it engenders produces another type of memory: heterogeneous, discontinuous, sensitive to events and bifurcations, and the plurality of fate.[14]

Knowledge and Technology: Thinking with the Hands

Hence, the hand (or hands): the hand is the method; it is that which connects us to the world and to others, the hand is a sign as well as a tool of intelligence. We can recall Aristotle's opposition to Anaxagore: "The most intelligent being is he who is capable of using the largest number of tools. Therefore, it is not because a man has hands that he is intelligent, but it is because he is intelligent that he has hands": a pair of hands, simultaneously identical and distinctive, inverted and complimentary, symmetrical and of different colours, ipseity and otherness, both the same and the other, the self and the other, but forming a pair; a new nod to the old master André Leroi-Gourhan and his expression *Ten-fingered civilisation*, which reminds us that, if the hand reaches towards the brain, it is perhaps because the brain is imminently "manual". We sometimes use the expression " to think with the hands" : Leroi-Gourhan asserts that "Not having to think with ten fingers any longer is equal to missing a part of one's normal, phytogenetically human thought".

So! The gesture and the word? The hand and the face? Both form a pair.

Many have delved into this issue and, aside from Aristotle and Leroi-Gourhan, the scope of this article prevents us from summoning forth the works of Paul Valéry, Charlotte Wolf, Martin Heidegger, Kant, Hegel, Montessori, Henri Focillon, Isabelle Létourneau and even Charles Darwin. The fact remains that hands are necessary for instituting language, but they are also necessary for miming, writing and, more generally, for doing.

[14] This issue is further developed *in* S. Poirot-Delpech, *Mémoires et histories de l'automatisation du contrôle aérien, Sociobiographie du CAUTRA*, Paris : L'Harmattan, 2010.

Hands are not organs, but tools; however, they are not simply *a* tool, but multiple ones (pluridisciplinary and polytechnic). Universal agents of action, the hand is the *organ of the possible* within which the power of humanity resides. Transformed by the intellect (which is *the Form of forms*), it is *the Instrument of instruments, the tool of tools*; the hand opposes nature, from whence it emerges and, owing to its sense of touch, its grasp of knowing and thinking, it permits us to love (and it is, for this reason, the most erotic organ, along with the face). Like the tongue, it is not tied to determined things, nor limited to particular functions, but finds fulfilment in keeping the diversity of its *possible* practical uses at arm's length and, in so doing, enjoys the freedom of *indetermination.*

The hand is essentially, to borrow Claude Lévi-Strauss's anthropological formula, that specifically human thing which reconciles and links the sensible with the intelligible.

The hand is both Man's actions and the condition in which he realizes his *Anthropos*; it is equally the manifestation of the latter, that of his intelligence and of his feelings, in other words, an *anthropophany.* We not only need hands in order to create, but also to conceptualize invention in its intuitive form. We need them in order to experience resistance and, in so doing, to enable us to interrogate, to question that which will become "the constructed real (*i.e.* a reality)". Deleuze highlights the importance of the hand in the films of Robert Bresson. In Bresson's work, it is the hand that connects, that connects fragments, disconnected fragments, fragments of the world. Bresson needs hands, notes Deleuze, in order to facilitate the connection of one part of space to another.[15]

The hand is always present in technical activity, as well as in memory, if we refer to Aristotle"s formula: memory is time. (Ricoeur, 2000: 7). This invites us to reconsider technology as a mode of memory. Let us briefly recall Bergson's distinction between a *mémoire pure*, spiritual memory in which an ensemble of recollections are stored and a *mémoire habitude,* linked to action, and which "deposits a series of pre-assembled mechanisms in the body". However, as Deleuze has rightly shown, these bergsonian dualities always precede a synthesis, and pure memory and habit memory produce immediate memory through action. In this case, memory allows for a relearning and, therefore, for representation, whereas in the other, memory is action without representation.

When Marcel Mauss formulated his concept of "techniques of the body" (Mauss, 1950), he borrowed "the old notion of habitus" from

[15] Cf Deleuze regarding the face: Deleuze G., *Cinéma tome I. L'image-mouvement* (1983), Paris: Minuit, 1999. Deleuze qualifies the function of the individuate face in a homologous manner to that of the hand ( ).

Aristotle. Even if he is not cited, we can also assume that Bergson is also there in the background, particularly when Mauss affirms that after having disassembled the real, "it must be "sewed up". Indeed, one can interpret Mauss's work as a transition in sociological work from habit- memory to memory representation. While walking, swimming, or loving, I am acted upon by an embodied tradition. I am acted upon by the past without recognizing it as such. It is memory without recognition. By restoring historicity, sociological work is, in this respect, memory work. It is not, or is rarely, a concrete recognition, as when, according to Bergson"s famous example, we remember the moment in which we have learned the lesson (pure memory) that we know by heart (habit-memory). However, we restore its historicity in a naturalized manner of behaving. It is indeed both a recognition and a relearning. The sociologist reminds us that there is time, that there is historicity. Yet, this embodied habit memory manifests itself through techniques.

This is an opportunity for Mauss to propose a definition of techniques that appears strange at first glance: techniques are effective, standardized and traditional "acts", as are effective, standardized and traditional ritual or magical acts. He specifies that the techniques would not inspire belief if they were not effective, or even efficient. Technical acts differ in that they are "experienced by their instigator on a physicochemical level". This assertion is surprising coming from a historian of categories such as Mauss. How is it possible to speak of an "experienced action" – such as throwing a boomerang, for example – as physicochemical?

Halbwachs helps us to shed some light on this anachronism when he asserts that "out of all social influences, those that take the form of technique are particular in that they function as if they were not social" (Halbwachs, 1927 : 27). Body techniques emerge from a collective memory that is characterized by the forgetting of historicity. These effective actions, which are transmitted through tradition, are returned to the realm of nature, meaning, in other words, to the realm of spontaneity and oblivion. What is "oblivion" in this sense? It is a "feeling", a presence without representation, an "unconscious representation" in the Bergsonian rather than Freudian sense.

This question relates back to the mode of presence of social facts. Social efficiency possesses a certain automaticity, the efficiency of which increases the more unconscious it becomes. However, according to Durkheim, the difference between social facts and facts of another order (physicochemical, for example) is their plasticity, their malleability. Mauss addresses the limits of social facts through the concept of body technique, in which the plasticity of the social intermingles with the durability of nature. By transforming itself into a "tool" or an "instrument" concerned with efficiency, our body allows us to forget the cultural dimension of techniques that can only be recalled through the gaze of an observer, or a socio-anthropologist. In this context, body techniques may be viewed as ways of using the body – standardized, transmissible, efficient – but may equally be understood as unconscious representations, most particularly when they are "experienced" on physicochemical, natural, or material level, despite the fact that they are nature and culture, material and representation...They function, as Halbwachs suggests, "as if they were not social".

Where are we going with this? Quite simply:

Is it possible to place this rereading of already ancient texts at the service of a socio-anthropology of technical societies? Is not this "oblivion", inherent and necessary to a certain type of efficiency, already at the heart of technoscience? More broadly, is not a society that regards efficiency as a central value (and not only as a necessity for an effective human action) a society that lauds oblivion, or abolishes time?

If it is memory that enables us to relearn and, therefore, to re-present an absent object belonging to the past (Ricoeur, 2000), it does not abolish time, but gives it a form. In making time exist, does not the socio-anthropologist's work create the possibility of distance and, in so doing, the possibility of critical space? Or, perhaps, the possibility to think about the space of politics, *or of its "lucidity"*, in which the fate of the real can be questioned?[16] What happens in particular when the body, as the instrument or tool described by Mauss (who, of course, did not reduce the body to this dimension), finds itself repaired, modified, even improved upon or changed by the incorporation of technical objects? In what particular "manners" will these new hybridizations cause us to forget, in the perpetual and imperative urgency of efficiency, that we are inscribed in history? Finally, what happens when technological scientific knowledge, which intervenes in the immediacy of our "manners" of thinking, acting and feeling, is developed to an unrepresentable scale (bordering on the *quantum scale* in nanotechnology)? What happens when automatisms become automats, when live in us, when they form, with us, a symbiotic relationship[17]? How is it possible to conceptualize politics in a world where social obligations are "incorporated" (and therefore forgotten in their historicity) in the form of material or immaterial techniques?

---

[16] 'La blessure la plus rapprochée du soleil', according to René Char (*Feuillets d'Hypnose*, 169).

[17] For more on this subjetc : Gras A., Moricot C., Poirot-Delpech S., Scardigli V., *Faced with automation* , Publications de la Sorbonne, 1994.

From this perspective, it is perhaps useful to recall that body techniques are not always automaticity and forgetting, but can engender rememoration and conscience. It is said that those gymnastics that are practiced in Europe come from China. However, practices that are more common (for example, body building or simple gym training) are techniques of bodily adaptation to social norms. What is sought after, of course, is the well-being of the practitioner, a concern for his or her body, but this well-being occurs through a submission to norms without necessarily invoking a consciousness of these norms, or any sort of reflexivity. Yet, other types of "body techniques" aim for consciousness rather than forgetting. Oftentimes emanating from the orient (yoga) and occasionally linked to psychoanalysis (as in *Rebirthing*), these body techniques aim to engender consciousness. Recalling or revealing events that have marked a person or habits and automatisms to which they are subjected through their corporeal imprint, these body techniques remind us that "I am a body".

Our body is the product of a social and personal history, and these body techniques aim to raise awareness of this history in order to open us up to another type of presence or existence in the world. As regards this "self-concern" through body techniques, we can, of course, refer to the later work of Foucault. This does not entail ceasing action and solely focusing on the act of remembering (which inhibits action), but to open spaces in which, through techniques, we can become aware of this forgetting and this automaticity. It is about opening a space, a space of consciousness which, by passing by way of a return to *self* understood as *ipseity* rather than *identity*, may potentially lead to the opening up of a space of judgment[18].

This is perhaps the central issue of what Georges Simondon referred to as a technical culture. Unlike in Gaston Bachelard, where culture is pushed into the arena of techno-scientific culture, or is fighting against common sense to secure the emergence of a *new man* that is detached from tradition, this concept of technical culture refers to a culture in which technology has the possibility of being "memorable" (meaning retranslated in its sociality, discussed and discussable, interrogated and searchable) and, in this way, is reintroduced into common sense and into Politics. This could be the vocation of socio-anthropological work.

At a time when successive reforms have forced universities to

---

[18] Florent Gaudez, "La dialectique *Altérité VS Ipséité* comme parcours anthropologique de la Connaissance. Une mise en scène à travers un cas en Socio-anthropologie du Texte littéraire", in *Altérité et Société*, (dir. Chryssoula Constantopoulou), Athènes, EKKE, CD-Rom, 2004.

undertake more and more inert practical work, to place more emphasis on "professionalizing" and less on thought, more on method and less on the "Humanities", more on the pseudo-innovative, more on promising but conceptually empty subjects (sustainable development), more "stooge" fieldwork (as if we weren't already aware of the fact that the more we discuss "fieldwork", the less we do) and, in forgetting, above all, that if the head exists no longer, neither does the hand, we would like to insist upon Michel Verret's (Verret, M., in Préface to: Christine Gamba-Nasica, 1999) warning:

> "Ten-fingered civilization, said Leroi-Gourhan.
> Decontrification takes one hand away.
> Delabourisation removes the other.
> Only the head remains...
> But which head, when the head no longer welcomes the hand?
> The head without hands loses its footing, as well.

Which we can transpose thus:

> "Ten-fingered civilization, said Leroi-Gourhan.
> Limiting thought takes one hand away.
> Marginalizing the humanities removes the other.
> Only the famous fieldwork remains (innovative, preferably!)
> But which fieldwork, when it neither welcomes thought nor the humanities?
> Fieldwork without Method loses its footing, as well.

What will we be left with?

Ultimately: The pads of the fingers on the pulp of the lips.

References

Arendt H. (1958). *The Human Condition*, London, Chicago, University of Chicago Press.

Char R. (1946). *Feuillets d'Hypnos* Paris, Folio, (N° 169).

Cortazar J. (1973). *Les Armes secrètes* Folio..

Deleuze G. (1999). *Cinéma tome . L'image-mouvement* (1983), Paris: Minuit.

Durkheim E. (1927). *Les règles de la méthode sociologique,* Paris : PUF.

Gaudez F. (2004). 'La dialectique Altérité VS Ipséité comme parcours anthropologique de la Connaissance. Une mise en scène à travers un cas en Socio-anthropologie du Texte littéraire », *in Actes du Colloque "Alterité et Société"*, Athènes, CD-ROM AISLF.

Gaudez F.H. (1997). *Pour une Socio-Anthropologie du texte littéraire. Approche sociologique du Texte-acteur chez Julio Cortázar*, Paris, L'Harmattan, (collection "Logiques sociales").

Gaudez F. (dir). (2007). *Sociologie des arts, sociologie des sciences,* Paris : Collection "Logiques Sociales" in the series "Sociologie des Arts", 286 p. L'Harmattan.

Gras A., Moricot C., Poirot-Delpech S., Scardigli V. (1994). *Face à l'automate,* Publications de la Sorbonne.

Halbwachs M. (1927). *Les cadres sociaux de la mémoire,* Paris : Albin Michel,

—. (2008). *Topographie légendaire des évangiles en terre sainte*, Paris: PUF.

Leenhardt J. (1986). "Un chemin de la connaissance : Les autonautes de la cosmoroute", *in Lo lùdico y lo fantàstico en la obra de Cortázar*, Centre for Latin American Research, University of Poitiers, Espiral hispano-americana, Editorial fudamentos.

Levi-Strauss C. "Introduction à l'œuvre de Marcel Mauss » , *in* Mauss. M., (1950) *Sociologie et Anthropologie,* Paris, PUF.

Mauss.M. (1950). *Sociologie et anthropologie,* Paris : PUF.

Passeron J.C. (2006). *Le raisonnement sociologique. L'espace non-poppérien de l'argumentation*, Paris: Albin Michel.

Poirot-Delpech S. (2010). *Mémoires et histoires de l'automatisation du contrôle aérien, Sociobiographie du CAUTRA*, Paris : L'Harmattan. 0.

Poirot-Delpech S. (2010). *La nature comme problème, in* Moricot. (Dir.), *Multiples du social. Regards socioanthropologiques,* Paris: Harmattan.

Ricoeur P. (1983). *Temps et récit. Tome I : L'intrigue et le récit* , Le Seuil.

—. (2000). *La mémoire, l'histoire, l'oubli*, Paris : Seuil.

Ramos S. (2007). "A abordagem socioantropologica de Sophie Poirot-Delpech", *in Revista Laboreal,* Volume III, n°2.

Verret M. (1999). Préface to: Christine Gamba-Nasica *Socialisations, experiences et dynamique identitaire. L'épreuve de l'entrée dans la vie active,* Paris : L'Harmattan.

Warnier J.P. (1999). *Construire la culture matérielle. L'homme qui pensait avec ses doigts,* Paris: PUF.

Worms F. (2000). *La philosophie en France au XXe siècle*, Folio Essais Gallimard, p. 96-103.

CONTRIBUTORS

In alphabetic order

Roberta Bartoletti is Associate Professor of Sociology of Culture and Communication at the Department of Communication Sciences of the Urbino University "Carlo Bo" (Italy). Among her last publications:. *The Crisis of Collective Memory and the Commodification of Nostalgia*, in Packard Noel (Ed.), "Sociology of Memory: Papers from the Spectrum", Cambridge Scholars Publishing 2009; *Memoria e comunicazione. Una teoria comunicativa complessa per le cose del moderno*, Angeli, Milano, 2007.

Roberto Cipriani is Full Professor of Sociology and Head of the Department of Education at University of Roma Tre. He is author of articles and books translated into 8 languages, *Sociology of Religion. A Historical Introduction*, in particular, is available in English, Italian, Spanish, Portuguese and Chinese.

Jérémy Damian graduated from the Political Sciences Institute of Grenoble in 2005 and currently Phd student at EMC2-LSG laboratory, he works toward a pragmatical socio–anthropology of sensitive experience, exploring how do we share expe-riences through social institutions, conventional signs and "built together sense-tions" : interobjectivity ? And in return how, exploring that, does it oblige to *practice* sociology and anthropology with different habits, tools and sensitivy, that is to imagine other mode of inquiry?

Alan Dunning is a founding member of the art/science collaboration, *The Einstein's Brain Project*. Over the last 15 years he has been included in more than 100 exhibitions, and received major awards from the Canada Council, La fondation Daniel Langlois, and the Social Sciences and Humanities Research Council of Canada.

Damayanthie Eluwawalage is an Assistant Professor at the Department of Human Ecology at the State University of New York, Oneonta, United States. She is a Professional Historian who specializes in costume history. Her research interests include design, design theory, costume history and fashion theory, focusing on the broader application and interpretation of dress, body and fashion theory in the sociological, historical and cultural contexts.

Florent Gaudez is a socio-anthropologist. He is Professor of Sociology at the Université Pierre-Mendès-France (UPMF – Grenoble, France) where he is Head of the Laboratory of Sociological Research *EMC*²-LSG (*Emotion-Mediation-Culture-Knowledge)* and oversees the Masters in '*Productions and Mediations of Cultural Forms*' (see http://sociologie. upmf-grenoble.fr/EMC2). Some of Gaudez's recent publications include *For a Socio-anthropology of literary texts. A Sociological Approach to the Text-actor in the work of Julio* Cortázar. He has also edited a number of collective works, including *Sociology of Arts, Sociology of Sciences; Middle-brow Arts Today; The Culture of Text; Knowledge of the Text; Figures of Otherness. How can one be a socio-anthropologist today?* (see http://www.harmattan.fr). He also edits the international journal *Sociology of Art – OPuS* (see *http://sociologieart.free.fr)* and is Vice President of the French Association of Sociology (see http://www.afs-socio-fr).

Gabriel Jderu holds a Ph.D. in Sociology. He is lecturer at Department of Sociology, University of Bucharest, Romania and he teaches Research Methods and Sociology of the Body. His current, post-doctoral research is project analyzes the body, in the context of automobility and seeks to understanding the relation between body and social emotions.

Angelina Karpovich is a Lecturer in Multimedia in the School of Engineering and Design, Brunel University. She is an anthropologist and filmmaker whose research focuses on the social impacts of media content and technologies.

Umberto Melotti is senior professor of Political Sociology at "La Sapienza" University, Rome. His fields of research include general sociology, cultural anthropology, sociology of change, international migration and security. Among his published works: *Revolución y Sociedad* (Fondo de Cultura Económica, México D.F.,1971), *Marx and the Third World* (Macmillan, London, 1978, also available in Spanish and

Chinese), *Sociología del hambre* (Fondo de Cultura Económica, México D.F.,1980²), *El hombre entre la naturaleza y la historia* (Península, Barcelona, 1982), *Oltre la sociobiologia* (Centro Studi Terzo Mondo, Milano, 1988), *Migrazioni internazionali, globalizzazione e culture politiche* (Bruno Mondadori, Milano, 2004), *Le banlieues* (Meltemi, Roma, 2007), *Perché la sociologia?* (Mondadori Università, Milano, 2010), *Migrazioni e sicurezza* (Chieti, Solfanelli, 2011).

Alireza Moula teaches sociology and social work and is ass. professor of social work at Karlstad University, Sweden. Its research is located at the intersection of sociology, social work, pragmatist philosophy and neuroscience. Moula has published several articles in the international journals and is the editor of a book in Swedish about empowerment oriented social work (2009). He is the initiator and coordinator of several research projects about empowering some groups of adolescents and women in Iran and Sweden. These projects aim to create a neuropragmatist knowledge bank that integrates knowledge from pragmatist philosophy with neuroscience. Moula holds his BA and MA in sociology and his PhD in social medicine from Linköping University in Sweden.

Matteo Negro is associate professor of Theoretical Philosophy at the niversity of Catania, Italy. He obtained his PhD in Philosophy in Fribourg, Switzerland, and specialized in Cognitive Neuroscience in Geneva. His philosophical research and his Italian and international publications focuse mainly on normativity, action, embodiment, cognitive neuroscience. Currently, he is leading, in collaboration with other Italian universities, a national research program on epistemology and subjectivity. He is a founding member of the International Society for Legal and Moral Philosophy (Cambridge, UK). He is a board member of the Research Committee on "The Body in the Social Sciences" of the International Sociological Association. Forthcoming: Perception, action et normativité, Bern-Berlin 2011.

Guido Nicolosi holds a research permanent position in Sociology of Culture and Communication at the DAPPSI, University of Catania, Italy. He teaches, as aggregate professor, Sociology of Communication and New Media. Member of several research teams in national and international research programmes, he has been working for the last ten years on the issue of "Body, Technology and Society". He has been visiting scholar in several European universities and cooperates with the Critical Technology Construction (CTC) of the Wageningen University and Research (NL).

Sophie Poirot-Delpech is a socio-antropologist. She is a Lecturer at the Université Paris 1 Panthéon-Sorbonne. Director of the Department of Social Sciences (SGSS), she also teaches sociology of technics in the Department of Philosophy. She is Researcher at the CETCOPRA (Centre d'Etude des Techniques, des Connaissances et des Pratiques) and associate researcher at the Laboratory of Sociology of Grenoble *EMC*^{2}-LSG (Emotion-Médiation-Culture-Connaissance). After years of research and investigation in the aeronautic world, she now works about issues on technic in the French sociological tradition. Her publications include *Mémoire et histoires de l'automatisation du contrôle aérien. Sociobiographie du CAUTRA (*L'Harmattan 2009), some collective work such as *L'imaginaire des techniques de pointe. Au doigt et à l'œil* (L'Harmattan, 1988), *Face à l'automate. Le pilote, le contrôleur et l'ingénieur* (Publications de la Sorbonne, 1996) et (avec Alain Gras) : *Grandeur et dépendance, sociologie des macrosystèmes techniques (PUF).*

Shinsuke Sakumichi is Professor of Social Psychology in the faculty of humanities of Hrisaki University, Japan. Born in Tokyo Japan, he received the B.S., M.S and Ph.D. degrees (literature) from Tohoku University, Sendai, Japan, in 1981, 1983 and 2001, respectively. During 1986-1989, he worked as a counselor for Tohoku University. In 1989, he joined the Department of Information and Behavior, Hirosaki University, where he is currently Professor of Social Psychology. He is Visiting Adjunct Professor in the Department of Social Welfare, Iwate Prefectural University of Morioka, Japan and Visiting Adjunct Professor in Nursing & Social Services, Health Sciences University of Hokkaido. His research interestes include health-care system in modernization and he has practiced ethnographic research in the Turkana, pastoralist of semi-arid Kenya since 1993.

Chris Speed is Reader in Digital Spaces at the Edinburgh School of Architecture and Landscape Architecture where he extends a practical and critical enquiry into how digital technology can engage with the field of architecture and human geography.

Thomas S. Smith (B.S., Johns Hopkins University; M.A., PhD. University of Chicago) is Professor of Sociology at the University of Rochester. He is the author of numerous scholarly articles and several books, including his much praised theoretical monograph *Strong Interaction* (Chicago 1992). Smith's more recent work makes use of computational methods to explore innate mechanisms active in attachment behavior and social networks.

Andreas Georg Staschelt trained as musician (violin performance) and received his academic education in the social sciences and humanities (Ph.D. in Sociology, Bielefeld University 1989). He is Senior Fellow of the Institute for Advanced Study in the Humanities, Essen and Professor of Media Studies, Aesthetics and Communication at Dortmund University of Technology (Institute of Music and Musicology) and at Dortmund University of Applied Sciences and Arts (Department of Social Sciences).

Paul Woodrow is a co-founder of *Einsteins Brain Project* in 1996 and has exhibited, lectured and published extensively in France , Argentina, Italy, Sweden. Russia, England, the USA, China, Mexico, Spain and Canada. He has received numerous distinguished awards and is presently Professor of Art Theory at the University of Calgary.

Ivan Varga is Emeritus Professor of Sociology, Queen's University, Kingsyon (Canada) and Honorary President of ISA Research Committee 22 (Sociology of religion). Is also a Member of the Editorial Board of International Sociology. Varga is author or editor of three books, and has contributed nearly 30 chapters in book published in Hungary, Poland, Czechoslovakia, Yougoslavia, France, the UK, and Canada.